COMPUTER APPLICATIONS
in
GENETICS

Lawrence Hasbrouck Snyder

[International Conference on Computer Applications in Genetics, University of Hawaii, 1969

COMPUTER APPLICATIONS
in
GENETICS

edited by

Newton E. Morton

Proceedings of a conference sponsored by the
University of Hawaii and the Genetics Study Section,
National Institutes of General Medical Sciences,
and dedicated to Lawrence Hasbrouck Snyder

1969

Library of Congress
Catalog Card Number 70-627391

Distributed by the
University of Hawaii Press
Honolulu, Hawaii
1969

Printed in the U.S.A.

FOREWORD

Within a very few years the high-speed computer has become a research tool of enormous value in many areas of science. One of these is genetics. The genetic analysis of natural populations, of quantitative traits, and of evolutionary relationships often involves elaborate computational procedures. This is conspicuously true of many kinds of human genetic research. Problems of segregation and linkage analysis, usually no problem in experimental organisms, present formidable difficulties to the human geneticist. Human data rarely appear in the neat and balanced packages that emerge from carefully designed laboratory studies. Elaborate analytical methods are required. Furthermore, the data are often very expensive to replicate and are sometimes unique, so that it becomes important to extract every possible bit of information from them; for this, computer methods are needed. Many of these have been developed by Dr. Morton and his associates at the University of Hawaii. Finally, the computer is useful in the development of theoretical models that are sufficiently complex to serve as realistic views of nature. The hope is that the publication of the various methods in this volume will make them more widely available to research workers.

The International Conference on Computer Applications in Genetics was held on September 2-4 at the University of Hawaii. It was jointly sponsored and supported by the University of Hawaii, the Population Genetics Laboratory, and the National Institutes of Health through its Genetics Study Section. The Study Section, in addition to its main task of reviewing and advising on research grants, has a broad general interest in genetic research. Several years ago it sponsored a series of three conferences on Methodology in Genetics which subsequently appeared in book form.* The present volume is in the same spirit and presents new methods in an area where these are changing most rapidly.

It was fortunate that Dr. L. H. Snyder, to whom the conference and this volume are dedicated, could be present at the conference. Dr. Snyder was a student of W. E. Castle and has had a distinguished career as scientist, teacher, and administrator (including the Presidency of the University of Hawaii). He was working in human genetics long before this became fashionable. At a time when most discussions of human genetics consisted of deploring man's shortcomings when compared to fruit flies and maize he was stimulating interest and developing methods for human genetic analysis. I should like to mention two aspects of his early work which are intellectual antecedents of ideas that appear in this volume. One is his early work with gene frequency methods and the discovery of what are now known as "Snyder's ratios." The other is his interesting Charles Cotterman in human genetics, with well known consequences, one of which appears in this volume.

James F. Crow, Chairman
Genetics Study Section
National Institutes of Health

*Methodology in Human Genetics (1962), Methodology in Mammalian Genetics (1963), and Methodology in Basic Genetics (1963). Ed. by W. J. Burdette. Holden-Day, San Francisco.

CONTENTS

FACTOR-UNION PHENOTYPE SYSTEMS

C. W. Cotterman

University of Wisconsin
Department of Medical Genetics

1. Introduction

An earlier publication (Cotterman, 1953) dealt with regular phenotype systems, defining these as many-one correspondence relations between a set of genotypes and a set of phenotypes. Taking "phenotypic identity" as an essentially undefined equivalence relation, one is assured of the broadest possible combinatorial classification, since an equivalence relation partitions a set into one or more subsets without restriction. Enumerations were made of the "essentially different" or anisomorphic systems for 2 and 3 alleles, defining these as systems which are distinct under permutations of the gene symbols; these were called phenograms.

A catalogue of regular phenotype systems (rps) naturally includes some rather bizarre systems, demanding for their explanation either unusual modes of gene expression or interaction or unusual behavior of the diagnostic reagents or test criteria. It is interesting, however, that Nature has contrived examples of many of these, and the remainder pose some interesting puzzles for both theoretical and experimental geneticists.

Here we wish to consider a special subset of rps's which Dr. Morton and I have come to know as factor-union systems (fU). These are systems which find the greatest appeal, demanding only the very simplest assumptions, yet often flexible enough on matters of interpretation to leave room for some vigorous scientific disputes. They are best exemplified by the inherited blood-type and serum-type systems of man and other vertebrates, varying from simple systems based upon two or three alleles to very complex ones involving 100 or more alleles. The term "factors" is employed here in deference to blood-group workers, who commonly refer to their antigenic units as factors.

Informally, we could define fU systems as those in which it is possible to assign to each allele a set of properties, which we call factors, in such a way that the phenotypes of all genotypes are then specified by the unions of two such sets. There is no difficulty in extending this principle to more complex cases (polyploids, multiple loci, sex-linkage, etc.), but here, as in the earlier paper, I shall strictly limit the discussion to fU systems with m alleles at one autosomal locus in a diploid species.

To some, this definition may seem to be only a rather cumbersome way of saying that the "genes are acting in a dominant or additive fashion in respect to all factors." I much prefer the longer definition for the following reasons. It is not how the genes are acting, but only under what conditions we can discern their presence, that really matters, at least to the formal geneticist. The shorter definition seems to commit us on questions of interpretation that might better be left open. Secondly, I have a definite need here, as in the earlier paper, for the term "dominance" in its original (mendelian) sense, i.e. as a dyadic relation on alleles: $A \rightarrow B$ if, and only if, AA and AB are phenotypically indistinguishable, with BB distinguishable from AA and AB. To also speak of dominance in reference to the factors would invite confusion. Lastly, the operation which combines two factor sets to spell out the phenotypes for each diploid genotype is, of course, not addition, but union.

The preferred definition also makes it a little clearer that we intend to restrict "factors" to 2-valued variables. A set of factors can therefore always be represented by a vector of 0's and 1's. It was the consistent use of these by Dr. Morton and his colleagues that led me to realize that (0, 1)-matrices are ideally suited to display the many charms about fU systems. In fact, without matrices I couldn't even define fU systems in a manner sufficiently explicit to suggest ways in which enumerations might be undertaken.

To prove theorems in relation to fU systems, however, one is apt to require a set-theoretical terminology and notation. Both graphs and matrices are very instructive, but the graphs (at least those I have devised) are highly specialized, and the operations which we perform on the matrices, though doubtless easy for a computer, are not the standard ones of matrix algebra. I wish therefore to include in this introduction a set-theoretical statement of the essentials of our problem, purged of all genetic terminology.

Let $G = \{a_1, a_2, \ldots a_m\}$ be a set of m elements $(m \geq 2)$ and let G_2 be its diplo-set, i.e. the set of all unordered pairs of elements of G, with duplications allowed. Let X_α and X_β be disjoint subsets of G_2; then we say that a proper subset $C \subset G$ cuts X_α from X_β or C is a cut-set for (X_α, X_β) and write

$$C /\!/ (X_\alpha, X_\beta) \text{ iff for all } (a_i a_j) \in X_\alpha, \{(a_i \in C) \cup (a_j \in C)\}$$

$$\underline{\text{and}} \text{ for all } (a_k a_l) \in X_\beta, \{(a_k \notin C) \cap (a_l \notin C)\}.$$

Here i and j are not necessarily distinct, nor k and l. Since the pair (X_α, X_β) is unordered, we need scarcely add "or vice versa" to the definition. Note that '//' defines an incidence relation on subsets of G to unordered pairs of disjoint subsets of G_2. We also need a relation on subsets of G to subsets of G_2 with the meaning C does not cut through X, with definition

$$C \not\frown X \text{ iff for all } (a_i a_j) \in X, \{(a_i \in C) \cup (a_j \in C)\}$$

$$\underline{\text{or}} \text{ for all } (a_i a_j) \in X, \{(a_i \notin C) \cap (a_j \notin C)\}.$$

Suppressing the a's and elevating their subscripts, we may illustrate the above definitions with

$$X_1 = \{11, 12, 13, 23\},$$

$$X_2 = \{14, 34, 35, 45\},$$

$$X_3 = \{22, 33\}.$$

Note that the simple subsets $\{1\}, \{2\}, \ldots \{5\}$ are not cut-sets for X_1 and X_2, but $\{4, 5\}$ qualifies. For X_2 and X_3 there are two cut-sets, $\{4, 5\}$ and $\{1, 4, 5\}$. There is no cut-set for X_1 and X_3. Note further that $\{1, 4, 5\}$ cuts through X_1, but $\{4, 5\}$ does not.

Let Π be a ϕ-partition of G_2. As the word is here used, a partition cannot have empty parcels ($X_\alpha \neq \emptyset$, for all α), so $\phi \leq m(m+1)/2$. We call Π a factor-union partition ($\mathfrak{P}$) and define

$$\Pi = \{X_1 | X_2 | \ldots | X_m || \ldots | X_\phi\} = \mathfrak{P} \qquad \text{iff}$$

(1) $(a_\alpha a_\alpha) \in X_\alpha$ for all $\alpha = 1, 2, \ldots m$, so that $\phi \geq m$,

and

(2) for all $(X_\alpha, X_\beta) \in \Pi$, $\exists\, C \subset G$ such that $C /\!/ (X_\alpha, X_\beta)$ and $C \not\frown X_\gamma$ for all $X_\gamma \in \Pi$ $(\gamma \neq \alpha, \gamma \neq \beta)$.

Let $f_\alpha \subseteq G$ $(f_\alpha \neq \emptyset)$ and let $X_\alpha \subset G_2$ $(X_\alpha \neq \emptyset)$; then f_α is called the field of X_α if f_α contains just those elements of G which are "represented" in X_α. More exactly, f_α is field of X_α iff

$$(a_i a_j) \in X_\alpha \text{ implies } (a_i \in f_\alpha) \cap (a_j \in f_\alpha)$$

$$\underline{\text{and}}\ a_i \in f_\alpha \text{ implies } \exists (a_i a_j) \in X_\alpha,$$

where i and j are not necessarily distinct.

The set $F = \{f_1; f_2; \ldots f_\phi\}$, in which f_α is field of X_α for all $X_\alpha \in \Pi$, is called the field-set of Π. Note that f and F are defined for all partitions of G_2. If $\mathfrak{P}$ is a factor-union partition, we substitute $\mathfrak{f}$ for f and $\mathfrak{F}$ for F. But, in the definition of $\mathfrak{P}$, Rule (1) requires that the fields $\mathfrak{f}_1, \mathfrak{f}_2, \ldots \mathfrak{f}_m$ be distinguished from the remaining $(\phi - m)$ fields, and this may be accomplished by the notation

$$\mathfrak{F} = \{\mathfrak{f}_1 \mid \mathfrak{f}_2 \mid \ldots \mid \mathfrak{f}_m \| \ldots \mid \mathfrak{f}_\phi\}.$$

Isomorphisms ($\simeq$) for partitions and their F-sets are defined in the usual way:

$$\Pi_i \simeq \Pi_j \text{ iff } \exists\ \xi \in |G| \text{ such that } \Pi_i = \xi(\Pi_j),$$

where ξ is an element of $|G|$, the group of permutations of elements of G. Similarly for $F_i \simeq F_j$. If the partitions are factor-union, we again substitute $\mathfrak{P}$ for Π and $\mathfrak{F}$ for F.

THEOREM 1.1 In every factor-union partition $\mathfrak{P}$, the fields of all X-subsets are distinct, so that the field-set is a subset of the power-set of G: $\mathfrak{F} \subset P(G)$.

Proof: For all (X_α, X_β), there exists $C \subset G$ such that $C /\!/ (X_\alpha, X_\beta)$, which implies that there exists $a_i \in G$ such that $a_i \in X_\alpha$ and $a_i \notin X_\beta$, or $a_i \in \mathfrak{f}_\alpha$ and $a_i \notin \mathfrak{f}_\beta$. Hence $\mathfrak{f}_\alpha \neq \mathfrak{f}_\beta$, for all α, β.

CONJECTURE If $\mathfrak{P}_i$ and $\mathfrak{P}_j$ are factor-union partitions and $\mathfrak{F}_i$ and $\mathfrak{F}_j$ are their corresponding field-sets, then $\mathfrak{F}_i \simeq \mathfrak{F}_j$ if, and only if, $\mathfrak{P}_i \simeq \mathfrak{P}_j$.

Note that the theorem and the conjecture refer just to factor-union partitions. The property of having ϕ distinct fields is not unique for factor-union partitions. For example, the following are not factor-union partitions,

$$\Pi_1 = \{11, 12 \mid 22, 23 \mid 33 \mid 13\}$$

$$\Pi_2 = \{11 \mid 22 \mid 33, 12 \mid 13, 23\},$$

but the four fields in Π_1 are distinct, while those of Π_2 are not:

$$F_1 = \{1, 2 \mid 2, 3 \mid 3 \mid 1, 3\}$$

$$F_2 = \{1 \mid 2 \mid 1, 2, 3 \mid 1, 2, 3\}.$$

Without the theorem, the conjecture would be untenable. But there is additional support for the conjecture in that it is known to hold in all cases with m = 2, 3 and 4. The field-set is seen to be a kind of abbreviated representation of the partition of G_2. As such, it may provide a convenient means of counting anisomorphic $\mathfrak{P}$-partitions, provided the conjecture is true.

2. Definitions and Notation

For convenience, almost all definitions specifically relating to our genetic problem are brought together here, with some notation. This will bridge the gap between the abstract definitions of Sections 1 and 3 and the discussion that follows.

- G — the set of m alleles; for purposes of illustration, we generally denote this set as $G = \{A, B, C, \ldots\}$
- H_1 — the set of m homozygous genotypes: {AA, BB, CC, ...}
- H_2 — the set of $m(m-1)/2$ heterozygotes: {AB, AC, AD, ...}
- G_2 — the "diplo-set" of $g = m(m+1)/2$ genotypes: $G_2 = H_1 \cup H_2$ (Mathematicians do not seem to have coined a word for the set of all unordered pairs of elements of a set. Genetics is concerned with many kinds of diplo-sets, G_2 being one.)
- $P(G)$ — the power-set of G, i.e. the set of all (2^m) subsets of G
- $|G|$ — the symmetric group, i.e. all (m!) permutations of G
- X_i — a phenoset, i.e. any subset of G_2 all of the genotypes of which correspond to the same phenotype (p_i). When matrices are used, p_i is represented by a (0,1) row vector.
- f_i — the field of X_i (or p_i), a non-empty subset of G containing just those alleles each of which is present in at least one genotype of the phenoset X_i
- k_i — the kernel of X_i (or p_i), a possibly empty subset of f_i containing just those alleles that are present in every genotype of the phenoset X_i. If $k_i = \emptyset$, X_i is said to have a null kernel.
- h_i — the number of genotypes in X_i. A phenoset is said to be simple if h = 1, complex if $h \geq 2$.
- $n(f_i)$ — the field number, i.e. the number of alleles in f_i; in general, $n(f_i) = 1, 2, \ldots m$.
- $n(k_i)$ — the kernel number, i.e. the number of alleles in k_i; in general, $n(k_i) = 0$, 1 or 2.
- $X^\bullet_i$ — a homophenoset, i.e. a phenoset containing one or more homozygous genotypes
- X^o_i — a heterophenoset, i.e. a phenoset containing heterozygous genotypes only
- $S^\bullet$ — a homophenoset structure, i.e. a set of homophenosets which is invariant under permutations of $|G|$
- S^o — a heterophenoset structure, i.e. a set of heterophenosets which is invariant under permutations of $|G|$
- $\Pi m.\phi$ — any partition of G_2 into phenosets $\{X_1 | X_2 | \ldots | X_\phi\}$ is called a regular phenotype system of degree m and index ϕ
- $\bar{\Pi} m.\phi$ — a phenogram is a set of Π-partitions of G_2 which is invariant under permutations of $|G|$
- F — the field-set, i.e. the set of all fields corresponding to a ϕ-partition of G_2

Although the words phenoset, field and kernel may be new, the sets which they describe are very familiar things to anyone who has worked with complex blood type systems. When one lists all of the genotypes corresponding to one blood type, his list is a phenoset. If he abstracts from this a list of all alleles represented in the phenoset, his list is a field. And if, in the field, there is one allele (there may possibly be two or none) which is common to all genotypes, that allele constitutes the kernel. These are the sets which the geneticist must consider when he wishes to infer genotypes from phenotypes and to infer parental contributions to those genotypes. It is not surprising, then, that when one attempts to define and enumerate factor-union systems using matrices, these three kinds of sets come immediately into prominence, begging for some recognition in our genetic nomenclature.

All of the above definitions are applicable to regular phenotype systems in general. Factor-union systems are subject to two restrictions:

(1) each homophenoset contains just one homozygote, and

(2) there is a general restriction on the composition of phenosets of a kind that can be described — very loosely — as factorability. At present, I know of no way to describe this property precisely except in terms of cut-sets (Section 1) and by means of matrices.

3. Special (0,1)-Matrices

This section describes some manipulations on (0,1)-matrices, the purposes of which will be readily apparent to a geneticist with just one or two examples. The work is quite easy for the brain and eye if the matrices are not too large, and I suspect all of the operations could be computerized without difficulty. They are, however, somewhat unorthodox, so their formal description is rather cumbersome; nevertheless, I think the following definitions are explicit. Note that the definitions are made in the abstract; some more-descriptive terms for these matrices will be suggested when examples are introduced.

DEF. 3.1 Let $p = |x_1, x_2, \ldots x_n|$ and $q = |y_1, y_2, \ldots y_n|$ denote (0,1)-vectors, i.e. $x_i = 0$ or 1, $y_i = 0$ or 1, for all i. If, in a (0,1)-matrix, both p and q are row vectors or both are column vectors, their union is defined by

$$p \cup q = |x_1 \cup y_1, x_2 \cup y_2, \ldots x_n \cup y_n|,$$

and their intersection by

$$p \cap q = |x_1 \cap y_1, x_2 \cap y_2, \ldots x_n \cap y_n|,$$

and the complement of p by

$$\sim p = |\sim x_1, \sim x_2, \ldots \sim x_n|,$$

where '∪', '∩' and '∼' are the operations in Boolean algebra defined by

$1 \cup 1 = 1$	$1 \cap 1 = 1$	
$0 \cup 1 = 1 \cup 0 = 1$	$0 \cap 1 = 1 \cap 0 = 0$	$\sim 1 = 0$
$0 \cup 0 = 0$	$0 \cap 0 = 0$	$\sim 0 = 1$

DEF. 3.2 Let Z be a (0,1)-matrix of order u x v, and let t be the number of equivalence classes of its row vectors with respect to identity, so that $t \leq u$. We say Z is column-irreducible if Z contains no submatrix of order u x s, $s < v$, having this same property (t equivalence classes for rows).

DEF. 3.3 Let P be a (0,1)-matrix of order m x n, having row vectors $p_1, p_2, \ldots p_m$; from this we construct a matrix P′ of order k x n, $k = m(m-1)/2$, having row vectors

$$p'_1 = p_1 \cup p_2, \; p'_2 = p_1 \cup p_3, \; \ldots \; p'_k = p_{m-1} \cup p_m.$$

The diplo-matrix P_2 of P is then defined as the partitioned matrix

$$P_2 = \left|\begin{array}{c} P \\ \hdashline P' \end{array}\right| \quad \text{of order } m(m+1)/2 \text{ x } n.$$

DEF. 3.4 A (0,1)-matrix (P) of order m x n is called a (factor-union) generator matrix just in case

(1) its m row vectors are distinct, and

(2) its diplo-matrix (P_2) is column-irreducible.

DEF. 3.5 Let P be a generator matrix of order m x n, and let P_2 be its diplo-matrix, having row vectors r_i (i = 1, 2, ... g) where $g = m(m+1)/2$. By the condensed diplo-matrix of P is meant a matrix $\overline{P}_2$ derived from P_2 by elimination of any row r_j just in case $r_j = r_i$ and $i < j$.

Note that condensation of P_2 can only affect the submatrix P′ of P_2, since the rows of P must be distinct by Def. 3.4. The condensed matrix P′ is denoted by $\overline{P}'$. In general, then, $\overline{P}_2$ has order ϕ x n, with $m(m+1)/2 \geq \phi \geq m$; if $\overline{P}_2 = P$, then $\phi = m$, and, at the other extreme, if $\overline{P}_2 = P_2$, then $\phi = m(m+1)/2$.

DEF. 3.6 Let P;I denote a partitioned (0,1)-matrix of order m x (n + m), where P is a generator matrix of order m x n and I is an identity matrix of order m x m. This will be called an augmented generator matrix, and its diplo-matrix is

$$P_2; I_2 = \left|\begin{array}{c:c} P & I \\ \hdashline P' & I' \end{array}\right| \quad \text{of order } m(m+1)/2 \text{ x } (n+m).$$

DEF. 3.7 Denote the partitioned row vectors of $P_2;I_2$ by

$$p_1|q_1, p_2|q_2, \ldots p_g|q_g,$$

where $g = m(m+1)/2$. Now let $P_2;F_2$ denote a partitioned matrix of the same order as that of $P_2;I_2$, having row vectors

$$p_1|f_1, p_2|f_2, \ldots p_g|f_g,$$

such that

$$p_i|f_i = p_i|q_i \bigcup_{p_i = p_j} p_j|q_j, \qquad (i \neq j)$$

where union "over $p_i = p_j$" means that each row vector of $P_2;I_2$ is replaced by the union of itself with all other row vectors having identical p-subvectors. In the condensed matrix $\overline{P}_2;\overline{F}_2$ the submatrix $\overline{F}_2$ is called the field matrix; it has the structure

$$F_2 = \left|\begin{array}{c} F \\ \hdashline \overline{F}' \end{array}\right| \quad \text{of order } \phi \text{ x } m,$$

with the same limits for ϕ as described for $\overline{P}_2$ in Def. 3.5. The submatrix F, which again is unaffected by condensation, is called the dominance matrix.

DEF. 3.8 Let J be the sum of the 1-entries of the dominance matrix F; then $d = J - m$ is called the dominance index.

DEF. 3.9 Denoting the row vectors of $P_2;I_2$ as before (Def. 3.7), let $P_2;K_2$ denote a partitioned matrix of the same order as that of $P_2;I_2$, having row vectors

$$p_1|k_1, p_2|k_2, \ldots p_g|k_g,$$

such that

$$p_i|k_i = p_i|q_i \bigcap_{p_i = p_j} p_j|q_j, \qquad (i \neq j)$$

where intersection "over $p_i = p_j$" means that each row vector of $P_2;I_2$ is replaced by the intersection of itself with all other row vectors having identical p-subvectors. In the condensed matrix $\overline{P}_2;\overline{K}_2$ the submatrix $\overline{K}_2$ is called the kernel matrix; its partitioned structure and order are the same as for the matrix $\overline{F}_2$.

DEF. 3.10 A regular m-allele ϕ-phenotype system $(\Pi_{m.\phi})$ is a factor-union phenotype system $(\mathfrak{P}_{m.\phi})$ if its partition of genotypes is specified by the augmented diplo-matrix $(P_2;I_2)$ of at least one generator matrix (P).

DEF. 3.11 Two factor-union phenotype systems of the same degree (m) and index (ϕ) are isomorphic and belong to the same phenogram if they are equivalent under permutation of the set of m alleles.

DEF. 3.12 Let P and Q be generator matrices of orders m x n and m x n′, respectively. We define P and Q as simply isomorphic if n = n′ and there is a permutation π_r of rows of P and a permutation π_c of columns of P which together map P onto Q.

DEF. 3.13 If P and Q are not simply isomorphic (in which case we can have n = n′ or n ≠ n′) they are I-isomorphic if their augmented diplo-matrices, $P_2; I_2$ and $Q_2; I_2$, define the same phenogram.

DEF. 3.14 Let U be a (0,1)-matrix of order ϕ x n. If, under any two permutations of its rows, U contains two submatrices P and Q of orders m x n and m′ x n (m and m′ not necessarily distinct) such that

(1) P and Q are both generators of U, i.e. $\overline{P}_2 = \overline{Q}_2 = U$,

and (2) $P_2; I_2$ and $Q_2; I_2$ define different phenograms,

then P and Q are G-isomorphic matrices. In this case, we shall say that P exhibits G-isomorphy of degree m′ and Q exhibits G-isomorphy of degree m.

DEF. 3.15 Two factor-union phenograms are G-isomorphic if they are generated by P-matrices which are G-isomorphic.

DEF. 3.16 Let P be a generator matrix and let P'' denote the set of ϕ row vectors of $\overline{P}_2$. Then P is said to be a basic generator matrix if P'' is closed under the operation of pairwise union, i.e. if $P'' = \overline{P''_2}$. Representing trivial cases are those P-matrices having $\phi = m$, for we must then have $P'' = P = \overline{P}_2 = \overline{P''_2}$.

The first thirteen definitions in the above list are discussed in Section 4, and the last three are discussed in Section 8.

4. Some Examples and Theorems

The phenotype matrices: P, P′, P_2, $\overline{P}_2$

As a first example, we choose a P-matrix which every student of human heredity will quickly recognize as the generator matrix for the 3-allele ABO blood group system:

$$P = \begin{array}{c} \begin{array}{cc} a & b \end{array} \\ \left|\begin{array}{cc} 1 & 0 \\ 0 & 1 \\ 0 & 0 \end{array}\right| \end{array} \begin{array}{c} \\ A \\ B \\ O \end{array}$$

Here the rows specify the antigenic factors (a and b) which Bernstein's hypothesis attributes to the three alleles A, B, O. If we "expand" P into its diplo-matrix, taking unions of pairs of row vectors of P in the strict numerical (or lexicographic) order prescribed by Def. 3.3, we obtain a matrix (P_2) of six rows, but this condenses (cf. Def. 3.5) into $\overline{P}_2$, which exhibits the four distinct phenotypes:

$$P_2 = \left|\begin{array}{cc} 1 & 0 \\ 0 & 1 \\ 0 & 0 \\ \hline 1 & 1 \\ 1 & 0 \\ 0 & 1 \end{array}\right| \begin{array}{l} AA \\ BB \\ OO \\ AB \\ AO \\ BO \end{array} \qquad \begin{array}{r} AO, AA \\ BO, BB \\ OO \\ AB \end{array} \left|\begin{array}{cc} 1 & 0 \\ 0 & 1 \\ 0 & 0 \\ \hline 1 & 1 \end{array}\right| = \overline{P}_2$$

The phenotype-genotype matrices: P;I, P′;I′, $P_2;I_2$

An obvious defect of the matrices described above is that they display the phenotypes without any specific genetic interpretation. They do, of course, make use of the factor-union principle and suggest that we are somehow interested in pairs of factor-sets contained in P. We added gene symbols to make the exposition clear, but a computer can't operate with such symbols. To remedy the defect, we attach to P a 3 x 3 identity matrix (I) and proceed exactly as before:

$$P;I = \begin{array}{c} \begin{array}{cc:ccc} a & b & A & B & O \end{array} \\ \left|\begin{array}{cc:ccc} 1 & 0 & 1 & 0 & 0 \\ 0 & 1 & 0 & 1 & 0 \\ 0 & 0 & 0 & 0 & 1 \end{array}\right| \end{array} \qquad P_2;I_2 = \begin{array}{c} \begin{array}{cc:ccc} a & b & A & B & O \end{array} \\ \left|\begin{array}{cc:ccc} 1 & 0 & 1 & 0 & 0 \\ 0 & 1 & 0 & 1 & 0 \\ 0 & 0 & 0 & 0 & 1 \\ \hdashline 1 & 1 & 1 & 1 & 0 \\ 1 & 0 & 1 & 0 & 1 \\ 0 & 1 & 0 & 1 & 1 \end{array}\right| \end{array}$$

The matrix $P_2;I_2$ now exhibits the phenotypes in alignment with the six genotypes. This clearly tells the "whole story" about Bernstein's hypothesis and defines phenogram 3-4-1 (cf. Table 1; Cotterman, 1953, Figs. 3, 11). We might wish that the matrix somehow showed more explicitly that there are just four phenotypes with two phenosets consisting of (AA, AO) and (BB, BO). But we cannot compress the matrix $P_2;I_2$ as was done with P_2, nor would we want to drop out any genotypes. Two means of compressing the information in $P_2;I_2$ are described in Defs. 3.7 and 3.9; these give rise to

The field and kernel matrices: $\overline{F}_2$, $\overline{K}_2$

$$\overline{P}_2;\overline{F}_2 = \begin{array}{c} \begin{array}{cc:ccc} a & b & A & B & O \end{array} \\ \left|\begin{array}{cc:ccc} 1 & 0 & 1 & 0 & 1 \\ 0 & 1 & 0 & 1 & 1 \\ 0 & 0 & 0 & 0 & 1 \\ \hdashline 1 & 1 & 1 & 1 & 0 \end{array}\right| \end{array} \qquad \overline{P}_2;\overline{K}_2 = \begin{array}{c} \begin{array}{cc:ccc} a & b & A & B & O \end{array} \\ \left|\begin{array}{cc:ccc} 1 & 0 & 1 & 0 & 0 \\ 0 & 1 & 0 & 1 & 0 \\ 0 & 0 & 0 & 0 & 1 \\ \hdashline 1 & 1 & 1 & 1 & 0 \end{array}\right| \end{array}$$

The first row vector in $\overline{P}_2;\overline{F}_2$ simply says that individuals who have factor a without b can possess gene A and gene O, but not B. The first row in $\overline{P}_2;\overline{K}_2$ says, in addition, that these same individuals must possess gene A. If it be asked why we say can in the first instance and must in the second, the answer lies in the definitions of field and kernel (Section 2) and in the meaning of union and intersection (Defs. 3.7, 3.9 and Section 9). Together, the p-, f- and k-vectors assert that

$$a \cap \text{not-}b \quad \text{implies} \quad AA \cup AO.$$

Similarly, the other three rows of $\overline{F}_2$ and $\overline{K}_2$ define the phenosets without ambiguity.

Will the above principle hold for any factor-union phenotype system? Or, could the process of consolidating phenosets into fields sometimes lead to a confounding of two or more "essentially different" systems? The conjecture of Section 1 answers the first question with "yes" and the second with "no." Moreover, the conjecture states that the $\overline{F}_2$ matrix alone is sufficient for deciphering the phenogram. If the conjecture holds, then we could describe $\overline{F}_2$ as the field matrix if the interest centers on the use of fields in genetic computer analysis or parentage tests, or, alternatively, as the phenogram matrix if interest centers on the system as a whole. The reader will note, incidentally, that Definitions 3.10, 3.13 and 3.14 place no reliance on the conjecture, using instead the phenotype-genotype matrix ($P_2;I_2$) as a means of identifying the phenogram.

The deciphering of the phenogram from $\overline{F}_2$ is by no means always easy, but it is of considerable interest that this matrix seems always to contain all of the essential information. In this connection, the partition of $\overline{F}_2$ into the two submatrices, F and $\overline{F}'$, is of vital significance. The matrix F contains the fields of the m homophenosets, while $\overline{F}'$ specifies the fields of the ϕ-m heterophenosets, and, since no permutation of the set G can transform a homophenoset into a heterophenoset, the distinction between F and $\overline{F}'$ dare not be ignored. Otherwise, it can happen that two phenograms of the same degree and index might be regarded as having identical field-sets. This is illustrated by a comparison of the $\overline{F}_2$-matrices of two 4-8 phenograms:

$$4\text{-}8\text{-}4, \overline{F}_2 = \left|\begin{array}{cccc} 1 & 0 & 0 & 1 \\ 0 & 1 & 0 & 0 \\ 0 & 0 & 1 & 0 \\ 0 & 0 & 0 & 1 \\ \hline 1 & 1 & 1 & 0 \\ 0 & 1 & 1 & 0 \\ 0 & 1 & 0 & 1 \\ 0 & 0 & 1 & 1 \end{array}\right| \begin{array}{l} AA, AD \\ BB \\ CC \\ DD \\ AB, AC \\ BC \\ BD \\ CD \end{array} \qquad 4\text{-}8\text{-}9, \overline{F}_2 = \left|\begin{array}{cccc} 1 & 1 & 1 & 0 \\ 0 & 1 & 0 & 0 \\ 0 & 0 & 1 & 0 \\ 0 & 0 & 0 & 1 \\ \hline 1 & 0 & 0 & 1 \\ 0 & 1 & 1 & 0 \\ 0 & 1 & 0 & 1 \\ 0 & 0 & 1 & 1 \end{array}\right| \begin{array}{l} AA, AB, AC \\ BB \\ CC \\ DD \\ AD \\ BC \\ BD \\ CD \end{array}$$

When a linear formula is used for a field-set, a double-bar indicates the partition into F and $\overline{F}'$. Thus, for phenogram 4-8-4 we write $\mathfrak{F}$ = {A,D | B | C | D || A,B,C | B,C | B,D | C,D} and for 4-8-9 we write $\mathfrak{F}$ = {A,B,C | B | C | D || A,D | B,C | B,D | C,D}.

The conjecture of Section 1 does not imply that the permutation group will always effect a 1:1 mapping of a set of isomorphic $\mathfrak{P}$-partitions onto their $\mathfrak{F}$-sets, and, in fact, there is one 4-allele phenogram which shows that the correspondence may be a many-one relation or epimorphism. The genotype partition

4-7-15,$\mathfrak{P}$ = {AA | BB | CC | DD || AB, AC | AD | BC, BD, CD}

has 12 distinct images under |G|, these forming a set isomorphic with $|G|/H_{BC}$, where H_{BC} is the subgroup consisting of the identity permutation I and the transposition (BC). But the corresponding field-set

4-7-15,$\mathfrak{F}$ = {A | B | C | D || A, B, C | A, D | B, C, D}

has only 6 images, forming a set isomorphic with |G|/V, where V is a Klein four-group consisting of I, (AD), (BC), (AD)(BC). Thus, the field-set or $\overline{F}_2$ matrix may sometimes identify the phenogram without particularizing one image. A necessary condition here is the presence of two or more heterophenosets having the same field number (cf. Section 2) but differing in structure (cf. Section 10). Inspection of phenogram 4-6-11 will show why this is not a sufficient condition.

I-isomorphic P-matrices

The principle of I-isomorphy (Def. 3.13) can be expressed quite simply. Some fU phenotype systems can arise from essentially different P-matrices; i.e. different sets of diagnostic reagents or criteria can sometimes accomplish the same over-all partition of the set of genotypes. The first opportunity for this arises in the case of phenogram 3-6-1. Its two I-isomorphic generator matrices (cf. Table 1) and their diplo-matrices are these:

P_2 = (P-1 above the dotted line)

1	0	0
0	1	0
0	0	1
1	1	0
1	0	1
0	1	1

(P-2 above the dotted line) = P_2

1	0	1	0
0	1	0	1
0	0	1	1
1	1	1	1
1	0	1	1
0	1	1	1

Since all six phenotypes are distinct in this example, we need hardly attach I_2-matrices to observe that both P_2-matrices define the same phenogram. A more interesting example is afforded by phenogram 4-7-9. Here we will write out the $P_2;I_2$ matrices, using the matrices P-1 and P-3 of Table 1:

$P_2;I_2$ =

4-7-9, P-1

1	0	0	1	1	0	0	0
0	1	0	1	0	1	0	0
0	0	1	1	0	0	1	0
0	0	1	0	0	0	0	1
1	1	0	1	1	1	0	0
1	0	1	1	1	0	1	0
1	0	1	1	1	0	0	1
0	1	1	1	0	1	1	0
0	1	1	1	0	1	0	1
0	0	1	1	0	0	1	1

4-7-9, P-3

1	1	0	1	0	1	0	0	0
0	1	0	1	1	0	1	0	0
0	0	1	1	1	0	0	1	0
0	0	1	0	1	0	0	0	1
1	1	0	1	1	1	1	0	0
1	1	1	1	1	1	0	1	0
1	1	1	1	1	1	0	0	1
0	1	1	1	1	0	1	1	0
0	1	1	1	1	0	1	0	1
0	0	1	1	1	0	0	1	1

The genotype partition is {AA | BB | CC, CD | DD || AB | AC, AD | BC, BD} in both cases. As an exercise, the reader may wish to verify that a third matrix, 4-7-9, P-2 of Table 1, will also yield this system, and show that the matrix $\overline{F}_2$ is the same in all three cases.

I-isomorphy in the above example can be viewed graphically, using a tetrahedral diagram (cf. Cotterman, 1953). The cuts through the graphs shown below can be identified with the column vectors of the matrices P-1 and P-3. This analogy explains the choice of the term cut-set in Section 1.

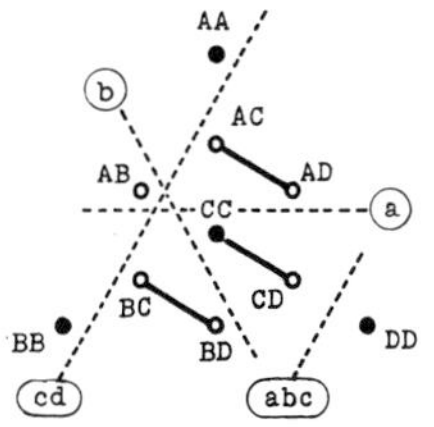

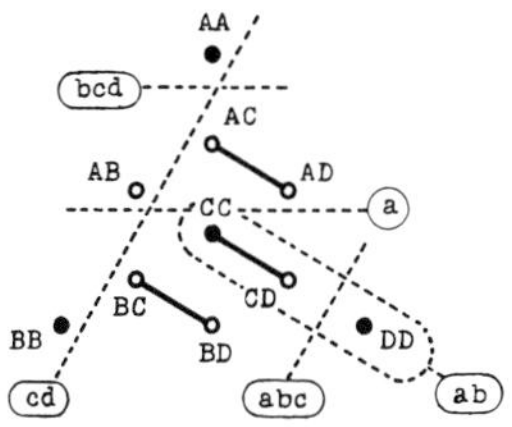

Irreducibility of matrix P_2

Two questions may be raised relative to Definition 3.4. Why is it required that the row vectors of P be distinct, and why must the diplo-matrix P_2 be column-irreducible? Would not any (0,1)-matrix yield a factor-union system?

It was an arbitrary decision to have the class fU not include degenerate phenotype systems, i.e. systems that are not discoverable except as systems of m-1 or fewer alleles (cf. Cotterman, 1953). This necessitates m distinct rows in P. Secondly, some limit must be placed on the number of columns if a combinatorial problem is desired. Geneticists may add any number of redundant columns they please, but their matrices will then reduce to one of the P-matrices of Table 1.

It should be carefully noted that the generator matrix, P, is not required to be column-irreducible. (Note that one column could be deleted from 3-6-1, P-1 and two columns from 3-6-1, P-2 without reducing the number of distinct rows of P.) It is rather the diplo-matrix, P_2, which must be column-irreducible. Can this characteristic of P be ascertained without examining P_2? Perhaps so, but I have not succeeded in finding a sufficient set of criteria. The following rule, however, is helpful in eliminating useless or redundant columns.

The matrix P_2 is not column-irreducible if P contains a column vector c_i such that

(1) $c_i = \text{col } |0,0,\ldots 0|$,

(2) $c_i = \text{col } |1,1,\ldots 1|$,

(3) $c_i = c_j$, where c_j is also a column vector of P,

(4) $c_i = c_j \cup c_k \cup \ldots$, where c_j, c_k, ... are additional column vectors of P.

As an example of the application of this rule, we may show phenogram 3-4-1 resulting from a matrix with four superfluous columns:

a	b	3	1	2	4
1	0	1	0	1	1
0	1	0	0	1	1
0	0	0	0	1	0
1	1	1	0	1	1
1	0	1	0	1	1
0	1	0	0	1	1

In this example we might identify column-3 as a duplicate anti-A reagent, column-1 as a serum lacking ABO-specificity, column-2 as a "panagglutinating" serum, and column-4 as an "anti-A+B" reagent, possibly an O-serum.

TABLE 1. Factor-union phenograms and P-matrices (m = 2, 3, 4)

Explanation of table

Column 1: Phenogram identification number, m-ϕ-s, where m = no. of alleles, ϕ = no. phenotypes, s = serial number.

Column 2: P-matrix serial number, used only when two or more I-isomorphic matrices generate the same phenogram.

Column 3: Column vectors of P or cut-sets. These subsets locate the 1-entries of the P-matrix when its rows are lettered A, B, ... Example:

4-9-1, P-3 =

	a	b	ac	bd	acd
A	1	0	1	0	1
B	0	1	0	1	0
C	0	0	1	0	1
D	0	0	0	1	1

Column 4: Ω = number of permutational images of phenogram

Column 5: d-x = dominance index and structure (cf. Fig. 1)

Column 6: m′ = degrees of G-isomorphic P-matrices

m-ϕ-s	P	Column vectors or cut-sets	Ω	d-x	m′
2-3-1	-	a b	1	0	3
2-2-1	-	a	2	1	-
3-6-1	P-1	a b c	1	0	-
	P-2	a b ac bc			4, 5, 6
3-5-1	-	a b ac	6	1	4, 5
3-5-2	-	a ab ac bc	3	0	4, 5
3-4-1	-	a b	3	2-1	4
3-4-2*	-	a ab ac	3	2-2	4
3-4-3*	-	a ab bc	6	1	4
3-4-4*	-	ab ac bc	1	0	4
3-3-1*	-	a ab	6	3	-
3-3-2*	-	ab ac	3	2-2	2

4-10-1	P-1	a b c d	1	0	-
	P-2	a b c ad bd			-
	P-3	a b c ad bcd			5
	P-4	a b c abd acd bcd			5, 6, 7
	P-5	a b ac bd acd bcd			5, 6, . . . 10
	P-6	a b ad bd abc acd bcd			5, 6, . . . 10
	P-7	a b ac bd cd			5
	P-8	a b ad bd cd abc			-
	P-9	a b ac bc bd acd			5, 6, 7
	P-10	a ab ac bd cd			5, 6, . . . 10
	P-11	a ab ac bc bd acd			5, 6, . . . 10
	P-12	ab ac ad bc bd			5, 6, . . . 10
4-9-1	P-1	a b c ad	12	1	-
	P-2	a b c abd acd			5, 6
	P-3	a b ac bd acd			5, 6, 7, 8, 9
	P-4	a b ac bc abd acd			5, 6, 7, 8, 9
	P-5	a b ac bc acd bcd			5, 6, 7, 8, 9
	P-6	a b ac bc cd			-
	P-7	a b ac bc ad			5, 6
	P-8	a ab ac bc bd			5, 6, 7, 8, 9
4-9-2	P-1	a b ac ad cd	12	0	-
	P-2	a b ac cd abd			5
	P-3	a b ac ad bcd			5, 6
	P-4	a b ac abd acd bcd			5, 6, 7, 8, 9
	P-5	a ab bc bd cd			-
	P-6	a ab bc cd abd			5, 6, 7, 8, 9
	P-7	a ab bc bd acd			5, 6, 7, 8, 9
	P-8	a ab ac bd acd bcd			5, 6, 7, 8, 9
	P-9	a ab ac bc abd acd bcd			5, 6, 7, 8, 9
	P-10	ab ac bc cd abd			5, 6, 7, 8, 9
4-9-3	-	ab ac bd cd	3	0	5, 6, 7, 8, 9
4-8-1	P-1	a b c abd	12	2-2	5
	P-2	a b ac bc abd			5, 6, 7, 8
	P-3	a b ac bc acd			5, 6, 7, 8
4-8-2	-	a b ac bd	12	2-3	5, 6, 7, 8
4-8-3	P-1	a b ac cd	24	1	
	P-2	a b ac acd bcd			5, 6, 7, 8
	P-3	a ab bc cd			5, 6, 7, 8
	P-4	a ab ac bd bcd			5, 6, 7, 8

Table 1 continued

m-ϕ-s	P	Column vectors or cut sets	Ω	d-x	m′
4-8-4	P-1	a ab ac bd acd	12	1	5, 6, 7, 8
	P-2	a ab ac bc abd acd			5, 6, 7, 8
	P-3	ab ac bc cd			5, 6, 7, 8
4-8-5	-	a ab ac bc abd bcd	24	1	5, 6, 7, 8
4-8-6	-	a b ac abd bcd	24	1	5, 6, 7, 8
4-8-7	P-1	a b cd abc abd	6	0	5
	P-2	a b abc abd acd bcd			5, 6, 7, 8
	P-3	a ab cd abc abd bcd			5, 6, 7, 8
4-8-8	-	a ab bc abd acd bcd	24	0	5, 6, 7, 8
4-8-9	P-1	a b ac ad	12	2-1	5
	P-2	a b ac abd acd			5, 6, 7, 8
	P-3	a ab bc bd			5, 6, 7, 8
4-8-10	P-1	a bc bd cd	4	0	-
	P-2	a bc cd abd			5, 6, 7, 8
	P-3	ab ac bc abd acd bcd			5, 6, 7, 8
4-8-11	P-1	a ab ac abd acd bcd	4	0	5, 6, 7, 8
	P-2	ab ac ad bcd			5, 6, 7, 8
4-8-12	-	ab bc cd abd acd	12	0	5, 6, 7, 8
4-7-1	P-1	a b c	4	3-3	-
	P-2	a b ac bc			5, 6, 7
4-7-2	-	a b ac bcd	24	2-2	5, 6, 7
4-7-3	-	a ab ac bd	24	2-3	5, 6, 7
4-7-4	-	a ab ac bc abd	24	2-2	5, 6, 7
4-7-5	-	a ab ac bc bcd	12	2-2	5, 6, 7
4-7-6	-	a ab bc acd bcd	24	1	5, 6, 7
4-7-7	-	a ab cd abc abd	12	1	5, 6, 7
4-7-8	-	a ab bc abd acd	24	1	5, 6, 7
4-7-9	P-1	a b cd abc	12	1	-
	P-2	a b abc acd bcd			5, 6, 7
	P-3	a ab cd abc bcd			5, 6, 7
4-7-10	-	a b ac acd	24	3-1	5, 6, 7
4-7-11	-	a b ac abd	24	3-4	5, 6, 7
4-7-12	-	a b abc abd acd	12	2-1	5, 6, 7
4-7-13	-	a ab bc abd bcd	24	2-1	5, 6, 7
4-7-14	-	ab ac bc abd acd	12	1	5, 6, 7
4-7-15	-	a bc abd acd bcd	12	0	5, 6, 7
4-7-16	-	a ab ac abd bcd	24	1	5, 6, 7
4-7-17	-	a ab abc abd acd bcd	12	0	5, 6, 7
4-7-18	-	ab ac cd abd	24	1	5, 6, 7
4-7-19	P-1	a ab ac abd acd	4	3-2	5, 6, 7
	P-2	ab ac ad			4, 5, 6, 7]†
4-7-20	-	a bc cd	12	2-1	4, 5, 6, 7]†
4-7-21	-	ab ac abd acd bcd	12	0	5, 6, 7
4-7-22	-	ab cd abc abd acd bcd	3	0	5, 6, 7
4-6-1	-	a b ac	24	4-3	5, 6
4-6-2	-	a ab ac bc	12	3-3	5, 6
4-6-3	-	a ab bc acd	24	2-2	5, 6
4-6-4	-	a ab cd abc	24	2-3	5, 6
4-6-5	-	a ab bc abd	24	3-4	5, 6
4-6-6	-	a b abc acd	24	3-1	5, 6
4-6-7	-	a ab bc bcd	24	3-1	5, 6
4-6-8	-	a ab ac bcd	12	2-2	5, 6
4-6-9	-	a ab abc acd bcd	24	1	5, 6
4-6-10	-	ab ac bc abd	12	2-2	5, 6
4-6-11	-	a bc abd acd	12	1	5, 6
4-6-12	-	ab ac cd	12	2-3	5, 6
4-6-13	-	a b abc abd	6	4-1	5, 6
4-6-14	-	a ab abc abd bcd	12	2-1	5, 6
4-6-15	-	a abc abd acd bcd	4	0	5, 6
4-6-16	-	a ab ac abd	24	4-2	5, 6
4 6-17	-	a ab abc abd acd	12	3-2	5, 6
4-5-18	-	ab ac abd bcd	24	1	5, 6
4-6-19	-	ab cd abc abd acd	12	1	5, 6
4-6-20	-	a bc abd bcd	24	2-1	3, 4, 5, 6]†
4-6-21	-	ab ac abd acd	12	3-2	3, 4, 5, 6]†
4-6-22	-	ab abc abd acd bcd	6	0	5, 6
4-5-1	-	a b abc	12	5-3	5
4-5-2	-	a ab ac	12	5-1	5
4-5-3	-	a ab bc	24	4-3	5
4-5-4	-	ab ac bc	4	3-3	5
4-5-5	-	a ab abc bcd	24	3-1	5
4-5-6	-	a ab abc acd	24	4-2	5
4-5-7	-	ab ac abd	24	4-2	3, 4, 5]†
4-5-8	-	a bc abd	24	3-4	3, 4, 5]†
4-5-9	-	ab ac bcd	12	2-2	5
4-5-10	-	ab cd abc acd	12	2-3	5
4-5-11	-	a ab abc abd	12	5-2	5

Table 1 concluded

m-ϕ-s	P	Column vectors or cut-sets	Ω	d-x	m′
4-5-12	-	a abc abd acd	4	3-2	5
4-5-13	-	ab abc acd bcd	12	1	5
4-5-14	-	a abc abd bcd	12	2-1	3, 4, 5 †
4-5-15	-	ab abc abd acd	12	3-2	3, 4, 5 †
4-5-16	-	abc abd acd bcd	1	0	5
4-4-1	-	a ab abc	24	6	-
4-4-2	-	ab ac	12	5-1	3
4-4-3	-	a abc abd	12	5-2	3
4-4-4	-	ab abc acd	24	4-2	3
4-4-5	-	abc abd acd	4	3-2	3

*In order to have factor-union phenograms numbered consecutively, it was necessary to assign new serial numbers to some three-allele phenograms (cf. Cotterman, 1953).

†G isomorphic phenograms of the same degree (m = m′)

TABLE 2. Numbers of phenograms of degree m, index ϕ, for factor-union systems, N($\mathfrak{P}$), and total rps systems, N(Π)

ϕ	m = 2 N($\mathfrak{P}$)	m = 2 N(Π)	m = 3 N($\mathfrak{P}$)	m = 3 N(Π)	m = 4 N($\mathfrak{P}$)	m = 4 *N(Π)
1	-	1	-	1	-	1
2	1	2	-	9	-	45
3	1	1	2	21	-	488
4	-	-	4	16	5	1604
5	-	-	2	4	16	1961
6	-	-	1	1	22	1071
7	-	-	-	-	22	301
8	-	-	-	-	12	48
9	-	-	-	-	3	5
10	-	-	-	-	1	1
	2	4	9	52	81	5525

*Data of Hartl and Maruyama (1968), the total confirming a calculation of Bennett (1957). Other data from Cotterman (1953, 1966).

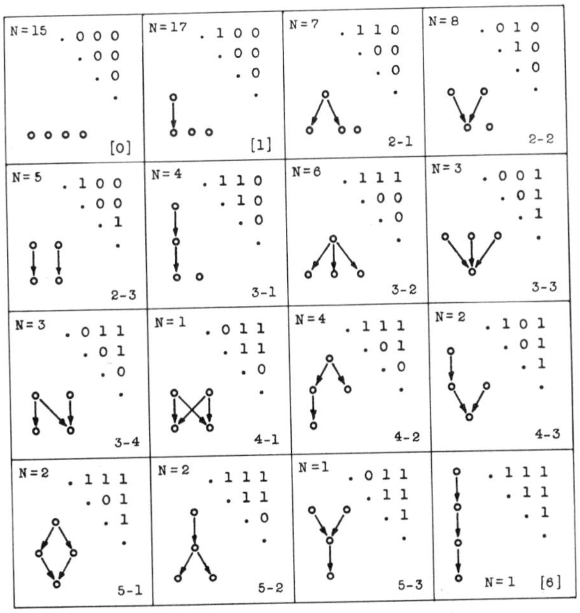

Figure 1. Dominance relation structures in four-allele factor-union systems. Structures are identified by numbers (d-x) at lower right, d being the dominance index, x being a serial number. N is the number of four-allele fU phenograms exhibiting the structure.

To show that the rule is insufficient, however, note that it does not disqualify either of the two following matrices, neither of which is a P-matrix. We leave it as an exercise to prove this by writing the diplo-matrix in each case. What should be done to each matrix to convert it into a P-matrix yielding a phenogram of the highest possible index (ϕ)? Verify by finding the matrix in Table 1.

1 1 1 1 0	1 0 1 0 1 0
0 1 0 0 1	1 1 0 0 0 1
0 0 1 0 1	0 0 0 1 1 1
0 0 0 1 1	0 1 1 1 0 0

Apart from the stipulation that the rows of P be distinct, there are no constraints upon the rows of the generator matrix. The m alleles are therefore free to dictate any combinations of the n phenotypic factors. Special interest attaches to those P-matrices which incorporate a null row vector (cf. Section 6) and a unit row vector (cf. Section 8). If P contains row vectors p_i, p_j, p_k such that $p_i = p_j \cup p_k$, then the phenotype system includes a null-kernel homophenoset (cf. Theorem 4.3).

Some theorems characterizing fU systems

From the definitions of Section 3, we can deduce some theorems concerning fU systems which, although they will not substitute for the Definition 3.10, are useful characterizations.

In every factor-union system:

THEOREM 4.1: All homozygotes have distinct phenotypes.

THEOREM 4.2: Dominance ($\rightarrow$) is a transitive relation.

THEOREM 4.3: If AA$\mathfrak{Z}$BC, then A $\rightarrow$ B and A $\rightarrow$ C.

THEOREM 4.4: The fields of all phenotypes are distinct.

Proof: The sign '$\mathfrak{Z}$' stands for the equivalence relation of phenotypic identity. Using p(A) and p(B) for the p-vectors of alleles A and B, we then note the following equivalences:

$$AA\mathfrak{Z}AB \equiv [p(A) \cup p(A) = p(A) \cup p(B)] \equiv p(B) \subseteq p(A)$$

But, from Defs. 3.10 and 3.4, $p(B) \neq p(A)$, therefore BB$\bar{\mathfrak{Z}}$AA, for all (A, B); this is Theorem 4.1. If $p(B) \subseteq p(A)$ and $p(B) \neq p(A)$, then $p(B) \subset p(A)$. The classical meaning of dominance is $(A \rightarrow B) \equiv (AA\mathfrak{Z}AB) \cap (AA\bar{\mathfrak{Z}}BB)$. Therefore $(A \rightarrow B) \equiv p(B) \subset p(A)$, and the transitivity of set inclusion ($\subset$) then assures the transitivity of dominance ($\rightarrow$); this is Theorem 4.2. Further, if AA$\mathfrak{Z}$BC, then $p(A) \cup p(A) = p(A) = p(B) \cup p(C)$, which implies $p(B) \subset p(A)$ and $p(C) \subset p(A)$, since $p(A) \neq p(B)$, $p(A) \neq p(C)$ and $p(B) \neq p(C)$. Therefore AA$\mathfrak{Z}$BC implies A $\rightarrow$ B and A $\rightarrow$ C; this is Theorem 4.3. Theorem 4.4 was proved in Section 1 (Theorem 1.1).

These rules enable us to identify many regular phenotype systems that are not factor-union systems. Theorem 4.1 alone disqualifies two of the two-allele phenograms (2-2-2 and 2-1-1):

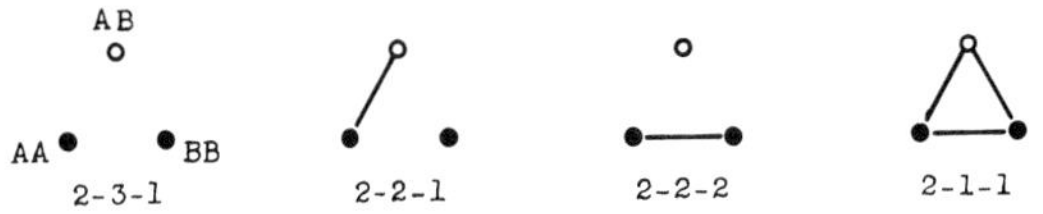

2-3-1 2-2-1 2-2-2 2-1-1

In fU systems, therefore, the relation between any two alleles is that of orthotaxy (A < > B) or dominance (A → B or B → A). Metataxy and parataxy (which includes iso-allelism) are excluded (cf. Cotterman, 1953, for definitions and explanation of the phenogram graphs).

Theorems 4.1, 4.2 and 4.3 together are sufficient for excluding all non-fU systems from the catalogue of 52 three-allele phenograms. Note which theorem excludes each of the following:

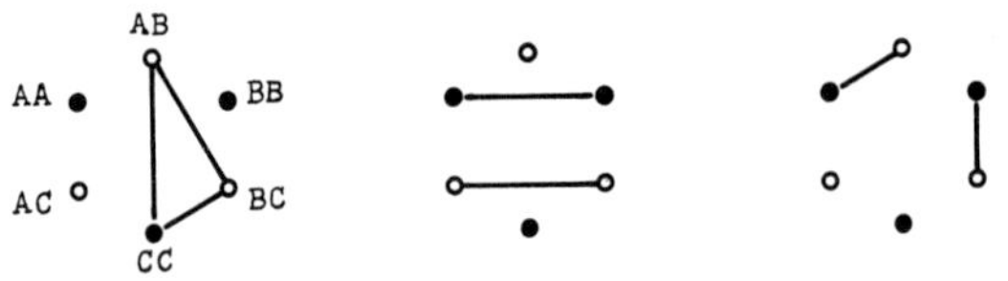

Passing to four-allele systems, we find non-fU systems that cannot be detected by the first three theorems, such as the following:

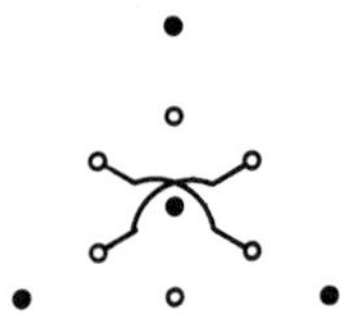

(I)

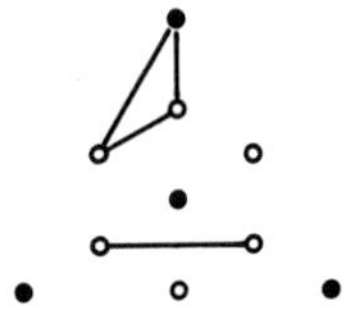

(I I)

System I is excluded by Theorem 4.4, since the two heterophenosets (AB, CD) and (AD, BC) have identical fields. But system II eludes all four theorems. Why is it, in fact, not factor-union? Because cuts through the tetrahedral diagram can only be of three kinds, those removing a vertex, an edge, or a face, and no combination of such will result in the partition shown. This is indeed the only sufficient rule. We could probably define fU systems in terms of cuts through an (m-1)-dimensional polytope, but this possibility will not be explored here.

An eight-allele system would hardly be considered complex by specialists in the field of cattle blood groups. Yet the case m = 8 is quite enough to show the complexity of our combinatorial problem. Do the following partitions represent factor-union systems? I know that one does, because it was generated by an 8 x 4 P-matrix (cf. Section 10). The other was concocted in such a way that it is not disqualified by Theorems 4.1-4.4. What would be your procedure for checking?

(1) { AA, AB, AF, AG, AH | BB, BG | CC, BC, CD, CE, CG, DE | DD | EE | FF, FG, FH | GG | HH | | DG | AC, AD, CF, DF, EF | AE | BD, BE | BF, BH | CH, DH, EG, EH, GH }

(2) { AA, AB, AD, AF, BD | BB, BF | CC, CE | DD, DF | EE | FF | GG, CG, CH, EG, GH | HH, EH | | EF | CD, CF, DE | AC, AE, AG, AH, BC, BG, DG, DH, FG | BE, BH, FH }

<u>Dominance relation structures</u>

It has been noted that very few regular phenotype systems can be uniquely described in terms of two-allele relations, such as dominance (Cotterman, 1953). Nevertheless, dominance relations tell part of the story about any fU system, and this part is contained in the submatrix F of the field matrix $\overline{F}_2$. Any relation on an m-set can be described by a (0, 1)-matrix of order m x m, and, if the relation is asymmetric and transitive, it is always possible to permute rows and columns simultaneously so as to have the matrix triangular. The P-matrices of Table 1 have their columns arrayed in a way that makes the dominance matrix (F) upper-triangular in every case.

Two relations on an m-set are said to have the same <u>structure</u> if they are equivalent under permutation of the rows and columns of the relation matrix (Davis, 1953). A dominance relation structure can also be portrayed by an unlabelled directed graph. Figure 1 shows the matrices and graphs of the 16 dominance structures possible in four-allele fU systems. Note that the dominance index (cf. Def. 3.8) cannot adequately summarize the dominance relations, and the dominance structure seldom uniquely describes the phenogram; for the latter we also need information concerning the heterophenosets, which is contained in $\overline{F}'$.

5. <u>P-Matrix Series</u>

There are a number of P-matrices which, because of certain symmetries that they display, deserve special consideration. For successive values of m, these matrices define series of phenograms for which the beginning member in each case is either phenogram 2-3-1 (two alleles without dominance) or 2-2-1 (two alleles with dominance). For each matrix series, $P^{(m)}$, we can write the phenogram identification number, m-ϕ-s, as a "general formula" comparable to those that characterize series of organic chemical compounds. And, as is frequently true of chemical series and of some algebraic series, the beginning member is often seen to be rather atypical.

Six such series will be considered here, and we illustrate each matrix for the case m = 4:

1 0 0 0	0 1 1 1	1 1 1	1 0 0	1 1	1 1 1
0 1 0 0	1 0 1 1	0 1 1	0 1 0	1 0	0 1 1
0 0 1 0	1 1 0 1	1 0 1	0 0 1	0 1	0 0 1
0 0 0 1	1 1 1 0	1 1 0	0 0 0	0 0	0 0 0
$I^{(4)}$	$J^{(4)}$	$L^{(4)}$	$A^{(4)}$	$C^{(4)}$	$D^{(4)}$

<u>Series $I^{(m)}$</u>

Here the generator matrix, P, is an identity matrix, $I^{(m)}$, of order m x m, which obviously makes the correspondences between genotypes and phenotypes 1:1. Ordinarily, this system might be described as one showing "complete lack of dominance," but for m = 4 there are no less than 15 fU phenograms which fit this description, having d = 0 (cf. Fig. 1 and Table 1). The phenogram series includes 2-3-1, 3-6-1, 4-10-1, etc., the general formula being m-g-1, where g = m(m+1)/2.

The beginning matrix in the series, $I^{(2)}$, is atypical in the following respects:

(1) $I^{(2)}$ is simply isomorphic with its complement

$$I^{(2)} = \begin{vmatrix} 1 & 0 \\ 0 & 1 \end{vmatrix} \simeq \begin{vmatrix} 0 & 1 \\ 1 & 0 \end{vmatrix} = \sim I^{(2)},$$

but this is not true for m $\geq$ 3; cf. series $J^{(m)}$.

(2) $I^{(2)}$ has no I-isomorphic P-matrix. For all m $\geq$ 3, however, there is always at least one I-isomorphic matrix for $I^{(m)}$ (Table 1).

(3) $I^{(2)}$ has a G-isomorphic P-matrix, namely $L^{(3)}$, which defines phenogram 3-3-2 (cf. Section 8). But for $m \geq 3$, $I^{(m)}$ possesses no G-isomorph.

Series $J^{(m)}$

This matrix is simply the complement of $I^{(m)}$, of order m x m. The phenogram series includes 2-3-1, 3-4-4, 4-5-16, etc., the general formula being m-(m+1)-s, where s is the number of m-allele fU phenograms of index (m + 1). The distinguishing feature is a heterophenoset incorporating all m(m − 1)/2 heterozygotes; this is the opposite extreme for the case of "no dominance" (d = 0). Unlike the case for $I^{(m)}$, the matrix $J^{(m)}$ never has an I-isomorph, but always has a G-isomorph (cf. Table 1).

Series $L^{(m)}$

The matrix $L^{(m)}$ is equivalent to $J^{(m-1)}$ augmented with a unit row vector $|1,1,\ldots 1|$; the order is m x (m − 1). The phenogram series includes 2-2-1, 3-3-2, 4-4-3, etc., the general formula being m-m-(m − 1). The distinguishing feature is a single complex homophenoset containing all heterozygotes. $L^{(m)}$ naturally has no I-isomorph, but $L^{(m)}$ is G-isomorphic with $J^{(m-1)}$ for all $m \geq 3$. The case of G-isomorphy for 2-3-1 and 3-3-2 is discussed in Section 8.

Series $A^{(m)}$

This matrix is the complement of $L^{(m)}$, of order m x (m-1). The phenogram series includes 2-2-1, 3-4-1, 4-7-1, 5-11-1, etc., the general formula being m-(h + 1)-1, where h = m(m − 1)/2. Because $A^{(3)}$ generates phenogram 3-4-1, the higher systems with $m \geq 4$ have been designated as ABO-like systems by Yasuda and Kimura (1968). In some respects, however, neither $A^{(2)}$ nor $A^{(3)}$ are typical members of the series. I-isomorphy is exhibited only by $A^{(m)}$ matrices with $m \geq 4$. For m = 2 and $m \geq 4$, $A^{(m)}$ has no G-isomorph, but $A^{(3)}$ is G-isomorphic with $C^{(4)}$.

Series $C^{(m)}$

The matrix $C^{(m)}$ is a P-matrix of order m x n, where $m = 2^n$; this completely specifies $C^{(m)}$, since Def. 3.4 demands that all rows be distinct and therefore these exhibit all combinations of the n factors. We have called this the power-set matrix or allele-complete matrix (Cotterman, 1966). Note that m can take only the values m = 2, 4, 8, etc., and that $C^{(m)}$ is simply isomorphic with its complement. The phenogram series includes 2-2-1, 4-4-2, 8-8-3, etc., the general formula being m-m-s, where $s = n = \log_2(m)$. $C^{(m)}$ has no I-isomorph, but G-isomorphy is a feature of all $C^{(m)}$ matrices for $m \geq 4$. The G-isomorphic systems 3-4-1 and 4-4-2 have some historical significance in human genetics, as will be related in Section 8.

Series $D^{(m)}$

This is a matrix of order m x (m − 1) in which element $x_{ij} = 1$ for all $i < j$ and $x_{ij} = 0$ otherwise. The phenogram series includes 2-2-1, 3-3-1, 4-4-1, etc., the general formula being m-m-1. Here the relation between any pair of alleles is one of dominance, so the dominance index takes its maximum value of d = m(m − 1)/2. For regular phenotype systems in general, a system of this kind was called a dominantly-connected system (Cotterman, 1953), and, since dominance in general is non-transitive, there can be many dominantly-connected phenograms for $m \geq 3$. But, owing to transitivity of dominance in fU systems (cf. Theorem 4.2), there is, for each m, only one dominantly-connected fU system, m-m-1.

The matrix $D^{(m)}$ is isomorphic with its complement and has no I-isomorph and no G-isomorph for any value of m. It is truly a "unique" system, and so we assign to it the serial number s = 1. The system is commonly observed in Drosophila and in mammalian and plant genetics, where the polymorphisms studied are chiefly concerned with morphological and pigmentary variations which are truly phenotypic in the sense of being observed by eye. Since the alleles are totally ordered by dominance in the system m-m-1, one may here refer to "multiple-allelic series" using the sign '>' for dominance. Workers in these fields are less apt to become intrigued by the combinatorics of phenotype systems than are those concerned with serological or biochemical systems.

When we regard m as fixed and ask how a given fU system may evolve into systems of higher index (ϕ) through acquisition of new diagnostic tests or "factors," we are led to consider a relation of P-matrix inclusion defined as follows.

DEF. 5.1 Let T(m) denote the set of all anisomorphic m-allele P-matrices, let P_i and P_j be elements (matrices) in T(m), and let C_i and C_j be their corresponding sets of column vectors. We define P_i as a generator submatrix of P_j and write $P_i \lhd P_j$ if, allowing permutations of rows and columns of P_i, C_i is a proper subset of C_j.

DEF. 5.2 A generator matrix P_i is a minimal generator matrix if there does not exist $P_j \in T(m)$ such that $P_j \lhd P_i$; similarly, P_i is a maximal generator matrix if there does not exist $P_j \in T(m)$ such that $P_i \lhd P_j$.

Not every column-submatrix or supermatrix of a P-matrix is a P-matrix, since the submatrix may have fewer than m distinct rows and supermatrix $\overline{E}_2$ may not be column-irreducible. So the relation '$\lhd$' is defined on a set of P-matrices. This is clearly an asymmetric and transitive relation, defining a hierarchy or partial order in T(m). The following diagram illustrates the relation in T(3), the set of ten 3-allele P-matrices (cf. Table 1). Of these, 3 are maximal and 3 are minimal P-matrices.

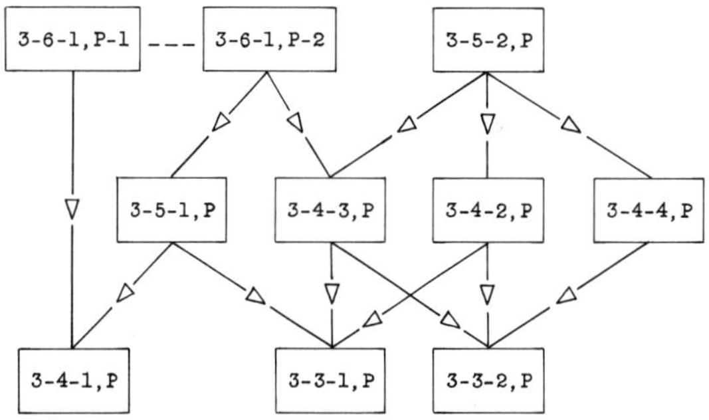

Since $\overline{P}_2$ is always column-irreducible, elimination of any column (factor) necessarily yields a system of lower index, i.e., a reduced number of phenotypes. But P itself may be column-irreducible, in which case it is a minimal m-allele generator matrix; elimination of any column here results in a system of lower degree, i.e., a reduced number of alleles. The following are minimal P-matrices:

2-2-1, P	3-3-1, P	4-4-1, P	4-4-4, P	4-5-8, P
	3-3-2, P	4-4-2, P	4-4-5, P	4-6-1, P
	3-4-1, P	4-4-3, P	4-5-1, P	4-7-1, P-1

Such matrices are either self-complementary (i.e. isomorphic with their complements) or else they come in dual pairs with respect to complementarity. For example, 3-3-1, P is self-complementary, but 3-3-2, P and 3-4-1, P are complements of each other. Having observed why this is so, the reader may wish to consider two related questions: If P and its complement are both P-matrices, are they necessarily minimal matrices? Does every P-matrix have a complement which is a P-matrix?

6. Enumeration of Phenograms and P-Matrices

Regular m-allele phenograms of the unrestricted class have been shown by Bennett (1957) and by Hartl and Maruyama (1968) to be countable with the aid of rather complex counting theorems. With the added restriction of the factor-union principle one might expect the enumeration of fU phenograms to be even more difficult. There are also more parameters of interest in the restricted case. For example, one can count fU phenograms by m, by ϕ, by dominance index and dominance structure, by G-isomorphy classes, by numbers of factors, numbers of I-isomorphic matrices, etc. Some tabulations of these kinds are presented elsewhere (Cotterman, 1966), along with a catalogue of fU phenogram graphs for $m \leq 4$.

Here we present a compilation of the P-matrices for fU systems with two, three and four alleles (Table 1). From P, one can easily write $P_2; I_2$, which defines the phenogram. There are 2 + 9 + 81 phenograms in the table, and 2 + 10 + 126 P-matrices, the latter being more numerous because of I-isomorphy.

Table 1 was prepared by a method that should be exhaustive. The columns of a P-matrix can be viewed as an n-subset of $\mathcal{P}'(G) = \{\mathcal{P}(G) - G - \emptyset\}$, where $\mathcal{P}(G)$ is the power-set of the m-set G. We therefore start with a catalogue of graphs representing the permutationally distinct subsets of $\mathcal{P}'(G)$. These bear a close analogy to the "simplicial complexes" of topology, for which there are apparently no known enumerative formulas. But the enumeration of P-matrices entails two further complications: we must eliminate matrices which are not row-distinct and those whose diplo-matrices are not column-irreducible. Finally, we obtain the fU phenograms after classification of the P-matrices into I-isomorphy sets.

Table 2 compares the numbers of fU phenograms of degree m and index ϕ with the corresponding numbers of regular (unrestricted) phenograms for two, three and four alleles.

A completely systematic ordering of phenograms could not be attempted, but some conventions for the assignment of serial numbers were mentioned in Section 5. We now call attention to a further principle utilized in Table 1.

DEF. 6.1 A factor-union phenogram is a <u>Θ-phenogram</u> if G contains a "silent" or "bottom recessive" allele, i.e. one having a null p-vector, $\check{p} = |0,0,\ldots 0|$.

THEOREM 6.1 The number of Θ-phenograms of degree m + 1, index $\phi + 1$ and dominance index d + m is equal to the number of fU phenograms of degree m, index ϕ and dominance index d.

<u>Proof:</u> Let P be any m x n generator matrix defining a fU phenogram of degree m, index ϕ and dominance index d. We will write $P = P_\theta$ if $\check{p} \in P$ and $P = P_1$ if $\check{p} \notin P$. Augmenting P_1 with a null row vector of order 1 x n gives a matrix $P_1/\check{p}$ of order (m + 1) x n. Also, after bordering P_θ with a unit column vector $|1,1,\ldots 1|$ of order m x 1, we may construct an augmented matrix $P_\theta/\check{p}$ of order (m + 1) x (n + 1). In both cases, the augmented matrix defines a Θ-phenogram of degree m + 1, index $\phi + 1$ and dominance index d + m. Two such augmented matrices are simply isomorphic (cf. Def. 3.12) if, and only if, they are derived from simply isomorphic P-matrices, and this is equally true with respect to I-isomorphy (cf. Def. 3.13). There is therefore a 1:1 correspondence between the set of m-allele fU phenograms and the set of (m + 1)-allele Θ-phenograms. This principle was utilized in the assignment of phenogram serial numbers, and the reader will find it instructive to compare the $\overline{P}_2; \overline{F}_2$ matrices for the following phenograms:

fU phenograms	Θ-phenograms
3-6-1, d = 0	4-7-1, d = 3
3-5-1, d = 1	4-6-1, d = 4
3-5-2, d = 0	4-6-2, d = 3
3-4-1, d = 2	4-5-1, d = 5
3-4-2, d = 2	4-5-2, d = 5
3-4-3, d = 1	4-5-3, d = 4
3-4-4, d = 0	4-5-4, d = 3
3-3-1, d = 3	4-4-1, d = 6
3-3-2, d = 2	4-4-2, d = 5

7. Interpretation of Phenotype Matrices

Matrices figure so prominently in this subject that some general remarks seem desirable on the interpretation of phenotypes that are expressible in matrix form. We shall comment briefly on the following points:

1. Postulation of a factor-union phenotype system does not require the presence of a phenotype matrix of the (0,1) type or its equivalent.

2. A phenotype matrix consistent with the factor-union principle does not logically exclude the possibility of "hybrid substances" or other properties possessed exclusively by heterozygotes.

3. Even the P-matrix itself is generally capable of a multiplicity of interpretations, according to whether one ascribes complexity in the matrix to the genes or gene-products, to the diagnostic reagents or test criteria, or to both.

One reason for the alternative definition presented in Section 1 was to show that a matrix of phenotypic "factors" is not essential to the definition of a factor-union system. A fU partition can be defined either in terms of sets of observed factors or in terms of subsets of the gene-set G itself. The latter, which we called cut-sets, bear a 1:1 correspondence to the column vectors of a P-matrix, and could be regarded as constituting a kind of phantom matrix.

Although there are no "factors" in evidence, the following systems are factor-union. One consists of four leaf-pigmentation types in Coleus, studied by Boye (1941); the phenogram is 3-4-1. The other is a hypothetical system which we could suppose has been found to account for six flower colors in some plant species.

PP, Pp	purple	R^1R^1, R^1R^2	violet
p^Gp^G, p^Gp	green	R^2R^2	blue
pp	pattern	R^3R^3, R^3R^4	red
Pp^G	grey	R^4R^4	pink
		R^1R^3, R^1R^4, R^2R^3	plum
		R^2R^4	lilac

The hypothetical system conforms to the rules stated in Theorems 4.1-4.4, but this does not guarantee that the system is factor-union. To settle the issue, we first note the dominance relations: $R^1 \rightarrow R^2$, $R^3 \rightarrow R^4$. This corresponds to dominance structure 2-3 of Figure 1, and Table 1 shows that only two 4-6 fU systems have this structure, namely 4-6-4 and 4-6-12. The latter fits the genotype partition shown, as we see from the $\bar{P}_2; \bar{F}_2$ matrix:

$$4\text{-}6\text{-}12, \bar{P}_2; \bar{F}_2 = \left|\begin{array}{ccc:cccc|l} 1 & 1 & 0 & 1 & 1 & 0 & 0 & R^1R^1, R^1R^2 \\ 1 & 0 & 0 & 0 & 1 & 0 & 0 & R^2R^2 \\ 0 & 1 & 1 & 0 & 0 & 1 & 1 & R^3R^3, R^3R^4 \\ 0 & 0 & 1 & 0 & 0 & 0 & 1 & R^4R^4 \\ \hdashline 1 & 1 & 1 & 1 & 1 & 1 & 1 & R^1R^3, R^1R^4, R^2R^3 \\ 1 & 0 & 1 & 0 & 1 & 0 & 1 & R^2R^4 \end{array}\right.$$

Although the matrix $\bar{P}_2$ is artificial, it is likely to interest the plant geneticist who may wish to supplement his subjective classification of flower colors with histological or biochemical studies. The matrix could conceivably serve as a scoring system for certain statistical studies. It also suggests an alternative notation for the four alleles which has the advantage that the phenotype system is completely revealed in the gene symbols themselves. This is shown in the second scheme below.

	ab	ac	cd
R^A	1	1	0
R^B	1	0	0
R^C	0	1	1
R^D	0	0	1

(i)

	a	b	c
R^{AB}	1	1	0
R^A	1	0	0
R^{BC}	0	1	1
R^C	0	0	1

(ii)

	ab	ac	d
R^A	1	1	0
R^B	1	0	0
R^{CD}	0	1	1
R^D	0	0	1

(iii)

In the construction of Table 1 we needed a linear formula for each P-matrix, which would allow the compact tabulation of many such matrices. Two simple alternatives are shown in (i) and (ii) above. In (i) the columns are described in terms of the rows, the latter being distinguished by single-letter designations. In (ii) the reverse procedure is followed. There are additional schemes, illustrated by (iii), in which compound designations are allowed in both rows and columns. We chose scheme (i) not because it suggests that the genes are "simple," but for reasons of convenience. This standardized the designations for alleles in Table 1 and more clearly reveals the number of columns, n, for each matrix. It also agrees with the notation for cut-sets in Section 1 and the method of matrix enumeration described in Section 6.

These notations can be regarded as equivalent codes for reconstructing the information in the phenotype matrix; they need not carry any implications about the nature of the gene-functions or gene-products nor about the nature of the diagnostic criteria or reagents. Frequently there is insufficient knowledge to choose between these notations along theoretical lines. In serological studies, for example, the matrix of agglutination or hemolysis reactions may constitute the whole of the information available. In such cases the author feels that it is wise to avoid references to imagined antigenic "substances" and antibody "components" and to select whichever notation for genes and reagents seems most convenient.

If a phenotype-genotype matrix is consistent with the factor-union principle, this does not necessarily mean that the genes are "acting independently" in heterozygotes, at least at the level of the observed phenotypic factors. The matrix of phenogram 2-3-1 will suffice for our illustration. Suppose that a chemist were told that three solutions (A, B, C) when tested with two reagents (X, Y) produced the following "phenotypes":

	X	Y
A	1	0
B	0	1
C	1	1

1 = white precipitate
0 = no precipitate

Asked for his interpretation, we imagine that the chemist would say that, without further information, several easy interpretations are possible, including the following.

	A	B	C	X	Y
(1)	α	β	α, β	anti-α	anti-β
(2)	α	β	γ	{anti-α, anti-γ}	{anti-β, anti-γ}
(3)	α	β	γ	anti-$\alpha\cup\gamma$	anti-$\beta\cup\gamma$

According to (1), we can assume that C shares substance α with A and substance β with B, these reacting to form precipitates with anti-α of X and anti-β of Y. Alternatively, if C contains a unique substance γ, we could suppose X and Y to have compositions shown in (2) or in (3), among other possibilities. Here "anti-$\alpha\cup\gamma$" signifies a substance yielding white precipitates with both α and γ. Dr. O. Smithies suggested a set of solutions which, at suitable concentrations, illustrate case (3) very nicely:

$$\alpha = Na_2SO_4 \qquad \text{anti-}\alpha\cup\gamma = Ba(NO_3)_2$$
$$\beta = NaOH$$
$$\gamma = Na_2CO_3 \qquad \text{anti-}\beta\cup\gamma = Zn(NO_3)_2$$

The situation is, I believe, in no significant way altered if we are given the additional information that A, B and C represent products (e.g. urines, leukocytes, tissue extracts) derived from organisms belonging to genotypes AA, BB and AB. If the possibility of a unique "hybrid substance" γ is to be excluded, this will require additional evidence concerning the reagents or reaction products. In keeping with Srb and Owen's definition of codominance we could surely say in this instance that AB "shares the properties of both homozygotes," but, if we are cautious to say just what the matrix entitles us to say, we will add "namely, the properties of forming white precipitates with X and with Y."

8. G-isomorphy and M-isomorphy

In 1924, almost a quarter of a century after Landsteiner's discovery, Bernstein proposed the generally accepted genetic theory for the ABO blood groups (cf. Section 4). Shortly thereafter, Wiener, Lederer and Polayes (1930) considered an alternative hypothesis. Dr. Wiener explains the problem in a most lucid manner in his book, Blood Groups and Transfusion:

"A theory to explain 'Bernstein exceptions' could be constructed on the hypothetical assumption of four completely linked pairs of genes (Ab), (aB), (ab) and (AB), or what amounts to practically the same thing, by assuming the existence of four allelic genes, of which three genes are the genes A, B, and O of Bernstein's theory, and the fourth allelic gene (C) acts like the genes A and B together and determines the presence of both agglutinogens, A and B. Naturally, the fourth allelic gene, if it exists at all, would have to be extremely rare. This possibility, which was first considered by the author (1930), and independently by Edwards and Etherington (1935), would imply the existence of four different genotypes in group AB, namely AB, CO, CA and CB. This would call for the existence, for example, of certain families O x AB (OO x CO) in which half of the children belong to group O and half to group AB, none to group A or to group B. Since such families have not been found, the existence of a fourth allelic gene seems very improbable."

This is probably the first instance in the literature of human genetics where two G-isomorphic phenotype systems were pitted against one another as rival genetic hypotheses. The question of the existence of a fourth allele is one of great interest from many points of view. Here, however, I wish to consider just the formality of genetic-hypothesis fitting. Why do we characterize the three-allele and the four-allele proposals as G-isomorphic?

In previous discussions of phenotype systems, we have considered the set of alleles (G) as fixed, and we asked: in how many ways can G_2 be partitioned into ϕ phenotypes? Now that our phenotypes are not wholly arbitrary or amorphous, having imposed a definite structural restriction upon them, we can invert the question: in how many ways can we seek to explain a set of ϕ observed phenotypes? We shall suppose that the phenotypes can be adequately described by a (0, 1)-matrix U. But we also need to specify what will be acceptable as an "explanation," otherwise the "ways" are hardly denumerable. Here, naturally, we wish to consider just factor-union systems of index ϕ, and the chosen phenotype system must have a P-matrix such that $\overline{P}_2$ exactly coincides with U. In other words, the proposed system must account for all observed phenotypes and "predict" no additional ones (cf. Def. 3.14).

Very simply, then, we ask: Given a (0, 1)-matrix U, how many factor-union phenograms can be fitted to the matrix? Here the set of alleles G becomes an adjustable parameter, and this explains the choice of the term G-isomorphy. When, as in Dr. Wiener's case, the geneticist considers two or more G-isomorphic phenotype systems simultaneously, we could say that he is fiddling on his G-string. A glance at the last column of Table 1 will show that there is an abundance of tunes for such fiddling. But let us first fix the general idea using matrices. A comparison of the two ABO hypotheses is most compactly made in the form of the $\overline{P}_2; \overline{F}_2$ matrices:

3-4-1			4-4-2		
1 0	1 0 1	AA, AO	1 1	1 1 1 1	CC, CA, CB, CO, AB
0 1	0 1 1	BB, BO	1 0	0 1 0 1	AA, AO
0 0	0 0 1	OO	0 1	0 0 1 1	BB, BO
1 1	1 1 0	AB	0 0	0 0 0 1	OO

It will be noted that Wiener neglected to mention a fifth genotype which belongs to the phenoset of group AB under the 4-allele theory, namely the homozygote CC. This omission was perhaps intentional since, if C is "extremely rare," CC would represent a rarity of the second order. Upon supplying the missing genotype, however, we note that the two theories differ in just one respect: one phenotype is made to correspond to a heterophenoset (in this case a simple heterophenoset AB) or, alternatively, to a complex homophenoset (CC, CA, CB, CO, AB). This is the essential distinction between G-isomorphic systems.

To show the full range of possibilities for G-isomorphy, we may compare the $\overline{P}_2$-matrices arising from three I-isomorphic

4-9-1, P-1	4-9-1, P-2	4-9-1, P-3
1 0 0 1	1 0 0 1 1	1 0 1 0 1
0 1 0 0	0 1 0 1 0	0 1 0 1 0
0 0 1 0	0 0 1 0 1	0 0 1 0 1
0 0 0 1	0 0 0 1 1	0 0 0 1 1
1 1 0 1	1 1 0 1 1	1 1 1 1 1*
1 0 1 1	1 0 1 1 1	1 0 1 1 1*
0 1 1 0	0 1 1 1 1	0 1 1 1 1*
0 1 0 1	0 1 0 1 1*	0 1 0 1 1*
0 0 1 1	0 0 1 1 1*	0 0 1 1 1*

5-9-a	6-9-(a, b)	5-9-b
0 1 0 1 1*	0 1 0 1 1*	0 0 1 1 1*
1 0 0 1 1	0 0 1 1 1*	1 0 0 1 1
0 1 0 1 0	1 0 0 1 1	0 1 0 1 0
0 0 1 0 1	0 1 0 1 0	0 0 1 0 1
0 0 0 1 1	0 0 1 0 1	0 0 0 1 1
1 1 0 1 1	0 0 0 1 1	1 0 1 1 1
0 1 1 1 1	0 1 1 1 1	0 1 1 1 1
1 0 1 1 1	1 1 0 1 1	1 1 0 1 1
0 0 1 1 1*	1 0 1 1 1	0 1 0 1 1*

P-matrices for phenogram 4-9-1. If, for any $\overline{P}_2$-matrix, the submatrix $\overline{P}'$ contains a row vector, say p*, which, in union with each of the row vectors of P, gives no vector not already present in $\overline{P}_2$, then p* can be transferred to P, and the matrix P*, thus augmented, is G-isomorphic with P (cf. Def. 3.14). In the case of 4-9-1, P-1, no row vector of $\overline{P}'$ can be transferred to P without necessitating additional vectors in $\overline{P}_2$. Also, no vector in P can be transferred to $\overline{P}'$, since the reduced P-matrix would not have $\overline{P}_2$ as its condensed diplo-matrix. Finally, no row vector of P can be interchanged with one of $\overline{P}'$. Therefore, P-1 has no G-isomorph.

At the opposite extreme, matrix P-3 is a basic generator matrix for phenogram 4-9-1 (cf. Def. 3.16). Any combination of the five row vectors of $\overline{P}'$ can be transferred to P to yield a matrix G-isomorphic with P-3. No two of these are simply isomorphic and no two are I-isomorphic; hence 4-9-1, P-3 belongs to a set of 32 G-isomorphs. Matrix 4-9-1, P-2 illustrates an intermediate condition. Its $\overline{P}'$ matrix contains two row vectors which can be shifted to P, but the two matrices thus formed are simply isomorphic (cf. Def. 3.12), as can be seen by performing on the P-matrix of 5-9-a the permutation (34) of rows and the permutation (23)(45) of columns. Thus, 4-9-1, P-2 belongs to a set of three G-isomorphs, one of degree 4, one of degree 5, and one of degree 6.

Phenogram graphs are used in Figure 2 to illustrate all three-allele factor-union phenograms and their two- and four-allele G-isomorphs. Table 1 reveals that G-isomorphy is a nearly universal characteristic of fU systems. In particular, the table shows that

(A) For every fU phenogram with $m \leq 4$, there is at least one basic generator matrix. (This implies that if P is not a basic generator matrix, then P possesses at least one I-isomorph which is a basic generator matrix.)

(B) For every fU phenogram with $m \leq 4$, with the sole exception of the dominantly-connected systems (m-m-1), there is at least one G-isomorphic phenogram of degree m', with $m'>m$, $m'=m$, or $m'<m$.

We shall refer to the above statements as conjectures A and B when the proviso "$m \leq 4$" is removed, and we will show, in Theorem 8.2, that A implies B. If conjecture A could be established, this would provide yet another approach to the enumeration problems.

THEOREM 8.1 A necessary condition for a basic generator matrix is that the matrix $\bar{P}_2$ must contain, either in P or in $\bar{P}'$, a unit row vector $\hat{p} = |1, 1, \ldots 1|$.

Proof: Let p_i and p_j be any two row vectors in P'', the set of ϕ row vectors of $\bar{P}_2$. Closure of P'' under '$\cup$' (cf. Def. 3.16) means that for all (p_i, p_j) if $p_i \cup p_j = p_k$, then $p_k \in P''$. This requires that P'', which is partially ordered by set inclusion, must contain a "first element," i.e. a row vector $\hat{p}$, such that $p_i \subset \hat{p}$ for all $p_i \neq \hat{p}$. But the column-irreducible requirement for P_2 (cf. Def. 3.4) means that P cannot contain a null column vector $c_i = |0, 0, \ldots 0|$, and this requires that $\hat{p} = |1, 1, \ldots 1|$. This is not a sufficient condition for a basic generator matrix. Table 1 includes four P-matrices (4-10-1, P-9, 4-9-1, P-7, 4-9-2, P-3 and 4-8-9, P-1) which are not basic generator matrices, although $\hat{p} \in \bar{P}'$. Also, starting with $m = 5$, it is easy to construct P-matrices which have $\hat{p} \in P$ and which are not basic generator matrices.

THEOREM 8.2 Every fU phenogram of degree m which has a basic generator matrix, with the exception of the dominantly-connected system (m-m-1), is G-isomorphic with at least one phenogram of degree m', with $m' > m$, $m' = m$, or $m' < m$.

Proof: Let P be a basic generator matrix for an m-allele phenogram $\mathfrak{P}$. From Def. 3.5, we know that $\phi \geq m$. If $\phi > m$, then $\bar{P}'$ contains at least one row vector p^* which can be transferred to P, defining a G-isomorphic phenogram $\mathfrak{P}^*$ of degree $m' = m + 1$. If $\phi = m$, there are two possibilities. If P is totally ordered by set inclusion, then $\mathfrak{P}$ is phenogram m-m-1, which has no G-isomorph; since no two row vectors of P are non-comparable, none can be transferred to $\bar{P}'$. For all other systems having $\phi = m$, however, $\mathfrak{P}$ has a G-isomorph of degree $m' < m$. We know that P contains at least three row vectors, including $\hat{p}$ (cf. Theorem 8.1), and, if P is not totally ordered by '$\subset$', we must have $(p_i, p_j, p_k) \in P$ such that $p_i \nsubseteq p_j$, $p_j \nsubseteq p_i$, $p_i \cup p_j = p_k$, and $p_k \subseteq \hat{p}$. Therefore, p_k can be transferred to $\bar{P}'$, defining a G-isomorphic phenogram $^*\mathfrak{P}$ of degree $m' = m - 1$. If P has a G-isomorph of degree $m' = m$, we must have $\phi > m$ and, accordingly, there exists a third G-isomorph of degree $m'' = m'+1 = m + 1$; there may or may not exist a G-isomorph of degree $m''' = m - 1$. (cf. phenograms 4-6-20, 4-6-21, 4-7-19, 4-7-20).

Although G-isomorphy is an equivalence relation when the term is applied to P-matrices (Def. 3.14), this need not be the case for phenograms (Def. 3.15) if the latter are generated by I-isomorphic P-matrices. To illustrate, the I-isomorphic matrices 4-7-19, P-1 and 4-7-19, P-2 are both basic generator matrices; hence each has a 7-allele, 7-phenotype G-isomorph. The latter describe different phenograms, as may be noted from the dominance structures, but, since $\phi = m$, there cannot exist a pair of G-isomorphic P-matrices for these systems. Thus, 4-7-19 is G-isomorphic with two 7-allele phenograms which are not G-isomorphs of each other.

Returning to the comparison of phenograms 3-4-1 and 4-4-2, we note that the latter was brought into consideration for the ABO blood groups just becuase it permits certain mating results that are disallowed under the three-allele hypothesis, namely, parent-child pairs (AB, O) and (O, AB). We could ask whether any two G-isomorphic fU systems might be expected to differ in their genetic consequences in this way. We now introduce two definitions for the purpose of making this question fully explicit.

DEF. 8.1 Let $\sigma_{ij.k}$ denote the probability (averaged over all matings and assuming a Hardy-Weinberg equilibrium) of phenotype k appearing in a child whose parents have phenotypes i and j. If the pair $[i, j]$ is ordered and i, j, k stand for any three phenotypes in a set of ϕ, then $\sigma_{ij.k}$ is an element of a three-dimensional matrix S, which is stochastic in one dimension since, for all $[i, j]$

$$\sum_{k=1}^{\phi} \sigma_{ij.k} = 1.$$

DEF. 8.2 If we replace each probability σ in S by a (0, 1)-variable

$$w_{ij.k} = 1 \quad \text{if} \quad \sigma_{ij.k} > 0,$$
$$w_{ij.k} = 0 \quad \text{if} \quad \sigma_{ij.k} = 0,$$

then the matrix $M = \{w_{ij.k}\}$ could be called the mating table for any phenotype system, and its complement $\sim M = \{\sim w_{ij.k}\}$ could be called the exclusion table. We shall say that any two G-isomorphic phenotype systems are M-isomorphic if their M-matrices are identical.

The matrix S could appropriately be called a Snyderian matrix, since, for the common and simplest case of two alleles with dominance, Snyder (1932) showed that

$$(2\text{-}2\text{-}1),\ S = \begin{vmatrix} [(1-\zeta^2),\ \zeta^2] & [(1-\zeta),\ \zeta] \\ [(1-\zeta),\ \zeta] & [0,\ 1] \end{vmatrix}$$

where $\zeta = q/(1 + q)$ and q is the frequency of the recessive gene.

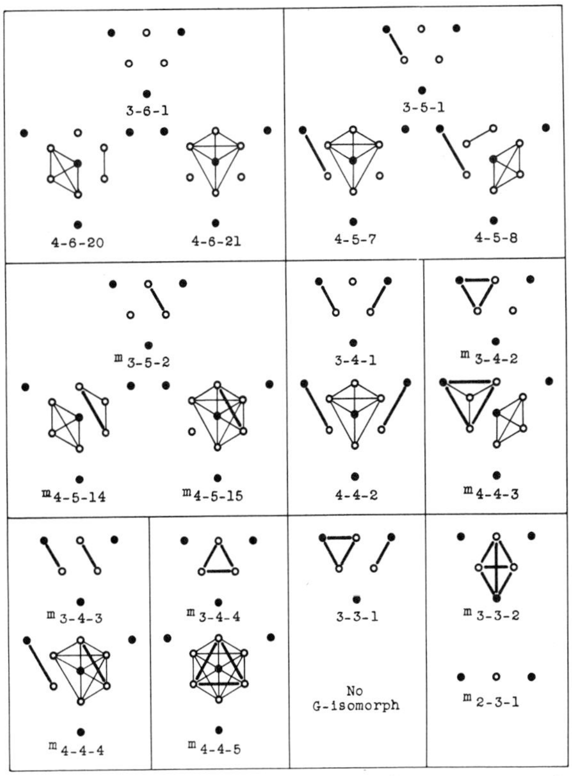

Figure 2. The nine three-allele factor-union phenograms compared with their two- and four-allele G-isomorphs. M-isomorphic sets are indicated by the symbol m.

The function ζ is frequently referred to as Snyder's ratio. By means of such functions, Dr. Snyder did much to dispel the once commonly held misconception that exact mathematical expectations could not be hoped for in the analysis of human heredity (cf. Snyder, 1941). With the exception of phenogram series m-g-1, every fU system will have an S-matrix in which some of the σ's are functions of the gene frequencies.

Of the nine three-allele fU phenograms, five possess G-isomorphic systems that are also M-isomorphic (cf. Fig. 2). These systems, 3-3-2, 3-4-2, 3-4-3, 3-4-4, 3-5-2, closely mimic their two- and four-allele counterparts with respect to their statistical manifestations. By definition, no trio of phenotypes in two parents and a child will serve to differentiate M-isomorphic systems. Discrimination must therefore be based upon a detailed analysis of heterogeneity in segregation ratios.

Phenograms 2-3-1 and 3-3-2 provide the simplest example of G-isomorphy and of M-isomorphy. Consider three alleles, M, N, C, having phenotype vectors $|1,0|$, $|0,1|$, $|1,1|$ and gene frequencies q_1, q_2, q_3, respectively. Denoting the three phenotypes by M, N and C, we can write the Snyderian matrix as

$$
(3\text{-}3\text{-}2),S = \left|\begin{array}{ccc} \overset{M}{[1,\ 0,\ 0]} & \overset{C}{[\lambda_1,\ 1-\lambda_1,\ 0]} & \overset{N}{[0,\ 1,\ 0]} \\ [\lambda_1,\ 1-\lambda_1,\ 0] & [\lambda_1^2,\ 1-\lambda_1^2-\lambda_2^2,\ \lambda_2^2] & [0,\ 1-\lambda_2,\ \lambda_2] \\ [0\ \ 1,\ 0] & [0,\ 1-\lambda_2,\ \lambda_2] & [0,\ 0,\ 1] \end{array}\right| \begin{array}{c} M \\ C \\ N \end{array}
$$

wherein $\lambda_1 = q_1(1 - q_1)/\bar{c}$, $\lambda_2 = q_2(1 - q_2)/\bar{c}$, and $\bar{c} = 1 - q_1^2 - q_2^2$ is the expected population frequency of phenotype C. If we put $q_3 = 0$, then $q_1 + q_2 = 1$, $\bar{c} = 2q_1q_2$, $\lambda_1 = \lambda_2 = 1/2$ and the above matrix reduces to (2-3-1),S, a familiar set of simple mendelian ratios. Obviously (3-3-2),M = (2-3-1),M.

Because hypothesis 3-3-2 differs from 2-3-1 only in its interpretation of phenotype C, we are interested in the expected values of $\bar{c}$, $\sigma_{CM.C}$, $\sigma_{CN.C}$, and $\sigma_{CC.C}$, each of which is 1/2 when $q_1 = q_2$ and $q_3 = 0$. Small values of q_3 will have but little effect upon these expectations. For example, with $q_3 = 0.01$ and $q_1 = q_2 = 0.495$, we have $\bar{c} = 0.510$, $\sigma_{CM.C} = \sigma_{CN.C} = 0.510$ and $\sigma_{CC.C} = 0.520$; with $q_3 = 0.05$ and $q_1 = q_2 = 0.475$, we have $\bar{c} = 0.549$, $\sigma_{CM.C} = \sigma_{CN.C} = 0.546$ and $\sigma_{CC.C} = 0.587$. In human genetical studies, it would certainly not be an easy matter to detect small departures of these magnitudes nor to demonstrate that the segregation ratios are heterogeneous when one or both parents has phenotype C. It is clear, however, that the demonstration of heterogeneity would be of far greater significance in establishing the existence of the "extra" allele postulated under system 3-3-2. Most convincing of all, perhaps, would be the demonstration that such heterogeneity disappears only when we select families in which the C-parents are derived from matings M x N. In this instance, it will be noticed that by reverting to the use of three-generation data, we are essentially reproducing an experiment of classical mendelian design.

In order to more easily deal with combinatorial principles, we found it convenient here to assume that one possesses a phenotype matrix which can be interpreted as a closed $\bar{P}_2$ matrix for one or more P-matrices. But this condition will frequently not be met in practice, since some alleles may be so rare that not all discernible combinations will be represented in the sample. This and other complications are discussed in detail by Morton and Miki (1968) in reference to the ever-expanding MN blood type system. For the population geneticist G-isomorphy presents a problem of concern lest the effects of unrecognized alleles be confused with genuine evidence of "heterozygote advantage" and other sources of discrepancy.

9. Parentage Exclusion Rules

In medico-legal investigations of disputed parentage, one generally is involved with tests on just three individuals, a child (z), its putative mother (X) and putative father (Y). The human geneticist who uses blood types as a means of detecting falsely represented parentage is, of course, not subject to this limitation. Having data on "whole families," he wishes to consider what we may call the general case: $X, Y, z_1, z_2, \ldots z_n$. We shall comment on this complex problem after considering in some detail the simpler

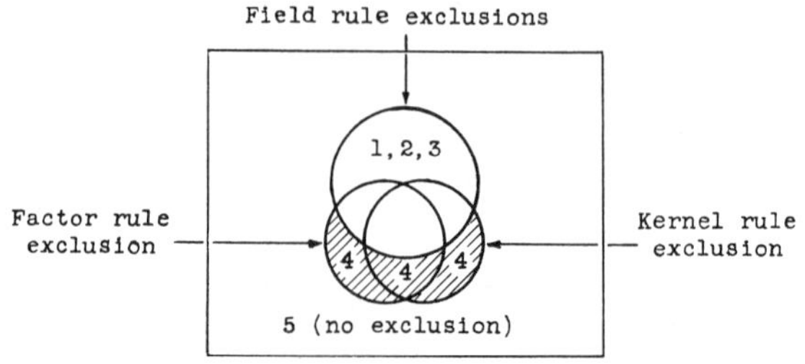

Figure 3. Venn-Euler diagram showing how the five exclusion classes are effected by the three exclusion rules.

case, i.e. three-way comparisons in a set (X, Y, z). Even for the geneticist the conclusions reached from such comparisons may be expected to be the most useful ones in his work.

Using the terminology of phenosets, fields and kernels, we can deduce some rules for parentage exclusion with fU systems that exhibit remarkable power, generality, efficiency, and discrimination.

By remarkable power I mean, in fact, infallibility. When used in combination, I believe that they will detect exclusion whenever and only when there is a basis for such under the phenotype system, provided this is fU. Neither errors of the first kind nor the second kind should be incurred, if my surmise is correct.

By generality I mean that the rules accomplish their purpose when used on any fU phenotype system. Moreover, the computer (human, electronic, or other) need not know the particular system or systems with which he, she or it is operating. Lists of genotypes for each phenotype, i.e. phenosets, need not be consulted. All that is needed is the p-f-k (phenotype-field-kernel) vector for each person involved. This is efficiency, or at least consistency.

By discrimination I mean that the exclusions are not thrown into one hat, but are parcelled into what I have called the "logical classes" of exclusion (Cotterman, 1951). Given tests on three individuals (X, Y, z), with the hypothesis

(1) X is mother of z, and

(2) Y is father of z,

there are, at worst, five different conclusions which may be reached. By "at worst" I mean under the supposition that both propositions (1) and (2) are false, a situation which the geneticist faces when he is confronted with adoptions, interchanged infants, falsely-claimed kidnapped persons, etc.

The first three categories (Table 3) include what I have

Table 3. Detection and Classification of Exclusions by Means of the Factor, Field and Kernel Rules

Class 1 Maternity exclusion Paternity exclusion $(X\ \overline{m}\ z) \cap (Y\ \overline{p}\ z)$	FIELD RULE excludes $\mathcal{F}(X, z) = \emptyset$	FIELD RULE excludes $\mathcal{F}(Y, z) = \emptyset$
Class 2 Maternity exclusion only $(X\ \overline{m}\ z) \cap (Y\ p\ z)$	FIELD RULE excludes $\mathcal{F}(X, z) = \emptyset$	$\mathcal{F}(Y, z) \neq \emptyset$
Class 3 Paternity exclusion only $(X\ m\ z) \cap (Y\ \overline{p}\ z)$	$\mathcal{F}(X, z) \neq \emptyset$	FIELD FULE excludes $\mathcal{F}(Y, z) = \emptyset$
Class 4 Joint-parentage exclusion $(XY)\ \overline{mp}\ z$	No field rule exclusions, but FACTOR RULE or KERNEL RULE excludes $[\mathcal{P}(XY/z) \neq \emptyset] \cup [\mathcal{K}(XY/z) \neq \emptyset]$	
Class 5 No exclusion	No field rule exclusions and No factor rule exclusion No kernel rule exclusion	

termed unconditional (Cotterman, 1951) or 1-link (Cotterman, 1966) exclusions. The symbol '$X\ \overline{m}\ z$' should be read: X is not mother of z, whoever the father may be. The father need not have been tested, and in fact wasn't if (2) is false. So the appropriate test here is a comparison of just two items, the fields of X and z. The symbol '$X\ m\ z$' reads: X is not excluded as mother of z.

Class 4 exclusion represents a "residue class." The symbol $XY(\overline{mp})z$ should be read: Neither X nor Y is unconditionally excluded as parent of z, but both can't be parents. This implies that tests for unconditional parentage exclusion have been applied with negative results, but that joint-parentage is excluded on some additional criterion. The hyphen in "joint-parentage exclusion" is intended to avoid confusion with class 1, which could perhaps be described as "joint parentage-exclusion."

The fifth and last category is also a residue class, which needs no explanation. It is noteworthy that all five exclusion classes are possible with one of the simplest of phenotype systems, phenogram 2-3-1 or the case of "two alleles without dominance." We shall illustrate this later.

The symbols X, Y, z have already been defined. Now let p(X), p(Y), p(z) denote the phenotypes of these individuals, let f(X), f(Y), f(z) be the corresponding phenotype fields, and let k(X), k(Y), k(z) be the phenotype kernels. All of these stand for (0,1) vectors, being either factor-sets (in the case of p) or allele-sets (in cases f and k). Further, let us agree to call the complement of f(X) the anti-field, $\sim$f(X), and the complement of k(X) the anti-kernel, $\sim$k(X).

Recalling that a field is a list of alleles, each of which is present in at least one genotype of the phenoset, and that a kernel is a list of alleles (never more than 2) each of which is present in every genotype of the phenoset, we may summarize as follows:

Given an individual z of phenotype p(z), then

k(z) includes alleles which	MUST BE,	(P = 1)
$\sim$k(z) includes alleles which	NEEDN'T BE,	(P < 1)
f(z) includes alleles which	CAN BE,	(P > 0)
$\sim$f(z) includes alleles which	CAN'T BE,	(P = 0)

in the genotype of z.

With this mnemonic scheme, our exclusion rules now almost write themselves. Consider first the field rule; this is our one (and only) criterion for unconditional exclusion of parentage: If there is no allele [$= \emptyset$] which can be in w [f(w)] and [$\cap$] can also be in z [f(z)], then parent-child relationship is excluded for w and z. Putting this more formally, we have:

FIELD RULE: Parent-child relationship is (unconditionally) excluded for w and z if, and only if, the phenotype fields of w and z are disjoint, i.e.

$$\mathcal{F}(w, z) \equiv f(w) \cap f(z) = \emptyset$$

In the case of the kernel rule, we argue: If there is at least one allele [$\neq \emptyset$] which must be in z [k(z)] and can't be in X [$\sim$f(X)] and can't be in Y [$\sim$f(Y)], then X and Y cannot both be parents of z.

KERNEL RULE: If $\mathcal{K}(XY/z) \equiv k(z) \cap \sim f(X) \cap \sim f(Y) \neq \emptyset$, then X and Y cannot both be parents of z.

An analogous statement can be made in reference to the phenotype factor-sets (p). In this case, however, we would probably prefer the verb forms 'is' and 'isn't', since the factors are observed properties rather than inferential ones.

FACTOR RULE: If $\mathcal{P}(XY/z) \equiv p(z) \cap \sim p(X) \cap \sim p(Y) \neq \emptyset$, then X and Y cannot both be parents of z.

Before considering the examples that follow, the reader should note the following principles.

1. The five classes of exclusion are discriminated only if a "complete test" is made. This consists of two applications of the field rule (X with z, Y with z), plus the factor and kernel rule tests. Each test involves a comparison, by intersection, of either two or three vectors.

2. Note that exclusion by the field rule requires a null intersection, but exclusions under the factor and kernel rules require non-null intersections.

3. Field rule exclusions are sometimes corroborated by the factor rule or kernel rule or both (e.g. cases 1,a, 2,a), and sometimes they are not (e.g. cases 1,b, 1,c). This, however, is immaterial, as only the field rule is empowered to make unconditional parentage exclusion; moreover, the conclusion $\overline{m} \cap (\overline{m} \cup \overline{p})$ is tautologically equivalent to $\overline{m}$ (cf. Fig. 3).

4. Class 4 exclusion requires a negative result (no exclusion) on both field rule tests. In most cases the factor and kernel rules will both detect exclusions of this class (e.g. cases 1,d, 3,a). In some phenotype systems having certain dominance structures, the factor rule can fail to exclude, but the kernel rule is then able to detect exclusion (e.g. case 2,b). Conversely, in systems having null-kernel phenotypes the kernel rule will always fail if $k(z) = \emptyset$, but the factor rule is then in position to make the exclusion (e.g. case 3,b). Although I have what I believe to be a formal proof of this proposition, I shall not detail this as it requires a tedious enumeration of various phenoset trios in fU systems. In any event, I have not discovered a case in which both kernel and factor rules would fail to provide an exclusion of class 4, and this is evidently the only point of doubt concerning the general power of these tests.

5. When we consider two children, z_1, z_2, with putative parents X and Y, the number of non-equivalent exclusion classes increases from 5 to 83 (Cotterman, 1966). And, rather surprisingly, at least 60 of these possibilities can be illustrated by results with a single four-allele system such as MNSs and A_1A_2BO. Using two such systems all 83 exclusion classes could easily occur. If it seemed desirable to distinguish all of these, this could certainly be accomplished by methods which take all phenoset compositions into account, e.g. by use of the $P_2;I_2$ matrix. But, in the interest of achieving a general method, invariant with respect to the phenotype system, the possibility should be explored that the p-f-k vectors alone might be sufficient, as in the simpler (X, Y, z) case.

6. From the field rule it is clear that a factor-union phenotype system affords no basis for unconditional parentage exclusion if, and only if, no two phenotype fields are disjoint. Theorem 9.1 delineates this class of fU systems.

THEOREM 9.1 A factor-union phenogram contains no two phenotype fields that are disjoint and therefore provides no basis for unconditional parentage exclusion if, and only if, it is a Θ-phenogram of index $\phi = m$.

Proof: Let $\check{p}$ denote a "last element" in the partial order of row vectors of P under set inclusion, i.e. $\check{p} \subset p_i$ for all $p_i \in P$, with $p_i \neq \check{p}$. Similarly, let $\check{f}$ be a last element in the set of row vectors of F. Then $\check{p} = |0,0,\ldots 0|$ since P cannot contain a unit column vector (cf. Def. 3.4 and Section 4), but $\check{f} = |0,0,\ldots 0,1|$ since F cannot contain a null row vector (cf. Def. 3.7). From the following biconditional statements,

(1) $\check{p} \in P$ iff $\check{f} \in F$,

(2) $\check{f} \in F$ iff $f_1 \cap f_j \neq \emptyset$ for all $(f_i, f_j) \in F$,

(3) $\phi = m$ iff $F = \overline{F}_2$,

we conclude that ($\check{p} \in P$ and $\phi = m$) implies $f_i \cap f_j \neq \emptyset$ for all $(f_i, f_j) \in \overline{F}_2$.

Conversely, $f_i \cap f_j \neq \emptyset$ for all $(f_i, f_j) \in \overline{F}_2$ implies (2) and (1). Suppose that $\check{p} \in P$ and $\phi > m$; then $\overline{P}'$ contains a row vector $p_k = p_i \cup p_j$, where p_i and p_j now represent non-comparable row vectors in P. If p_i and p_j are disjoint, then the corresponding row vectors of F, f_i and f_j, are disjoint. But if p_i and p_j are non-disjoint, then f_k and $\check{f}$ are disjoint. Therefore $f_i \cap f_j \neq \emptyset$ for all $(f_i, f_j) \in \overline{F}_2$ implies ($\check{p} \in P$ and $\phi = m$).

From Theorem 6.1, the number of m-allele m-phenotype Θ-phenograms is equal to the number of fU phenograms of degree m − 1 and index m − 1. Denoting this number by θ_m, we find

$$\theta_2 = 1,\ \theta_3 = 1,\ \theta_4 = 2,\ \theta_5 = 5,\ \theta_6 = 15,\ \theta_7 = 53.$$

The first four systems, having $\phi = m \leq 4$, have P-matrices belonging to the series $C^{(m)}$ and $D^{(m)}$ (cf. Section 5).

Example 1. Using the MN blood-types as an example of phenogram 2-3-1, the following $\overline{P}_2;\overline{F}_2;\overline{K}_2$ matrix serves as our key:

	p	f	k
"M"	1 0	1 0	1 0
"N"	0 1	0 1	0 1
"MN"	1 1	1 1	1 1

⊛ denotes exclusion under each rule

Case 1(a). Assume: X= "M", Y = "M", z = "N"

f(X) = 1 0	f(Y) = 1 0	~p(X) = 0 1	~f(X) = 0 1
f(z) = 0 1	f(z) = 0 1	~p(Y) = 0 1	~f(Y) = 0 1
⊛ ∩ = 0 0	⊛ ∩ = 0 0	p(z) = 0 1	k(z) = 0 1
EXCLUSION (Class 1)		⊛ ∩ = 0 1	⊛ ∩ = 0 1

Case 1(b). Assume: X = "M", Y = "MN", z = "N"

f(X) = 1 0	f(Y) = 1 1	~p(X) = 0 1	~f(X) = 0 1
f(z) = 0 1	f(z) = 0 1	~p(Y) = 0 0	~f(Y) = 0 0
⊛ ∩ = 0 0	∩ = 0 1	p(z) = 0 1	k(z) = 0 1
EXCLUSION (Class 2)		∩ = 0 0	∩ = 0 0

Case 1(c). Assume: X = "N", Y = "M", z = "N"

f(X) = 0 1	f(Y) = 1 0	~p(X) = 1 0	~f(X) = 1 0
f(z) = 0 1	f(z) = 0 1	~p(Y) = 0 1	~f(Y) = 0 1
∩ = 0 1	⊛ ∩ = 0 0	p(z) = 0 1	k(z) = 0 1
EXCLUSION (Class 3)		∩ = 0 0	∩ = 0 0

Case 1(d). Assume: X = "M", Y = "M", z = "MN"

f(X) = 1 0	f(Y) = 1 0	~p(X) = 0 1	~f(X) = 0 1
f(z) = 1 1	f(z) = 1 1	~p(Y) = 0 1	~f(Y) = 0 1
∩ = 1 0	∩ = 1 0	p(z) = 1 1	k(z) = 1 1
EXCLUSION (Class 4)		⊛ ∩ = 0 1	⊛ ∩ = 0 1

Case 1(e). Assume: X = "MN", Y = "N", z = "N"

f(X) = 1 1	f(Y) = 0 1	~p(X) = 0 0	~f(X) = 0 0
f(z) = 0 1	f(z) = 0 1	~p(Y) = 1 0	~f(Y) = 1 0
∩ = 0 1	∩ = 0 1	p(z) = 0 1	k(z) = 0 1
NO EXCLUSION (Class 5)		∩ = 0 0	∩ = 0 0

Example 2. Using the A_1A_2BO blood-types as an example of phenogram 4-6-1, the following $\overline{P}_2;\overline{F}_2;\overline{K}_2$ matrix serves as our key:

	p	f	k
"A_1"	1 1 0	1 1 0 1	1 0 0 0
"A_2"	0 1 0	0 1 0 1	0 1 0 0
"B"	0 0 1	0 0 1 1	0 0 1 0
"O"	0 0 0	0 0 0 1	0 0 0 1
"A_1B"	1 1 1	1 0 1 0	1 0 1 0
"A_2B"	0 1 1	0 1 1 0	0 1 1 0

Case 2(a). Assume: X = "A_1", Y = "A_2", z = "A_1B"

f(X) = 1101	f(Y) = 0101	~p(X) = 001	~f(X) = 0010
f(z) = 1010	f(z) = 1010	~p(Y) = 101	~f(Y) = 1010
∩ = 1000	⊛ ∩ = 0000	p(z) = 111	k(z) = 1010
EXCLUSION (Class 3)		⊛ ∩ = 001	⊛ ∩ = 0010

Case 2(b). Assume: X = "A_1B", Y = "B", z = "A_2B"

f(X) = 1010	f(Y) = 0011	~p(X) = 000	~f(X) = 0101
f(z) = 0110	f(z) = 0110	~p(Y) = 110	~f(Y) = 1100
∩ = 0010	∩ = 0010	p(z) = 011	k(z) = 0110
EXCLUSION (Class 4)		∩ = 000	⊛ ∩ = 0100

Example 3. Using the MNSs blood-types as an example of phenogram 4-9-3, the following $\bar{P}_2;\bar{F}_2;\bar{K}_2$ matrix serves as our key:

	p	f	k
"MS"	1 0 1 0	1 0 0 0	1 0 0 0
"Ms"	1 0 0 1	0 1 0 0	0 1 0 0
"NS"	0 1 1 0	0 0 1 0	0 0 1 0
"Ns"	0 1 0 1	0 0 0 1	0 0 0 1
"MSs"	1 0 1 1	1 1 0 0	1 1 0 0
"MNS"	1 1 1 0	1 0 1 0	1 0 1 0
"MNSs"	1 1 1 1	1 1 1 1	0 0 0 0
"MNs"	1 1 0 1	0 1 0 1	0 1 0 1
"NSs"	0 1 1 1	0 0 1 1	0 0 1 1

Case 3(a). Assume: X = "Ns", Y = "MNs", z = "NSs"

f(X) = 0001	f(Y) = 0101	$\sim$p(X) = 1010	$\sim$f(X) = 1110
f(z) = 0011	f(z) = 0011	$\sim$p(Y) = 0010	$\sim$f(Y) = 1010
$\cap$ = 0001	$\cap$ = 0001	p(z) = 0111	k(z) = 0011
EXCLUSION (Class 4)		$\circledast\cap$ = 0010	$\circledast\cap$ = 0010

Case 3(b). Assume: X = "MSs", Y = "Ms", z = "MNSs"

f(X) = 1100	f(Y) = 0100	$\sim$p(X) = 0100	$\sim$f(X) = 0011
f(z) = 1111	f(z) = 1111	$\sim$p(Y) = 0110	$\sim$f(Y) = 1011
$\cap$ = 1100	$\cap$ = 0100	p(z) = 1111	k(z) = 0000
EXCLUSION (Class 4)		$\circledast\cap$ = 0100	$\cap$ = 0000

10. Phenoset Structures

Just as we are interested in permutationally distinct phenotype systems or partitions of G_2, we are also interested in their permutationally distinct parts, i.e. phenosets. These have been designated as phenoset structures (Section 2). They are much simpler than phenograms and can be enumerated rather easily by graphical methods, both in the unrestricted case (rps) and for fU systems (Cotterman, 1966).

There are just eight homophenoset structures and ten heterophenoset structures exhibited by fU systems with $m \leq 4$ (Fig. 4). The phenogram number accompanying each graph in this figure refers to its first appearance in Table 1. Note that no phenogram, not even the simplest ones (2-3-1, 2-2-1), can incorporate all phenoset structures possible with a fixed value of m. The maximum number of structures occurring in the same phenogram is five in the case of four-allele fU systems, this condition being realized in phenograms 4-6-5, 4-6-6 and 4-6-7.

The points of phenoset-structure graphs represent genes (alleles) and the lines represent heterozygous genotypes. In only one respect do these graphs resemble phenogram graphs: homozygotes are again represented by filled points (•), which serve to distinguish the structures of homo- and heterophenosets. Graphs of heterophenosets differ in other ways from those of homophenosets, and the latter follow different rules of construction for fU systems and non-fU systems. Here we will consider only the specifications for graphs of phenoset structures in factor-union systems. The graph-theoretical terminology employed here follows that of Harary (1964); see also Busacker and Saaty (1965).

A fU homophenoset structure is represented by an undirected graph $S^\bullet$ of h lines and v points, of which one point, the "root" (•), is adjacent to all other points. An $S^\bullet$ graph is therefore never disconnected. Its v points denote the v alleles, one of them also denoting the homozygote. The number of genotypes in $S^\bullet$ is therefore h + 1. The peculiarities of this graph can be traced to two consequences of our definition of a factor-union system. Theorem 4.1 allows just one homozygote in the set of genotypes; this requires that the graph be "rooted" on one distinctive point. Theorem 4.3 allows $S^\bullet$ to contain lines not incident to the root, but such lines must be included in circuits of three lines, two of them incident to the root. Adjacency of the root to all other points means that this allele is dominant to all others. Transitivity of dominance (cf. Theorem 4.2) does not impose a restriction upon $S^\bullet$, although it does impose restrictions upon pairs of $S^\bullet$ graphs in the "dissected" graph described below.

A fU heterophenoset structure is represented by an undirected graph S° of h lines and v points with one restriction: the degree of each point is a positive integer, i.e. every point is incident to one or more lines. An S° graph can be connected or disconnected, but in either case no point stands alone. Again v indicates the number of alleles, but here the number of genotypes in S° is just h.

Since phenoset structures are, by definition, invariant under permutations of gene symbols, the points of $S^\bullet$ and S° graphs bear no labels. If labelled, the graph then represents one particular phenoset of the structure and the set of labels designates the field. A point represents an allele in the kernel if, and only if, it is incident with all lines of the graph.

Denote by $\Delta(v,h)$ the set of all topologically distinct or anisomorphic undirected graphs with v points and h lines. Consider first the set $\Delta(v-1, h-v+1)$. If to each of these graphs we append a v-th point (•) and v − 1 lines adjoining this point to all

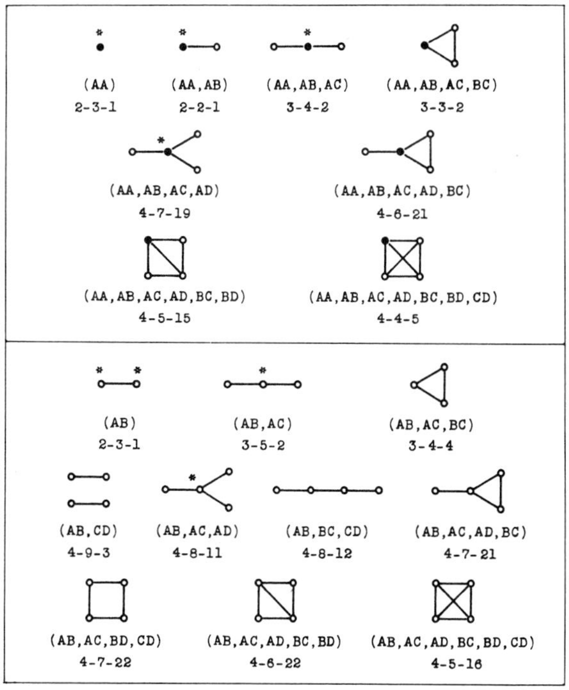

Figure 4. Phenoset structures in factor-union systems ($m \leq 4$). Alleles belonging to the kernel are marked by asterisks.

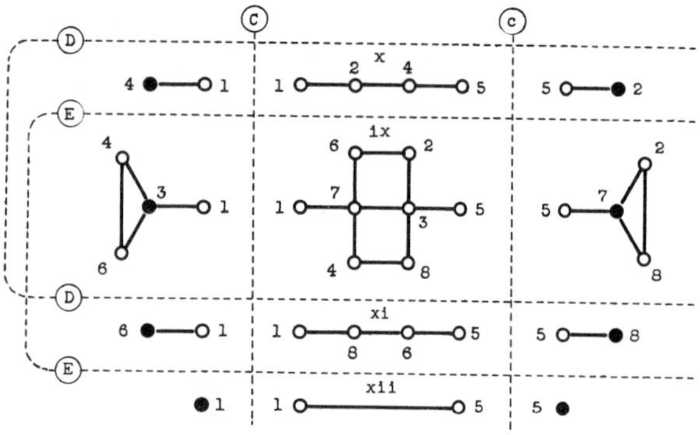

Figure 5. The anatomy of an 8-allele 12-phenotype factor-union system, the (CDEc) Rh-system.

others, we construct the set of all $S^{\bullet}$ graphs having v points and h lines. Consider next the set $\Delta(v-1, h)$. By adding to each of these graphs an isolated v-th point, we obtain a set which includes just those graphs of $\Delta(v, h)$ which are ineligible as S^{o} graphs. These transformations furnish the indication of proof of Theorem 10.1. Using these relations it is easy to determine phenoset structure numbers from a table of N(v, h), such as that provided by Riordan (Table 7, Chapter 6).

THEOREM 10.1 Let N(v, h) be the number of anisomorphic undirected graphs with v points and h lines. For phenosets of factor-union systems which contain v alleles and h heterozygous genotypes, $N(v-1, h-v+1)$ enumerates the homophenoset structures and $N(v, h) - N(v-1, h)$ enumerates heterophenoset structures.

If we represent the set of genotypes G_2 by a complete graph, i.e. one having m points and $m(m-1)/2$ lines connecting all pairs of points, then a fU phenogram can be depicted by assigning ϕ colors to both lines and points, each point having a distinct color. More conveniently, the complete graph can be dissected through its points to yield ϕ phenoset graphs. This representation lacks some of the advantages of the phenogram graphs used in earlier sections of this paper, but the dissected graph readily adapts to cases with five or more alleles. An example is shown in Figure 5, which illustrates a commonly employed Rh classification based upon the use of four antisera. The eight alleles are here assigned numbers corresponding to their ranked frequencies in most Caucasian populations. A key for the interpretation of this figure is provided by the following matrix:

$\bar{P}_2; \bar{F}_2$ =

C D E c	7 2 8 5 3 4 6 1		
1 1 1 0	1 1 1 1 0 0 0 0	7	vii
1 1 0 0	0 1 0 1 0 0 0 0	2	ii
1 0 1 0	0 0 1 1 0 0 0 0	8	viii
1 0 0 0	0 0 0 1 0 0 0 0	5	v
0 1 1 1	0 0 0 0 1 1 1 1	3	iii
0 1 0 1	0 0 0 0 0 1 0 1	4	iv
0 0 1 1	0 0 0 0 0 0 1 1	6	vi
0 0 0 1	0 0 0 0 0 0 0 1	1	i
1 1 1 1	1 1 1 1 1 1 1 1		ix
1 1 0 1	0 1 0 1 0 1 0 1		x
1 0 1 1	0 0 1 1 0 0 1 1		xi
1 0 0 1	0 0 0 1 0 0 0 1		xii

REFERENCES

Bennett, J. H. 1957. The enumeration of genotype-phenotype correspondences. Heredity, (Lond.) 11: 403-409.

Boye, C. L. 1941. An allelic series in Coleus. J. Genet. (Cambr.) 42: 191-196.

Busacker, R. G. and Saaty, T. L. 1965. Finite Graphs and Networks. McGraw-Hill, New York, Pp. xiv. + 294.

Cotterman, C. W. 1951. A note on the detection of interchanged children. Am. J. Human Genet. 3: 362-375.

Cotterman, C. W. 1953. Regular two-allele and three-allele phenotype systems. Part I. Am. J. Human Genet. 5: 193-235.

Cotterman, C. W. 1966. Combinatorial Problems in Genetics. Unpublished lecture notes. (Chaps. 10, 11).

Davis, R. L. 1953. The numbers of structures of finite relations. Proc. Am. Math. Soc. 4: 486-495.

Edwards, J. F., and Etherington, I. M. H. 1935. Blood group inheritance. Nature, (Lond.) 136: 297; ibid. 136:646.

Harary, F. 1964. Combinatorial problems in graphical enumeration. Chap. 6, pp. 185-217, in Applied Combinatorial Mathematics, ed. E. F. Beckenbach. John Wiley, New York.

Hartl, D. L. and Maruyama, T. 1968. Phenogram enumeration: the number of regular genotype-phenotype correspondences in genetic systems. J. Theoret. Biol. 20: 129-163.

Morton, N. E. and Miki, C. 1968. Estimation of gene frequencies in the MN system. Vox Sang. 15: 15-24.

Riordan, J. 1958. An Introduction to Combinatorial Analysis. John Wiley, New York. Pp. x. +244.

Snyder, L. H. 1932. The inheritance of taste deficiency in man. Ohio J. Sci. 32: 436-440.

Snyder, L. H. 1934. Modern analysis of human pedigrees. Eugen. News 19: 61-69.

Wiener, A. S. 1943. Blood Groups and Transfusion. Third Edition. C. C. Thomas, Springfield, Ill. Pp. xix. +438.

Wiener, A. S., Lederer, M, and Polayes, S. H. 1930. Studies in isohemagglutination, III. On the heredity of the Landsteiner blood groups. J. Immunol. 18: 218-221.

Yasuda, N. and Kimura, M. 1968. A gene-counting method of maximum likelihood for estimating gene frequencies in ABO and ABO-like systems. Ann. Human Genet. 31: 409-420.

Publication No. 1302 of the Laboratory of Genetics of the University of Wisconsin, Madison, Wisconsin, U.S.A.

DISCUSSION ON FACTOR-UNION SYSTEMS

N. E. Morton and J. H. Edwards

MORTON: The first step in an analysis of segregation, linkage, parentage exclusion, and gene frequency estimation is usually to represent a set of alleles as a factor-union system. I have been intrigued by the ease with which one factor-union system may be transformed into another to eliminate technical errors or to study particular alleles. For example, consider the sample of 275,664 ABO types of Swiss army personnel and air raid wardens reported by Rosin (1956). Preparatory to a study of isolation by distance among the 3101 communes into which this sample was partitioned, we obtained maximum likelihood estimates of gene frequencies by the ALLTYPE computer program, using the methods of Yasuda (1968). On the assumption of panmixia, group B is in excess and group AB is deficient, with $X_1^2 = 264$ (Table 1). When the gene frequencies and inbreeding coefficient F are estimated simultaneously, taking a high trial value of the inbreeding coefficient to avoid singularity as F approaches zero (Yasuda, 1968), we obtain a perfect fit to the observed frequencies with the incredible estimate of $F = .497 \pm .011$. This indicates a common error, whereby AB genotypes (and especially A_2B) are misclassified as B. Fisher and Taylor (1940) drew attention to systematic errors in AB typing and excluded this blood group with a technique that was published later by Dobson and Ikin (1946) and Roberts (1948). Unfortunately, exclusion of the AB phenotype does not correct for misclassification of AB as B, but it is easy to devise an appropriate method. If B and AB are not accurately distinguished they may be pooled, giving the factor-union system shown on the right side of Table 1. The first binary factor corresponds to anti-B, and the second to anti-A+B. In this system, the maximum-likelihood estimate of the A gene frequency (p) is significantly increased at the expense of the O gene (r). The recorded AB frequency is only 86 percent of its expected value. Among AB genotypes, 20 percent are A_2B in western Europe (Mourant et al., 1958), suggesting that many A_2B individuals were misclassified as group B in this sample. This error is eliminated by pooling B and AB, without appreciably increasing the standard errors. We therefore adopted this method in the distance analysis (Morton et al., 1968). Note that to study isolation by distance for each allele separately, we may focus attention on the A, B, and A+B factors alone.

Azevedo (1968) has used reduced factor-union systems to detect and minimize classification errors of isozymes after long-term storage.

Azevedo, Eliane. 1968. Erythrocyte isozymes, other polymorphisms, and the coefficient of kinship in northeastern Brazil. Ph.D. thesis, University of Hawaii.

Dobson, A. M., and Ikin, E. W. 1946. The ABO blood groups in the United Kingdom: frequencies based on a very large sample. J. Path. Bact. 48:221-227.

Fisher, R. A., and Taylor, G. L. 1940. Scandinavian influence in Scottish ethnology. Nature (London) 145:590.

Morton, N. E., Yasuda, N., Miki, C., and Yee S. 1968. Population structure of the ABO blood groups in Switzerland. Amer. J. Human Genet. 20:420-429.

Mourant, A. E., Kopec, C., and Domaniewska-Sobczak, K. 1958. The ABO blood groups: comprehensive tables and maps of world distribution. Thomas, Springfield, Ill. 276 pp.

Roberts, J. A. F. 1948. The frequencies of the ABO blood groups in southwestern England. Ann. Eugen. (London) 14:109-116.

Rosin, S. 1956. Die Verteilung der ABO-Blutgruppen in der Schweiz. Arch. Klaus-Stift. Vererbungsforsch. 31:1-127.

Yasuda, N. 1968. Estimation of the inbreeding coefficient from phenotype and mating type frequencies. Computer applications in genetics, pp.

EDWARDS: Although "factor-union" is the logically correct term for these systems, I think it unfortunate that the term should be used in genetics since the whole basis of gene replication and manifestation is the non-union of the genetic factors (Mendel's Erbfaktoren).

Mendelian systems are usually considered to be chromosomal: historically this cannot be a reasonable definition. Mendel demonstrated that genetic factors could be inferred whose expression was determined by the logical relations of "and" and "or", also termed "intersection" and "union", which, in this context, he termed "dominant" and "recessive."

I would like to see the term "Mendelian system" replace Dr. Cotterman's "factor-union system." Genes could then be chromosomal or non-chromosomal, expression could be Mendelian or non-Mendelian, and rare peculiarities, as the haptoglobin 2^m allele, irregular Mendelian.

If this use of Mendelian is objectionable, may I suggest Boolean; the factor-union system is a strictly Boolean system involving binary strings and relationships which are unchanged on simultaneous interchange of 0 and 1 and "and" and "or" throughout.

MORTON: Most geneticists would not want to call cytoplasmic particles Mendelian. Boolean is too general a term for factor-union systems, which permit only one of the four elementary Boolean operations to form phenotypes.

TABLE I

ABO BLOOD GROUPS IN SWITZERLAND

Phenotypes	Separating B and AB				Pooling B and AB			
Phenotypes	Binary Code	Observed Number	Expected Number (F=0)	X^2	Phenotypes	Binary Code	Observed Number	Expected Number (F=0)
A	10	130,201	129,105.53	9.30	A	01	130,201	130,201.00
B	01	23,263	22,018.21	70.37	B+AB	11	31,590	31,590.00
AB	11	8,327	9,626.30	175.37	. . .	. . .	. . .	. . .
O	00	113,873	114,913.95	9.43	O	00	113,873	113,873.00
Total	. . .	275,664	. . .	264.47	. . .	. . .	275,664	. . .
Genes	Binary Code	Frequency	Standard Error		Genes	Binary Code	Frequency	Standard Error
A	10	.295205	.000675		A	01	.298241	.000703
B	01	.059146	.000322		B	11	.059041	.000322
O	00	.645649	.000712		O	00	.642718	.000730

ESTIMATION OF BREEDING VALUES WITH SIMULTANEOUS ADJUSTMENT FOR FIXED ENVIRONMENTAL EFFECTS

Walter R. Harvey

Ohio State University

Techniques which adjust for important fixed environmental effects while simultaneously obtaining estimates of additively genetic breeding values are currently in use on a large scale in several applied animal breeding areas. For example, the National Sire Proving Program of the U.S.D.A. in dairy cattle makes use of this principle. A more direct application of least squares and maximum likelihood techniques is being used to estimate breeding values for the stocks entered in Poultry Random Sample Tests in the United States. (See Agricultural Research Service 44-79, 1960-1968 -- a USDA publication). Several studies are currently underway to determine the value of direct least squares and maximum likelihood estimation procedures in dairy sire proving programs. One report has recently been published (Miller et al., 1968).

Most of the theory and techniques to be discussed in this presentation have been published by Dr. C. R. Henderson of Cornell and his co-workers. A list of the pertinent literature is given at the end of the paper. The primary purpose of this paper is to discuss some of the basic problems involved in computing maximum likelihood estimates of breeding values (or producing abilities) when adjustment must be made from the data for fixed environmental effects. It will be assumed throughout that the primary goal is the selection of individuals so as to maximize subsequent performance.

ONE-WAY CLASSIFICATION

When no adjustments are required for fixed environmental effects or if adjustment factors are used from outside the data, the model will often reduce to the one-way classification model as follows:

$$y_{ij} = \mu + a_i + e_{ij}$$
$$i = 1, 2, \ldots, p$$
$$j = 1, 2, \ldots, n_i$$

Now if the intra-class correlation

$$r = \frac{\sigma_a^2}{\sigma_a^2 + \sigma_e^2},$$

or the ratio σ_e^2/σ_a^2, and μ are known, the maximum likelihood (ML) equations for the a_i are as follows:

	$\hat{\hat{a}}_1$	$\hat{\hat{a}}_2$	...	$\hat{\hat{a}}_p$	RHM
a_1:	$n_1 + \frac{1-r}{r}$	0	...	0	$Y_1 - n_1\mu$
a_2:	0	$n_2 + \frac{1-r}{r}$	...	0	$Y_2 - n_2\mu$
.	.	.		.	.
.	.	.		.	.
.	.	.		.	.
a_p:	0	0	...	$n_p + \frac{1-r}{r}$	$Y_p - n_p\mu$

From these equations it is seen that

$$\hat{\hat{a}}_i = \frac{Y_i - n_i\mu}{n_i + \frac{1-r}{r}}$$

$$= \frac{r(Y_i - n_i\mu)}{n_i r + 1 - r}$$

$$= \frac{n_i r}{1 + (n_i - 1)r} (\bar{y}_i - \mu)$$

and the maximum likelihood estimates of the breeding values or producing abilities are $\mu + \hat{\hat{a}}_i$. These estimates are unbiased in this case regardless of the manner in which the a_i are selected provided the records for each individual, whether they be different measurements of the same trait or measurements of the same trait in equally related family members, are randomly obtained.

If μ is unknown and must be estimated from the data it is essential that the a_i in the data be randomly drawn from the population to which inferences are to be made. The records on each animal or breeding group (the e_{ij}) must also be randomly obtained, of course. However, in practice the number of records for each a_i (the n_i) will often be correlated with the a_i themselves. In this case, one may still obtain an unbiased estimate of μ by weighting each animal or breeding group mean equally, i.e.,

$$\hat{\mu} = \frac{\sum_i \bar{y}_i}{p},$$

the least squares estimate of the population mean. If one uses this estimate of the mean in the ML equations above the ML estimates for the breeding values or producing abilities become

$$\hat{\mu} + \hat{\hat{a}}_i = \hat{\mu} + \frac{n_i r}{1 + (n_i - 1)r} (\hat{a}_i)$$

where $\hat{a}_i = \bar{y}_i - \hat{\mu}$, the least-squares (LS) estimate of a_i. Hence, the ML estimates of breeding values may be obtained by simply regressing the LS estimate of the a_i in this case.

When the a_i and the e_{ij} are randomly drawn and the number of records for each a_i (the n_i) are not correlated with the a_i, one may use the maximum likelihood estimate of μ, if desired. The ML estimate of μ is the mean of the ML estimates of breeding values, whereas the mean of the simple class averages is the LS estimate of μ. When both the mean and the breeding values are estimated simultaneously by ML the equations are as follows:

	$\hat{\hat{\mu}}$	$\hat{\hat{a}}_1$	$\hat{\hat{a}}_2$	...	$\hat{\hat{a}}_p$	RHM
μ:	n.	n_1	n_2	...	n_p	Y.
a_1:	n_1	$n_1 + k$	0	...	0	Y_1
a_2:	n_2	0	$n_2 + k$	...	0	Y_2
.	.	.	.		.	.
.	.	.	.		.	.
.	.	.	.		.	.
a_p:	n_p	0	0	...	$n_p + k$	Y_p

where $k = (1-r)/r$. Since $\sum_i \hat{\hat{a}}_i = 0$, one may impose this restriction in the usual manner in these equations and reduce the number of equations to be solved simultaneously by one, if desired.

When the a_i represented in the data are a random sample from the population of a_i, the e_{ij} are random and uncorrelated, and the frequencies are uncorrelated with the a_i, one may obtain an unbiased estimate of r from the data in the usual manner. However, if the data are not numerous r will be inaccurately estimated and it will often be preferable to use an estimate of r from outside the data.

Computational Example

To illustrate the computational procedures discussed above suppose the following data are available:

Sire	Progeny	y_{ij}
1	1	12
	2	14
2	1	10
	2	6
	3	8

Now

$$\hat{\mu} = (1/2)\left[\frac{12+14}{2} + \frac{10+6+8}{3}\right] = 10.5$$

and suppose r = .4, then

$$\hat{\mu} + \hat{\hat{a}}_1 = 10.5 + \frac{(2)(.4)}{1+(1)(.4)}(13 - 10.5)$$
$$= 10.5 + 1.43$$
$$= 11.93$$
$$\hat{\mu} + \hat{\hat{a}}_2 = 10.5 + \frac{(3)(.4)}{1+(2)(.4)}(8 - 10.5)$$
$$= 10.5 - 1.67$$
$$= 8.83$$

One may also obtain these same estimates of the breeding values by setting up the usual LS equations, imposing the restriction that $\sum_i \hat{a}_i = 0$, solving the remaining equations and then regressing the LS constant estimates for the a_i as follows:

LS Equations

$$\begin{pmatrix} 5 & 2 & 3 \\ 2 & 2 & 0 \\ 3 & 0 & 3 \end{pmatrix} \begin{pmatrix} \hat{\mu} \\ \hat{a}_1 \\ \hat{a}_2 \end{pmatrix} = \begin{pmatrix} 50 \\ 26 \\ 24 \end{pmatrix}$$

Reduced LS Equations

$$\begin{pmatrix} 5 \\ -1 \end{pmatrix} \begin{pmatrix} -1 \\ 5 \end{pmatrix} \begin{pmatrix} \hat{\mu} \\ \hat{a}_1 \end{pmatrix} = \begin{pmatrix} 50 \\ 2 \end{pmatrix}$$

Solution

$$\hat{\mu} = 10.5$$
$$\hat{a}_1 = 2.5$$

and

$$\hat{a}_2 = -\hat{a}_i = -2.5$$

ML Estimates of the a_i

$$\hat{\hat{a}}_1 = \frac{n_1 r}{1+(n_1-1)r}(\hat{a}_1)$$
$$= 1.43$$
$$\hat{\hat{a}}_2 = -1.67$$

as obtained above.

If one assumes that $\mu = \hat{\mu} = 10.5$ and then sets up the ML equations for the a_i with r = .4, the following equations result:

$$\begin{pmatrix} 3.5 & 0 \\ 0 & 4.5 \end{pmatrix} \begin{pmatrix} \hat{\hat{a}}_1 \\ \hat{\hat{a}}_2 \end{pmatrix} = \begin{pmatrix} 5.0 \\ -7.5 \end{pmatrix}$$

where:

$$3.5 = n_1 + \frac{1-r}{r} = 2 + 1.5$$
$$4.5 = n_2 + \frac{1-r}{r} = 3 + 1.5$$
$$5.0 = 26 - (2)(10.5)$$
$$-7.5 = 24 - (3)(10.5)$$

The solution of these two equations, of course, yields the same ML estimates of the a_i as obtained above. It is obvious that the right hand members of these equations would be different and hence the ML estimates of breeding values would also be different if a value for μ other than 10.5 were used. The estimate of μ which forces the $\hat{\hat{a}}_i$ to sum to zero is the ML estimate of μ and is obtained by solving the following equations:

$$\begin{pmatrix} 5 & 2 & 3 \\ 2 & 3.5 & 0 \\ 3 & 0 & 4.5 \end{pmatrix} \begin{pmatrix} \hat{\hat{\mu}} \\ \hat{\hat{a}}_1 \\ \hat{\hat{a}}_2 \end{pmatrix} = \begin{pmatrix} 50 \\ 26 \\ 24 \end{pmatrix}$$

One may first reduce these equations by one since $\sum_i \hat{\hat{a}}_i = 0$ as follows:

$$\begin{pmatrix} 5 & -1 \\ -1 & 8 \end{pmatrix} \begin{pmatrix} \hat{\hat{\mu}} \\ \hat{\hat{a}}_1 \end{pmatrix} = \begin{pmatrix} 50 \\ 2 \end{pmatrix}$$

The solution is

$$\hat{\hat{\mu}} = 10.31$$
$$\hat{\hat{a}}_1 = 1.54$$

and

$$\hat{\hat{a}}_2 = -\hat{\hat{a}}_1 = -1.54$$

The estimated breeding values are

$$\hat{\hat{\mu}} + \hat{\hat{a}}_1 = 11.85$$
$$\hat{\hat{\mu}} + \hat{\hat{a}}_2 = 8.77$$

as compared with 11.93 and 8.83 when the LS estimate of μ was used.

TWO-WAY CLASSIFICATION WITH NO INTERACTION

The model in this case is the same as for the incomplete block design with one replicate:

$$y_{ijk} = \mu + a_i + b_j + e_{ijk}$$
$$i = 1, 2, \ldots, p$$
$$j = 1, 2, \ldots, q$$
$$k = 1, 2, \ldots, n_{ij}$$

The computational procedure which should be followed in order to obtain the "best" estimates of the breeding values or producing abilities (the $\mu + a_i$) depends on the a priori information available and the characteristics of the data available. It will be assumed that the b_j are fixed environmental effects and that these must be estimated from the available data in all cases. In order for the estimates of the b_j to be unbiased the records available on each animal (a_i class) must be distributed at random with respect to the different b_j classes, or the expectation of the mean of the a_i available in each b_j class must be the same.

When μ is Unknown and r is Known

When this situation exists one must decide whether it is valid to assume that (i) the a_i were chosen at random from the population of a_i and that (ii) the expectation of the mean of the a_i available for each b_j class is the same; i.e., $E(\bar{a}_{.1}) = E(\bar{a}_{.2}) = \ldots = E(\bar{a}_{.q})$. If the first assumption is invalid, no unbiased estimate of μ exists from the data unless one can adjust for the nonrandomness in the selection of the a_i. However, it is possible to obtain unbiased estimates of μ and the b_j even though the second assumption is not true. A satisfactory estimation procedure to follow when μ is unknown, r is known, assumption (i) is true, and assumption (ii) is false is as follows:

1). Set up the LS equations in the usual manner for the $\mu + a_i$ and the b_j. If p is large the equations for the $\mu + a_i$ may easily be "swept out" by LS absorption procedures; i.e., the coefficients in the equations for the b_j and the corresponding right hand members are adjusted for the confounding of the b_j with the $\mu + a_i$ effects. In this case, the estimate of μ may be obtained from the mean of the $\hat{\mu} + \hat{a}_i$, which are computed with a "back solution" procedure, viz.,

$$\hat{\mu} + \hat{a}_i = \frac{1}{n_{i.}}(Y_{i.} - \sum_j n_{ij}\hat{b}_j).$$

However, an unbiased and often satisfactory estimate of μ may be more easily computed as

$$\bar{\mu} = \frac{1}{n_{..}} (Y_{..} - \sum_j n_{.j} \hat{b}_j)$$

since the a_i are random.

2). Compute the ML estimates of the breeding values or producing abilities

$$\hat{\mu} + \hat{\hat{a}}_i = \hat{\mu} + \frac{A_{ii}^{-1} r}{1 + (A_{ii}^{-1} - 1)r} (\hat{a}_i)$$

where the A_{ii} are the inverse diagonal elements for the $\hat{\mu} + \hat{a}_i$ and these may be computed during the "back solution" for the $\hat{\mu} + \hat{a}_i$ as follows:

$$A_{ii} = \frac{1}{n_{i.}} [1 + \frac{1}{n_{i.}} \sum_j^{q-1} \sum_{j'} (n_{ij} - n_{iq})(n_{ij'} - n_{iq}) C_{jj'}]$$

and the $C_{jj'}$ are the inverse elements involving the b_j when the restriction is imposed that $\sum_j \hat{b}_j = 0$ and the equation for b_q has been eliminated.

If $\bar{\mu}$ is used as the estimate of μ instead of $\hat{\mu}$, the $\bar{a}_i = \hat{\mu} + \hat{a}_i - \bar{\mu}$ would be used to compute the ML estimates of breeding values, $\bar{\mu} + \hat{\hat{a}}_i$, instead of the $\hat{a}_i$.

The regression coefficient in the formula for $\hat{\mu} + \hat{\hat{a}}_i$ may be written in terms of the variance components, if desired, viz.,

$$\frac{\sigma_a^2}{\sigma_a^2 + A_{ii}\,\sigma_e^2} .$$

When μ is unknown, r is known, and assumptions (i) and (ii) are both true, one can estimate the b_j a bit more accurately by making use of the differences among the a_i which are confounded with the b_j, as well as the differences among the b_j within the a_i classes (Cunningham and Henderson, 1966). In addition, the ML estimate of μ may be obtained from the data. The estimation procedure is given below.

1). The complete set of equations is as follows:

	$\hat{\mu}$	$\hat{a}_i$	$\hat{b}_j$	RHM
μ:	$n_{..}$	$n_{i.}$	$n_{.j}$	$Y_{..}$
a_i:	$n_{i.}$	$_0\; n_{i.} + k\; ^0$	n_{ij}	$Y_{i.}$
b_j:	$n_{.j}$	n_{ij}	$_0\; n_{.j}\; ^0$	$Y_{.j}$

where $k = \frac{1-r}{r}$, as before. Only one restriction or constraint is required in order to obtain a unique solution to these equations. A convenient restriction is that $\sum_j \hat{b}_j = 0$.

When the number of animals is large one may absorb the equations for the a_i in the usual manner. However, in this case, the equation for μ is not absorbed when the equations for the a_i are absorbed by ML since $r < 1.0$.

As was true with the one-way classification, the $\sum_i \hat{a}_i$ will equal zero since these constants will now be measured as a deviation from the ML estimate of the mean, $\hat{\hat{\mu}}$. Finally, the estimates of the breeding values of producing abilities will be the $\hat{\mu} + \hat{\hat{a}}_i$.

This direct ML procedure combines two sets of estimates of the b_j, namely (i) the set that may be computed by conventional LS procedures under the full model and (ii) the set that may be computed on an "inter-block" basis using weighted LS procedures. In the first case, one would normally set up the equations for the b_j on a "within A class basis," i.e., the equations for the $\mu + a_i$ would be absorbed in the usual manner. In the second case, one would set up the equations for the b_j on a "between A class basis." If the investigator suspects that these two sets of estimates for the b_j are different he should examine both sets prior to using the direct ML procedure. There has been one case reported where both sets of LS estimates of the fixed effects are biased but in opposite directions (Henderson et al., 1959) so that the ML procedure exactly accounts for both biases to yield unbiased estimates. This appears to be a special case, and in general, it seems more likely that the "inter-block" estimates will be biased and the "intra-block" estimates will be either unbiased or biased to a lesser extent.

When μ and r are Unknown

In this case, one will need to estimate σ_a^2 and σ_e^2 from the data. In order to obtain unbiased estimates of these variance components it is necessary that the a_i be randomly drawn. Although one may be able to assume that this is true for the entire sample of a_i it is not too uncommon to find that $E(\bar{a}_{.j}) \neq E(\bar{a}_{.j'})$, where $j \neq j'$, even though the e_{ij} are uncorrelated. For example, if the a_i represent stocks entered in random sample tests in a given year and the b_j represent the test location effects it is unreasonable to assume that a random sample of stocks were entered in each test, i.e., that $E(\bar{a}_{.j}) = E(\bar{a}_{.j'})$.

When these conditions exist one may use Method 3 of Henderson to obtain unbiased estimates of the b_j and the variance components. In the process of doing this analysis (see Henderson, 1953, and Harvey, 1960) one may, with little additional work, obtain the $\hat{a}_i$, $\hat{\mu}$ and the inverse diagonal elements for the $\hat{\mu} + \hat{a}_i$, the A_{ii}. The least squares constant estimates of the a_i are then regressed as indicated above to obtain the ML estimates of the breeding values or producing abilities.

If it seems valid to assume that the a_i are drawn at random and that they are distributed at random with respect to the b_j, one may complete a ML analysis (Cunningham and Henderson, 1968) to estimate simultaneously the variance components and the fixed effects, the b_j, by an iterative procedure. The procedure is as follows:

1). Complete a Method 3 analysis to estimate r_1.

2). Modify the diagonals of the LS equations for the a_i with $(1-r_1)/r_1$ (as indicated above) and re-estimate the variance components from the ML sum of squares for A and its expectation.

3). Repeat (2) until the estimate of r and the fixed effects, the b_j, remain constant. The ML estimates of μ and the a_i will be obtained directly from the last iteration.

Computational Example

Suppose the data given previously were also classified into b_j classes to yield the following set of LS equatons:

$$\begin{pmatrix} 5 & 2 & 3 & 4 & 1 \\ 2 & 2 & 0 & 1 & 1 \\ 3 & 0 & 3 & 3 & 0 \\ 4 & 1 & 3 & 4 & 0 \\ 1 & 1 & 0 & 0 & 1 \end{pmatrix} \begin{pmatrix} \hat{\mu} \\ \hat{a}_1 \\ \hat{a}_2 \\ \hat{b}_1 \\ \hat{b}_2 \end{pmatrix} = \begin{pmatrix} 50 \\ 26 \\ 24 \\ 36 \\ 14 \end{pmatrix}$$

When the restrictions are imposed that $\sum_i \hat{a}_i = \sum_j \hat{b}_j = 0$ and the a_2 and b_2 equations are eliminated the following reduced set of equations remain:

$$\begin{pmatrix} 5 \\ -1 \\ 3 \end{pmatrix} \begin{pmatrix} -1 & 3 \\ 5 & -3 \\ -3 & 5 \end{pmatrix} \begin{pmatrix} \hat{\mu} \\ \hat{a}_1 \\ \hat{b}_1 \end{pmatrix} = \begin{pmatrix} 50 \\ 2 \\ 22 \end{pmatrix}$$

The solution to these equations and $\hat{a}_2$ and $\hat{b}_2$ are

$$\hat{\mu} = 11 \qquad \hat{b}_1 = -1$$
$$\hat{a}_1 = 2 \qquad \hat{b}_2 = 1$$
$$\hat{a}_2 = -2$$

The inverse of the reduced set of equations is

$$\begin{pmatrix} 5 & -1 & 3 \\ -1 & 5 & -3 \\ 3 & -3 & 5 \end{pmatrix}^{-1} = \frac{1}{12}\begin{pmatrix} 4 & -1 & -3 \\ -1 & 4 & 3 \\ -3 & 3 & 6 \end{pmatrix}$$

and the inverse diagonal elements corresponding to $\hat{\mu} + \hat{a}_1$ and $\hat{\mu} + \hat{a}_2$ are

$$A_{11} = \frac{4}{12} + \frac{4}{12} + (2)(-\frac{1}{12}) = \frac{1}{2}$$
$$A_{22} = \frac{4}{12} + \frac{4}{12} + (2)(\frac{1}{12}) = \frac{5}{6}$$

Hence, if $r = .4$, the estimated breeding values or producing abilities would be

$$\hat{\mu} + \hat{\hat{a}}_1 = 11 + \frac{(2)(.4)}{1+(2-1)(.4)}\,(2)$$
$$= 11 + .571\,(2)$$
$$= 11 + 1.143$$
$$= 12.143$$

$$\hat{\mu} + \hat{\hat{a}}_2 = 11 + \frac{(6/5)(.4)}{1+[(6/5)-1](.4)}\,(-2)$$
$$= 11 + .444\,(-2)$$
$$= 11 - .889$$
$$= 10.111$$

Now if $r = .4$, the a_i were randomly chosen, and these effects are distributed at random with respect to the b_j effects, one could complete the ML analysis directly as follows:

ML Equations

$$\begin{pmatrix} 5 & 2 & 3 & 4 & 1 \\ 2 & 3.5 & 0 & 1 & 1 \\ 3 & 0 & 4.5 & 3 & 0 \\ 4 & 1 & 3 & 4 & 0 \\ 1 & 1 & 0 & 0 & 1 \end{pmatrix} \begin{pmatrix} \hat{\mu} \\ \hat{a}_1 \\ \hat{a}_2 \\ \hat{b}_1 \\ \hat{b}_2 \end{pmatrix} = \begin{pmatrix} 50 \\ 26 \\ 24 \\ 36 \\ 14 \end{pmatrix}$$

One restriction must be imposed on the $\hat{b}_j$ in order to obtain a unique solution. When the restriction is imposed that $\sum_j \hat{b}_j = 0$ the following equations result:

$$\begin{pmatrix} 5 & 2 & 3 & 3 \\ 2 & 3.5 & 0 & 0 \\ 3 & 0 & 4.5 & 3 \\ 3 & 0 & 3 & 5 \end{pmatrix} \begin{pmatrix} \hat{\hat{\mu}} \\ \hat{\hat{a}}_1 \\ \hat{\hat{a}}_2 \\ \hat{\hat{b}}_1 \end{pmatrix} = \begin{pmatrix} 50 \\ 26 \\ 24 \\ 22 \end{pmatrix}$$

One may further reduce these equations since $\sum_i \hat{\hat{a}}_i = 0$ to the following:

$$\begin{pmatrix} 5 & -1 & 3 \\ -1 & 8 & -3 \\ 3 & -3 & 5 \end{pmatrix} \begin{pmatrix} \hat{\hat{\mu}} \\ \hat{\hat{a}}_1 \\ \hat{\hat{b}}_1 \end{pmatrix} = \begin{pmatrix} 50 \\ 2 \\ 22 \end{pmatrix}$$

Estimates of all the constants are

$$\hat{\hat{\mu}} = 11.25 \qquad \hat{\hat{b}}_1 = -1.75$$
$$\hat{\hat{a}}_1 = 1.00 \qquad \hat{\hat{b}}_2 = 1.75$$
$$\hat{\hat{a}}_2 = -1.00$$

The estimates of the breeding values or producing abilities using this procedure are

$$\hat{\hat{\mu}} + \hat{\hat{a}}_1 = 11.25 + 1.00 = 12.25$$

$$\hat{\hat{\mu}} + \hat{\hat{a}}_2 = 11.25 - 1.00 = 10.25$$

compared with 12.14 and 10.11 when the regressed LS estimates of the a_i and μ were used.

It should be noted that even though the estimates of breeding values do not differ materially using these two methods, the estimates of b_j do differ materially. Of course, this is a very small example and such differences can therefore easily occur due to chance. However, these differences illustrate the fact that if the a_i effects are not distributed at random with respect to the b_j or if $E(\bar{a}_{.j}) \ne E(\bar{a}_{.j'})$ one will obtain biased estimates of the b_j using direct maximum likelihood procedures.

EXTENSIONS TO COMPLEX MODELS
No Interaction of Random With Fixed Effects

Regardless of the number of sets of fixed effects, whether discrete or continuous independent variables, the principles described above still apply, provided there is only one set of non-interacting random effects. Therefore, for these cases one may partition the general linear model in matrix algebra as follows:

$$y = WB_1 + XB_2 + e$$

where y is a column vector of the set of observations, W and X are the design matrices for the B_1 and B_2 sets of effects, respectively, and e is a column vector of the random errors.

<u>Regressed LS Means Procedure</u> If one plans to first obtain the LS estimates of the random effects, say the a_i, and then regress these making use of an a priori estimate of r, B_1 would contain the $\mu + a_i$ and B_2 would contain all sets of fixed effects. The computing procedures that would usually be most convenient in this case would be as follows:

1). Sort the data in sequence by A classes and absorb the $\mu + a_i(B_1)$ equations into the equations for the B_2 effects. If X_R is the X matrix after necessary restrictions have been imposed on the fixed effects, then the entire set of LS equations would be

$$\begin{pmatrix} W'W \\ X'_R W \end{pmatrix} \begin{pmatrix} W'X_R \\ X'_R X_R \end{pmatrix} \begin{pmatrix} B_1 \\ B_2 \end{pmatrix} = \begin{pmatrix} W'y \\ X'_R y \end{pmatrix}$$

and $W'W$ is a diagonal matrix. To simplify notation let

$$\begin{pmatrix} W'W \\ X'_R W \end{pmatrix} \begin{pmatrix} W'X_R \\ X'_R X_R \end{pmatrix} = \begin{pmatrix} D & N \\ N' & S \end{pmatrix}$$

and

$$\begin{pmatrix} W'y \\ X'_R y \end{pmatrix} = \begin{pmatrix} Y_1 \\ Y_2 \end{pmatrix}.$$

The reduced set of equations after absorption of the B_1 equations are

$$(S - N'D^{-1}N)B_2 = (Y_2 - N'D^{-1}Y_1)$$

and

$$\hat{B}_2 = (S - N'D^{-1}N)^{-1}(Y_2 - N'D^{-1}Y_1).$$

2). Compute the LS estimates of the $\mu + a_i$ (the $\hat{B}_1$) and the inverse diagonal elements for these constants. These may all be computed simultaneously with a "back solution" technique.

If

$$\begin{pmatrix} D & N \\ N' & S \end{pmatrix}^{-1} = \begin{pmatrix} A & G \\ G' & C \end{pmatrix}$$

then

$$A_{ii} = D_{ii}^{-1}\left[1 + D_{ii}^{-1}\sum_{j}^{q}\sum_{j'}^{q} N_{ij} N_{ij'} C_{jj'}\right]$$

where D_{ii} is the number of observations in the ith A class, q is the number of equations for the fixed effects in the reduced set of equations, the N_{ij} are elements in the N matrix, and the $C_{jj'}$ are the inverse elements of the $S - N'D^{-1}N$ matrix.

The $\hat{\mu} + \hat{a}_i$ may be computed at the same time as follows:

$$\hat{B}_1 = \hat{\mu} + \hat{a}_1 = D^{-1}(Y_1 - N\,\hat{B}_2).$$

3). If μ must be estimated from the data compute the estimate as the simple average of the $\hat{\mu} + \hat{a}_i$ or from

$$\frac{Y. - N.\hat{B}_2}{\mathrm{tr}\ D}$$

where Y. is the over-all total of the individual observations.

4). Compute the ML estimates of the breeding values or producing abilities:

$$\hat{\mu} + \hat{a}_i = \hat{\mu} + \frac{A_{ii}^{-1}\ r}{1+(A_{ii}^{-1}-1)r}\,(\hat{a}_i)\ .$$

Direct ML Procedure If one can assume that the recovery of the information regarding the fixed effects which is confounded with the random effects (the a_i) will not lead to biased estimates, the direct ML procedure may be used to improve slightly the accuracy of adjustments for the fixed environmental effects. In this case, the computing procedures are as follows for the general model being considered:

1). Absorb the equations for the a_i (the B_1) into the equations for μ and the other fixed effects (the B_2) by maximum likelihood. This absorption process is accomplished with a simple modification in the diagonal matrix D (Henderson et al., 1959). The quantity $\frac{1-r}{r}$ is added to each diagonal element, where r is the intra-class correlation as defined previously. The reduced set of equations for the fixed effects will now be

$$[S - N'(D + Ik)^{-1}N]\,B_2 = [Y_2 - N'(D + Ik)^{-1}Y_1]$$

where I is the identity matrix and $k = \frac{1-r}{r}$. The solution of these equations now includes an estimate of μ as well as all fixed effects and is given by

$$\hat{B}_2 = [S - N'(D + Ik)^{-1}N]^{-1}[Y_2 - N'(D + Ik)^{-1}Y_1]\,.$$

2). Compute the ML estimates of the a_i by a "back solution" procedure,

$$\hat{a}_i = (D + Ik)^{-1}(Y_1 - N\,\hat{B}_2).$$

If one desires to obtain the $\hat{\hat{\mu}} + \hat{\hat{a}}_i$, the ML estimates of breeding values or producing abilities, the column for μ is omitted from N, and $\hat{\hat{\mu}}$ is omitted from $\hat{B}_2$ in calculating $N\,\hat{B}_2$.

Iterative ML Procedure When r must be estimated from the data and it is valid to assume that the "recovery of inter-block information" does not result in biased estimates of the fixed effects, one should use the iterative ML procedure described recently by Cunningham and Henderson (1968). If this assumption cannot be made, the results are likely to be more useful if one stops after the first round of iteration, i.e., with the estimates of the variance components σ_a^2 and σ_e^2 that are obtained from a Method 3 analysis of Henderson and with the LS estimates of the constants.

Interactions of Random With Fixed Effects

A direct ML procedure has not yet been published (to the author's knowledge) to handle analyses which fall into this class. However, the extension of the "regressed LS means" procedure to models of this type seems to be straightforward. If the model contains only one set of random effects (say a_i), such as stocks, animals, sires, etc., and only one interaction of the random set of effects with some set of fixed effects (say b_j), the "regressed LS means" (the breeding values or producing abilities) are computed as follows:

$$\bar{\mu} + \hat{a}_i = \bar{\mu} + \hat{\beta}\ \hat{a}_i$$

where $\hat{\beta}$ is the regression coefficient, $\bar{\mu}$ is some estimate of the population mean and $\hat{a}_i$ is the LS estimate of a_i when the interaction $(ab)_{ij}$ is ignored. The regression coefficient may be written in terms of the variance components, the inverse elements for the $\hat{\mu} + \hat{a}_i$ and the numbers as

$$\hat{\beta} = \frac{\sigma_a^2 + A_{ii}k_i\sigma_{ab}^2}{\sigma_a^2 + A_{ii}(\sigma_e^2 + k_i\sigma_{ab}^2)}$$

and

$$k_i = \frac{\sum_j n_{ij}^2}{n_{i.}}\ .$$

If

$$r_1 = \frac{\sigma_a^2}{\sigma_a^2 + \sigma_{ab}^2 + \sigma_e^2}$$

$$r_2 = \frac{\sigma_a^2 + \sigma_{ab}^2}{\sigma_a^2 + \sigma_{ab}^2 + \sigma_e^2}$$

one may write the regression coefficient as

$$\hat{\beta} = \frac{A_{ii}^{-1}r_1 + k_i(r_2 - r_1)}{1 + (k_i - 1)r_2 + (A_{ii}^{-1} - k_i)r_1}\ .$$

The second term in the numerator of the regression coefficient in each form of $\hat{\beta}$ given above exists because of the assumption that the b_j effects are fixed and that these same effects are present under the environmental conditions for which the animals or breeding groups are to be selected. If the b_j represent effects such as years or locations and one really wants to select the animals or breeding groups for use in another set of years or locations, the second term in the numerator should be deleted in computing the "regressed means." Of course, this procedure then requires one to assume that the variance components (or correlations) used in computing the $\hat{\mu} + \hat{a}_i$ are unbiased estimates, even though other b_j classes are involved.

If satisfactory estimates of μ, r_1 and r_2 are available from outside the data, one will obtain unbiased estimates of the breeding values or producing abilities even though the a_i in the data were not chosen at random from the population of a_i. However, if these parameters must be estimated from the data the sample of a_i must have been randomly drawn. In this case, the variance components would be estimated by making use of Method 3 of Henderson (Searle and Henderson, 1961 and 1967, and Harvey, 1964).

If the set of random effects (the a_i) interacts with two sets of discrete fixed effects (say b_j and c_k) there will be four intra-class correlations (or four variance components) that must be known or that must be estimated from the data in order to compute the "regressed" LS means, namely

$$r_1 = \frac{\sigma_a^2}{\sigma_a^2 + \sigma_{ab}^2 + \sigma_{ac}^2 + \sigma_e^2}$$

$$r_2 = \frac{\sigma_a^2 + \sigma_{ab}^2}{\sigma_a^2 + \sigma_{ab}^2 + \sigma_{ac}^2 + \sigma_e^2}$$

$$r_3 = \frac{\sigma_a^2 + \sigma_{ac}^2}{\sigma_a^2 + \sigma_{ab}^2 + \sigma_{ac}^2 + \sigma_e^2}$$

$$r_4 = \frac{\sigma_a^2 + \sigma_{ab}^2 + \sigma_{ac}^2}{\sigma_a^2 + \sigma_{ab}^2 + \sigma_{ac}^2 + \sigma_e^2} .$$

In this case, the ML estimates of the A class means (the regressed means) may be computed as follows:

$$\hat{\mu} + \hat{a}_i = \hat{\mu} + \frac{\sigma_a^2 + A_{ii}(k_{1i}\,\sigma_{ab}^2 + k_{2i}\,\sigma_{ac}^2)}{\sigma_a^2 + A_{ii}(\sigma_e^2 + k_{1i}\,\sigma_{ab}^2 + k_{2i}\,\sigma_{ac}^2)} \quad (\hat{a}_i)$$

where

$$k_{1i} = \frac{\sum_j n_{ij.}^2}{n_{i..}}$$

$$k_{2i} = \frac{\sum_k n_{i.k}^2}{n_{i..}} .$$

Again, the use of the above regression coefficient assumes that the environmental conditions under which the selected animals or breeding groups are to be used will contain the same b_j and c_k classes as are present in these data. If new c_k classes will be present (such as new years) but the same b_j classes will be present, one would want to omit the term which contains σ_{ac}^2 from the numerator.

If new classes of both the b_j and c_k will be present, one would retain only σ_a^2 in the numerator of the regression coefficient.

The derivation of the appropriate regression coefficient for a particular set of data and for a particular set of subsequent environmental conditions will usually not be difficult. One must simply keep in mind that the regression coefficient desired is the regression of true breeding value (or true producing ability) for the particular set of environmental conditions specified on the available phenotypic measurement. When the variance components are estimated from the data this will always be the variance of $\hat{\mu} + \hat{a}_i$ divided into σ_a^2 if new classes of fixed effects will be present in the future environmental conditions. If the same fixed effect classes are to be present one must add terms to the numerator as shown in the above examples.

REFERENCES

Agricultural Research Service. Annual reports of random sample egg production tests -- Combined summary. ARS 44-79 series. 1960 - 1968.

Cunningham, E. P. and Henderson, C. R. 1966. Analytical techniques for incomplete block experiments. Biometrics 22:829-842.

Cunningham, E. P. and Henderson, C. R. 1968. An iterative procedure for estimating fixed effects and variance components in mixed model situations. Biometrics 24:13-25.

Harvey, Walter R. 1960. Least squares analysis of data with unequal subclass numbers. U. S. Dept. Agri., ARS 20-8. 157pp.

Harvey, Walter R. 1964. Computing procedures for a generalized least squares analysis program. Mimeograph paper presented at the Analysis of Variance Conference at Ft. Collins, Colorado, July.

Henderson, C. R. 1948. Estimation of general, specific and maternal combining abilities in crosses among inbred lines of swine. Unpublished Ph.D. Thesis. Iowa State University Library. (See pages 35-41, especially).

Henderson, C. R. 1953. Estimation of variance and covariance components. Biometrics 9:226-252.

Henderson, C. R., Kempthorne, O., Searle, S. R., and von Krosigk, C. M. 1959. The estimation of environmental and genetic trends from records subject to culling. Biometrics 15:192-218.

Henderson, C. R. 1963. Selection index and expected genetic advance. Statistical Genetics and Plant Breeding. NAS-NRC 982.

Miller, R. H., Harvey, W. R., Tabler, K. A., McDaniel, B. T., Corley, E. L. 1966. Maximum likelihood estimates of age effects. Jour. Dairy Sci. 49:65-73.

Miller, R. H., McDaniel, B. T. and Plowman, R. D. 1968. Comparison of three methods of sire evaluation. Jour. Dairy Sci. 51:782-791.

Searle, S. R. and Henderson, C. R. 1967. Computing procedures for estimating components of variance in the two-way classification, mixed model. Biometrics 17:607-616. 1961. Also see the correction published for this paper in Biometrics 23:852-853.

DISCUSSION ON ESTIMATION OF BREEDING VALUES

ELSTON: Dr. Harvey has presented in his paper many different and useful methods for computing breeding values. I think, however, that those who are not experts in this field may not be able to see the wood for the trees, and so I propose here to clarify the issues involved, especially with regard to the choice between least squares and maximum likelihood estimation. In trying to estimate any quantity or parameter three main issues are involved: (i) the underlying model, which includes the distributional properties to be assumed, (ii) the method of estimation and (iii) the computational method. It is desirable and helpful to consider each of these issues separately. As Dr. Harvey has very adequately dealt with the last of these issues, I shall concentrate on the first two. I shall use the same notation as Dr. Harvey, so that $\mu + a_i$ represents a breeding value that we wish to estimate.

ONE-WAY CLASSIFICATION

The simplest model for the observed values y_{ij} can be expressed as

$$\left.\begin{array}{l} y_{ij} = \mu + a_i + e_{ij}, \\ \mu \text{ and } a_i \text{ constants, } e_{ij} \text{ a random variable with } E(e_{ij}) = 0. \end{array}\right\} \quad (1)$$

In this model the form of the distribution of y_{ij} is not completely specified, and so maximum likelihood (ML) estimation is not possible. We can obtain least squares estimates of $(\mu + a_i)$, but neither μ nor the individual a_i are estimable - i.e., there is no function of the observations y_{ij} whose expected value is μ or a_i. (We can only estimate the a_i if either μ or one of them - or some function of them - is known).

If we modify (1) by adding the assumption that the e_{ij} are normally distributed with variance-covariance matrix $I\sigma^2$, i.e., the e_{ij} are independent $N(0, \sigma^2)$ random variables, then we can obtain ML estimates of $(\mu + a_i)$, and these are the same as the least squares estimates. The least squares estimates are the same whatever assumptions we make about the distribution of e_{ij}, so long as $E(e_{ij}) = 0$. The ML estimates depend very much upon the distribution of the e_{ij}: the particular case of independent $N(0, \sigma^2)$ random variables always makes least squares and ML estimates identical.

When Dr. Harvey talks about ML estimation, however, he is assuming a third, and quite different, model. One form of it is often expressed as follows:

$$\left.\begin{array}{c} y_{ij} = \mu + a_i + e_{ij}, \\ \mu \text{ constant, } a_i \text{ independent } N(0, \sigma_a^2), \; e_{ij} \text{ independent } N(0, \sigma_e^2), \\ a_i \text{ and } e_{ij} \text{ independent.} \end{array}\right\} \quad (2)$$

This is not quite the model Dr. Harvey is assuming, but I should like to consider it to clarify certain points. Using (2) we can write out the likelihood of the observed sample as a function of μ and a_i, and we find that, by the usual methods, ML estimates of $\mu + a_i$ can be obtained. This is a peculiar result, since (2) assumes that the a_i are random variables, not parameters. Rather than talking about estimating random variables, I think a better and more logical formulation of the problem is to call the a_i constant parameters, but to add that we know that the a_i in our sample are such as might have been obtained as a random sample from the distribution $N(0, \sigma_a^2)$. Using this knowledge ML estimation ends up with the same result as assuming (2), but I personally find (2) difficult to accept as a conceptual framework. Assuming (2) we have $E(\mu + a_i) = \mu$, and so any least squares estimate must be of μ alone. If we reformulate the problem as I have suggested, then we say that the a_i are constants and the least squares method simply ignores the further information about them: the least squares estimates are therefore the same as those found assuming (1).

Now the ML estimates $\mu + \hat{a}_i$ that Dr. Harvey gives are ML whether the a_i in our sample are such as might have come from a distribution with mean 0 or any other mean; and it is easy to see that these estimates must be unbiased if $E(y_{ij}) = \mu + a_i$. But they are only ML estimates if this distribution is assumed to be normal. If we are to suppose there has been non-random selection of the a_i from some normal distribution, then the a_i in our sample are not such as might have been obtained as a random sample from a normal distribution; and in this case, under a more appropriate model, the estimates Dr. Harvey gives are neither ML nor least squares estimates.

TWO-WAY AND HIGHER-WAY CLASSIFICATIONS

Provided the extra fixed environmental effects that are introduced do not interact with the a_i, all higher-way classifications can be considered simply as a two-way classification. For example, if we wish to estimate breeding values correcting for herd and year effects, both these latter being fixed, then we can let each separate herd-year effect be represented by b_j and write

$$y_{ijk} = \mu + a_i + b_j + e_{ijk} \quad (3)$$

The model analogous to (1) appends to (3)

μ, a_i and b_j constants, e_{ijk} a random variable with $E(e_{ijk}) = 0$.

Under this model, however, $(\mu + a_i)$ is not estimable: we must make some further assumptions in order to make any reasonable estimates of the breeding values.

The model analogous to (2) appends to (3)

μ and b_j constants, a_i independent $N(0, \sigma_a^2)$, e_{ijk} independent $N(0, \sigma_e^2)$, a_i and e_{ijk} independent.

As before, however, I think it is better to consider the a_i as constants, but such as might have been obtained as a random sample from the distribution $N(0, \sigma_a^2)$. On this assumption we cannot obtain least squares estimates of $(\mu + a_i)$, but we can obtain ML estimates. What these estimates are, and whether they are unbiased, depends, of course, on what is known about μ, σ_a^2 and σ_e^2. Dr. Harvey gives computational procedures for the various possibilities. But it is important to note, as he says in other words, that when fixed environmental effects are added to the model unbiased estimates of the breeding values are possible only when the a_i in our sample are such as might have come from a distribution with mean zero.

Finally, with regard to the case where there are interactions between the a_i and the b_j, I am not sure just what model underlies the methods that Dr. Harvey gives. I think it is very important that the model should be clearly specified, and in this case I believe that even definition of the breeding values is no trivial matter. In the usual mixed model, where the a_i are random variables, it is common to assume that any interaction effects are random variables that are independent of the a_i; I suspect this is one assumption that Dr. Harvey is making, but which is hardly defensible. The only logical way that I know of considering interactions in a mixed model is to consider the underlying distribution to be multivariate, in the way Scheffé (1959) does. But in this formulation the a_i are definitely random variables, and so not estimable parameters.

Scheffé, H. 1959. The analysis of variance. Wiley, New York.

GENETIC EXPERIENCE WITH A GENERAL MAXIMUM LIKELIHOOD ESTIMATION PROGRAM

T. Edward Reed
Departments of Zoology, Paediatrics, and Anthropology
University of Toronto

Maximum likelihood (ML) estimation computer programs for gene frequency estimation have been written by Balakrishnan and Sanghvi (1965), Kurczynski and Steinberg (1967), and MacCluer et al. (1967). A general program for ML estimation of any genetic parameter, gene frequency or not, would clearly be useful also but, when I sought such a program some two and a half years ago, I could find none. I therefore wrote my own, first in a version specifically for estimating certain parameters affecting segregation in the MN blood group system, and, about a year later, in a completely general form (MAXLIK). I have used this general form rather extensively in a variety of genetic situations and have found it very useful and, I believe, rather "robust."

MAXLIK is written in Fortran IV and has been used on the IBM 7094 computers of the University of California (Berkeley) and University of Toronto. Since Dr. W. J. Schull has successfully adapted MAXLIK for his IBM 1130 computer at the University of Michigan, we can say that MAXLIK is neither location- nor computer-bound (Reed and Schull, 1968). I would like to describe very briefly the theory behind MAXLIK and then list and discuss its applications to date.

THEORY

MAXLIK obtains ML estimates of parameters x_i which specify the probabilities f_j of specified classes. It uses Fisher's efficient scoring method as described by Rao (1952, pp. 168-172). Figure 1 gives the basic equations. The partial derivatives are obtained by numerical differentiation as shown in Figure 2. This differentiation has been shown to be adequate by direct comparison of results obtained by it with those obtained by algebraic differentiation. The program was checked initially, in ABO gene frequency estimation, with desk calculator results and also with various published calculations. Later, Rh gene frequency estimates (5 parameters) of MAXLIK were compared with those of the MAXIM program of Kurczynski and Steinberg (1967); agreement was excellent.

INPUT AND OUTPUT

The user must supply a subroutine, FREQ, relating the parameters x_i to the probabilities f_j for the problem in question. An example, A_1A_2BO gene frequency estimation with possible phenotype misclassification, is shown in Figure 3. In addition, the numbers of parameters, classes, and problems for FREQ must be specified and the number in each class and a set of trial estimates for the x_i must be given. Other input features are seen in a sample output, shown in Figure 4. These include labels, comments, and the degrees of freedom.

Output includes final estimates of the x_i (if iteration converged), their standard errors, and the goodness-of-fit of expected class numbers to the observed. Other output items are also to be seen on Figures 4 and 5.

Figure 1 - Maximum Likelihood Scoring Procedure in MAXLIK

Parameters: $x_1, x_2, \cdots x_i \cdots x_{imax}$ (E.g., gene frequencies)

Classes: $c_1, c_2, \cdots c_j \cdots c_{jmax}$ (E.g., phenotypes)

Class numbers: n_j . Class probabilities (expected): f_j .

Score for $x_i = S_i = \frac{\partial \log L}{\partial x_i} = \sum_j \frac{n_j}{f_j} \frac{\partial f_j}{\partial x_i}$ (L = likelihood)

Information (expected value per observation)

k, lth element of information matrix

$$= I_{kl} = \sum_j \frac{1}{f_j} \frac{\partial f_j}{\partial x_k} \frac{\partial f_j}{\partial x_l}$$

Correction for $x_i = \Delta x_i = \frac{1}{N} \sum_l C_{il} S_l$ ($N = \Sigma n_i$, C_{il} = i, lth element of covariance matrix)

Iteration: Continue until all $\Delta x_i < 10^{-6}$ or number of iterations = 100

Figure 2 - Numerical Differentiation in MAXLIK

$\frac{\partial f_j}{\partial x_i}$ is estimated by $\frac{F(x+h) - F(x-h)}{2h}$,

where $F(x + h)$ = expected probability of the jth class when the ith parameter = x + h ($h = 10^{-8}$)

Note: $\frac{F(x+h) - F(x)}{h}$ gives almost identical results

Figure 3 - Subroutine FREQ for Estimating Gene Frequencies and Possible Misclassification in the A_1A_2BO System.

```
SUBROUTINE FREQ (X, F, W, K)

Calculate F(J) from X(I) where X(I) is frequency of the I-th
gene or parameter and F(J) is the frequency of the J-th
phenotype in the A1A2BO blood group system. In this
example, misclassification of A2B as B is assumed to be
possible.
    I CODE -- 1 = A1, 2 = A2, 3 = B, 4 = proportion
                 of the A2B misclassified as B, 5 = 0

    J CODE -- 1 = A1, 2 = A2, 3 = B, 4 = A1B, 5 = A2B,
                 6 = 0

DOUBLE PRECISION X(15), F(10)
X(5)  = 1. - X(1) - X(2) - X(3)
F(1)  = X(1)**2 + 2.*X(1)*X(2) + 2.*X(1)*X(5)
F(2)  = X(2)**2 + 2.*X(2)*X(5)
F(3)  = X(2)**2 + 2.*X(3)*X(5) + 2.*X(4)*X(2)*X(3)
F(4)  = 2.*X(1)*X(3)
F(5)  = 2.X(2)*X(3)*(1. - X(4))
F(6)  = X(5)**2
RETURN
END
```

APPLICATIONS, RESULTS AND PROBLEMS

Gene Frequency Estimation: Pure gene frequency estimation is very simple when the assumptions made in FREQ are justified. This will be the case when the genes present in the population are known, grouping and sampling are reliable, and there are no gross departures from Hardy-Weinberg equilibrium. FREQ is simply all Hardy-Weinberg relations between phenotype frequencies and gene frequencies.

Blood group and serum group systems used with MAXLIK to date are: ABO, A_1A_2BO, MNSs, Rh (6 alleles), and Gm (2, 3, and 4 alleles). Some of the blood group data have been published (Reed, 1968). The maximum number of "independent" parameters estimated here is therefore five (Rh). For most estimations there appear to be no problems of any kind. The number of iterations required is usually four or five and the goodness-of-fit is good. The procedure seems "robust" in that it is quite insensitive, within wide limits, to the closeness of the trial estimates to the final estimates. This robustness is illustrated in Figure 6. A survey of 55 pairs of p(A), q(B) gene frequency trial estimates in the ABO system was made. As can be seen, each pair led (in four to six iterations) to the same consistent final estimates. Since the entire likelihood surface has been quite closely sampled, it appears that there is only one "peak" present. I would hesitate to generalize from this.

Also, in the Rh (6 allele) system, seven sets of trial estimates were chosen to vary around the correct values, each set having some parameters departing from the true values by 0.1 or 0.2 or 0.4. All converged, in five to seven iterations, to the same final correct values in a large (N=5056) Caucasian population. The same pattern of deviations in sets of trial estimates, applied to a small (N=335) "Mexican" population, also showed similar convergence to the correct final estimates in nine to eleven iterations (close trial estimates required eight).

The only "problems" to date are not really problems. Both have involved a large California Negro population (Reed, 1968) and in each case, Rh and MNSs, it was known before estimation that the data were deficient in one respect. D^u tests were not recorded for the Rh data, and the chi-square for fit was significant at the 0.01 level. In the MNSs data the S-s- phenotype was not recorded and iteration did not converge. There was no problem in Rh or MNSs estimation in Caucasians or in Mexicans. The "problems" were that for Negroes, the FREQ's did not provide for frequent D^u+ bloods in Rh estimation or for S-s- phenotypes in MNSs estimation. Obviously, the parameters cannot be estimated if they are not correctly related to phenotypes.

Non-gene Frequency Estimation: The following types of parameters have been estimated: probability of non-paternity, probability of serological grouping error, probability of "general" error in phenotype classification, degree of racial mixture, and relative fitness of genotype. Data on the first three parameters have been published for the MN system in Caucasian families (Reed and Milkovich, 1968). An example of estimation of racial mixture, using the Gm system (4 alleles) with California Caucasians and Negroes, is shown in Figure 7.

Usually one or more gene frequencies are also estimated at the same time that non-gene frequency parameters are estimated. The subroutine and output shown in Figures 3 and 4, as well as 6, show examples of such estimation. Algebraically and statistically, of course, there is no difference between the two types of parameters and, if the parameter is a probability or proportion, the resemblance is particularly close. Relative fitnesses, however, since they are not inherently limited to a 0-1 range, may offer

Figure 4 - Sample Output from FREQ of Fig. 3.
(Distributions and Estimates)

PROBLEM NO. 1 * * * RESULTS OF CALCULATIONS * * *

A1A2BO Gene Frequency Estimation. Data of Formaggio 1951 (Mourant Et Al., 1958, The ABO Blood Groups, PG. 222). M= Proportion of A2B Misclassified as B.

DISTRIBUTIONS IN THE J CLASSES

CLASS	OBS. NO.	EXP. NO.	OBS. PROP.	EXP. PROP.	CHI-SQUARE (CELL VALUES AND TOTAL)
A1	1951	0.195212D 04	0.390200D 00	0.390424D 00	0.644130D -03
A2	429	0.428806D 03	0.858000D -01	0.857612D -01	0.875481D -04
B	508	0.509188D 03	0.101600D 00	0.101838D 00	0.277176D -02
A1B	178	0.176710D 03	0.356000D -01	0.353420D -01	0.941627D -02
A2B	17	0.170398D 02	0.340000D -02	0.340795D -02	0.927504D -04
O	1917	0.191613D 04	0.383400D 00	0.383227D 00	0.390853D -03
TOTAL	5000	0.500000D 04	0.100000D 01	0.100000D 01	0.134033D -01

NUMBER OF ITERATIONS = 4 D.F. = 1 NOT SIGNIFICANT AT .05 LEVEL

PARAMETER ESTIMATES

PARAMETER	TRIAL EST.	FINAL EST.	VARIANCE	ST. ERROR	DELTA X (LAST CORRECTION)
A1	0.200000D 00	0.242218D 00	0.212514D -04	0.460993D -02	0.103104D -07
A2	0.700000D -01	0.657739D -01	0.937167D -05	0.306132D -02	0.971692D -08
B	0.700000D -01	0.729550D -01	0.702603D -05	0.265067D -02	0.101103D -07
M	0.500000D 00	0.644897D 00	0.742764D -02	0.861838D -01	0.346508D -07

Figure 5 - Sample Output from FREQ of Fig. 3.
(Information, Covariance, and Product Matrices)

INFORMATION MATRIX (PER OBS.)

X(I)	A1	A2	B	M
A1	0.981813D 01	0.217513D 01	0.260685D 01	-0.137504D -01
A2	0.217513D 01	0.226882D 02	0.213801D 01	-0.150793D 00
B	0.260685D 01	0.213801D 01	0.292693D 02	-0.687520D -02
M	-0.137504D -01	-0.150793D 00	-0.687520D -02	0.279305D -01

COVARIANCE MATRIX (PER OBS.)

X(I)	A1	A2	B	M
A1	0.106257D 00	-0.936214D -02	-0.877994D -02	-0.394905D -03
A2	-0.936214D -02	0.468584D -01	-0.253079D -02	0.247749D 00
B	-0.877994D -02	-0.253079D -02	0.351302D -01	-0.933840D -02
M	-0.394905D -03	0.247749D 00	-0.933840D -02	0.371382D 02

I*C PRODUCT MATRIX (TEST FOR UNIT MATRIX)

X(I)	A1	A2	B	M
A1	0.100000D 01	-0.281893D -17	-0.566496D -17	-0.000000D -38
A2	0.563853D -16	0.100000D 01	0.650521D -18	0.444089D -15
B	0.293829D -16	-0.105168D -16	0.100000D 01	0.277556D -16
M	-0.194818 D -18	-0.867362D -18	0.271051 D -19	0.100000D 01

certain problems, especially when more than one is estimated.

The dependence of successful estimation on the closeness of the trial estimates to the final correct estimates has not been extensively explored. Experience suggests, however, that when the non-gene frequency parameters are also probabilities or proportions, estimation is still rather "robust." For example, sets of trial estimates used with the data shown on Figures 3 and 4 show that single-parameter deviations of +0.2 or -0.2 from the correct estimate permitted estimation to proceed correctly. Also, initial values of 0.01 for all four parameters gave correct estimates. However, when all trial estimates were +0.2 high, an error message ($\sqrt{-x}$) was given and grossly impossible final estimates were calculated. Limited experience in estimating multiple relative fitnesses suggests that close initial estimates may be required here.

POSSIBLE IMPROVEMENTS

Dr. R. C. Elston has suggested to me that use of observed values instead of expected values in the information matrix would improve the convergence. It would be interesting to try this. However, since MAXLIK has performed very satisfactorily in almost all situations to date and usually requires only four to six iterations, this would seem like a rather minor possible improvement.

A potentially valuable improvement might be the addition of instructions to cause MAXLIK to "explore" the space around the i_{max}-dimensional point specified by the final estimates. It could report the "area" over which the final estimate vector represented a likelihood maximum. This could increase our confidence in the results in certain cases.

Other basic improvements are undoubtedly possible. I believe that the results to date, however, demonstrate that MAXLIK is already a useful program.

REFERENCES

Balakrishnan, V. and Sanghvi, L. D. 1965. Use of digital computer in the estimation of blood group gene frequencies. Acta. Genet. Stat. Med. 15: 345-357.

Kurczynski, T. W. and Steinberg, A. G. 1967. A general program for maximum likelihood estimation of gene frequencies. Amer. J. Hum. Genet. 19: 178-179.

MacCluer, J. W., Griffith, R., Sing, C. F., and Schull, W. J. 1967. Some genetic programs to supplement self-instruction in FORTRAN. Amer. J. Hum. Genet. 19: 189-221.

Rao, C. R. 1952. Advanced Statistical Methods in Biometric Research. Wiley: New York.

Reed, T. E. 1968. Distribution and tests of independence of seven blood group systems in a large multiracial sample from California. Amer. J. Hum. Genet. 20: 142-150.

Reed, T. E. and Milkovitch, L. 1968. The accuracy of blood grouping of cord blood specimens with special reference to the MN system. Vox. Sang. 14: 9-17.

Reed, T. E. and Schull, W. J. 1968. A general maximum likelihood estimation program. Amer. J. Hum. Genet. 20:579-580.

Figure 6 - Results of 55 Trial Estimates in ABO Gene Frequency Estimation

Number of iterations, for each indicated pair of **trial p, q** values, to reach correct final estimate. (Best pair in box.)

P_A \ q_B	.05	.1	.2	.3	.4	.5	.6	.7	.8	.9
.9	5									
.8	5	5								
.7	5	5	6							
.6	5	5	5	6						
.5	5	5	5	5	6					
.4	4	4	5	5	5	6				
.3	4	4	4	5	5	5	6			
.2	4	4	4	5	5	5	5	5		
.1	4	4	4	4	5	5	5	5	5	
.05	4	4	4	4	5	5	5	5	5	5

Data: California Caucasians (Reed, 1968)

Group	O	A	B	AB	Σ
Number	2361	2033	507	155	5056

Final estimates for all pairs:

$p = .2471 \pm .0046$

$q = .0678 \pm .0025$

$r = .6851 \pm .0050$

$X_1^2 = 1.764$

Figure 7 - Estimation of the Proportion of Caucasian Ancestry (M) in California Negroes Using Gm (1, 2, 5) Data. (C = Caucasian, N = Negro, + - + = Gm (1, -5, 2), e.g.)

PROBLEM NO. 1 * * * RESULTS OF CALCULATIONS * * *

ESTIMATION OF M FROM Gm(1, 2, 5) DATA. CAUCASIANS AND NEGROES OF OAKLAND, CALIF.

DISTRIBUTIONS IN THE J CLASSES

CLASS	OBS. NO.	EXP. NO.	OBS. PROP.	EXP. PROP.	CHI-SQUARES
C. +--	27	0.319644D 02	0.365854D-01	0.433122D-01	0.771033D 00
C. -+-	183	0.194018D 03	0.247967D 00	0.262897D 00	0.625730D 00
C. ++-	168	0.157502D 03	0.227642D 00	0.213417D 00	0.699784D 00
C. +-+	25	0.309865D 02	0.338753D-01	0.419872D-01	0.115659D 01
C. +++	75	0.635291D 02	0.101626D 00	0.860828D-01	0.207119D 01
N. +--	3	0.129721D 01	0.406504D-02	0.175773D-02	0.223519D 01
N. -+-	8	0.787380D 01	0.108401D-01	0.106691D-01	0.202261D-02
N. ++-	239	0.236225D 03	0.323848D 00	0.320088D 00	0.326038D-01
N. +-+	2	0.125752D 01	0.271003D-02	0.170396D-02	0.438383D 00
N. +++	8	0.133467D 02	0.108401D-01	0.180849D-01	0.214188D 01
	738	0.738000D 03	0.100000D 01	0.100000D 01	0.101744D 02

NUMBER OF ITERATIONS = 5 D.F. = 5

PARAMETER ESTIMATES

PARAMETER	TRIAL EST.	FINAL EST.	VARIANCE	ST. ERROR	DELTA X
M	0.100000D 00	0.273148D 00	0.134365D-02	0.366558D-01	-0.100511D-06
C. GM-1	0.250000D 00	0.258595D 00	0.199504D-03	0.141246D-01	-0.770575D-07
C. GM-5	0.650000D 00	0.637100D 00	0.229175D-03	0.151385D-01	0.793935D-07

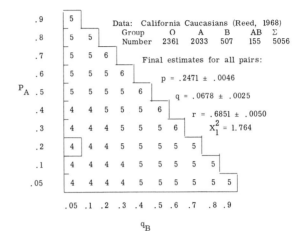

ITERATION PROBLEMS

R. C. Elston
University of North Carolina

In genetics we are frequently interested in obtaining maximum likelihood estimates, and I believe that it is in this area that most, if not all, of our iteration problems lie. I shall therefore confine myself to this particular topic here. In section 1 I shall introduce the problem by discussing several general methods for solving nonlinear equations, and then in section 2 we shall see the implications for maximum likelihood estimation of one unknown genetic parameter; the extension to several unknown parameters will be left to section 3.

1. ITERATIVE SOLUTIONS FOR NONLINEAR EQUATIONS

Suppose $\hat{\theta}$ is that value of θ that satisfies the equation $s(\theta) = 0$. What methods can we use to determine $\hat{\theta}$, to a specified degree of precision? We shall assume throughout this discussion that there is one and only one such value $\hat{\theta}$.

1.1. GRAPHICAL METHOD

One of the simplest methods is simply to evaluate $s(\theta)$ for various values of θ and then use these points to plot a graph of $s(\theta)$ against θ. $\hat{\theta}$ is that point at which the graph crosses the θ-axis. Any desired degree of precision can be obtained by taking points close enough together in the region where $s(\theta)$ is approximately zero, and this can be done in an iterative manner. This method can reasonably be applied in those cases where $\hat{\theta}$ is known to lie within a specified interval. But, if $\hat{\theta}$ may lie anywhere from $-\infty$ to $+\infty$, there may be difficulty in approximately locating $\hat{\theta}$ in the first place.

1.2. BISECTION METHOD

The bisection method is based upon the following theorem:

> If $s(\theta)$ is continuous from $\theta = \theta_0$ to $\theta = \theta_1$, and if $s(\theta_0)$ and $s(\theta_1)$ have opposite signs, then there is at least one real root of $s(\theta) = 0$ between θ_0 and θ_1.

It is first necessary to find appropriate θ_0 and θ_1 that fulfill the conditions of the theorem. We then let $\theta_2 = (\theta_0 + \theta_1)/2$ and calculate $s(\theta_2)$. If $s(\theta_2) = 0$, θ_2 is the desired solution; if not, let j (j = 0 or 1) be such that θ_2 and θ_j fulfill the conditions of the theorem. The process is then repeated with θ_2 and θ_j, which bracket $\hat{\theta}$ within an interval half as large as the previous one. Clearly this process can be continued until $\hat{\theta}$ is known to any desired degree of precision, and the process must converge.

1.3. STRAIGHT-LINE FITTING

From now on it will be convenient to use the notation $s_j = s(\theta_j)$, so that (s_j, θ_j) specifies a point on the curve. Given any two points on the curve, say (s_0, θ_0) and (s_1, θ_1), we can construct a straight line that passes through them and determine the point at which this line crosses the θ-axis; it is given by

$$\theta_2 = (s_1\theta_0 - s_0\theta_1)/(s_1 - s_0) .$$

This can equivalently be written

$$\theta_2 = \theta_1 - s_1/s(\theta_1, \theta_0) \tag{1}$$

where

$$s(\theta_1, \theta_0) = (s_1 - s_0)/(\theta_1 - \theta_0) . \tag{2}$$

Provided $s(\theta)$ is approximately linear, at least in the region we are concerned with, θ_2 will be an approximation to $\hat{\theta}$. The process can be repeated with the two points (s_1, θ_1) and (s_2, θ_2) to obtain θ_3, and so on analogously. Eventually, provided there is convergence, $\hat{\theta}$ can be obtained to any desired degree of precision. This method is sometimes called the secant method. (Depending upon the configuration of the points it may be better to obtain θ_3 from (s_0, θ_0) and (s_2, θ_2), but it is hardly worth bothering to allow for this. The problem disappears as soon as we get to θ_4; for $j \geq 4$ we always compute θ_j from (s_{j-2}, θ_{j-2}) and (s_{j-1}, θ_{j-1}).)

We can also fit a straight line from one point alone, if we know the derivative of the curve at that point. Denote the derivative at the j-th point by $s_j' = s'(\theta_j)$. Then the straight line passing through (s_1, θ_1) and with slope s_1' crosses the θ-axis at

$$\theta_2 = \theta_1 - s_1/s_1' . \tag{3}$$

If we now calculate s_2 and s_2' from θ_2, we can use the results to obtain θ_3, and so on, fitting a succession of straight lines just as before. This is the Newton-Raphson method, and, provided it converges, it will eventually give $\hat{\theta}$ to any desired degree of precision It is clear that this method is identical with the secant method if we differentiate s numerically using the simple formula (2) instead of s'.

The advantages of straight-line fitting, as opposed to the bisection method, are two-fold. It is not necessary at the start to hunt for two values $s(\theta_0)$ and $s(\theta_1)$ that are of opposite sign, and convergence is much faster in the final stages of the iteration. The great disadvantage, however, is that there may not be convergence, and so $\hat{\theta}$ is never obtained. Conditions that ensure convergence of the Newton-Raphson method have been given (see e.g. Isaacson and Keller, 1966, or Saaty and Bram, 1964), but it will be more instructive here to consider some cases where the method does not converge.

(1) $s(\theta)$ may have a point of inflection near θ, and so the process will diverge. This is illustrated in Figure 1a. If, furthermore, $s(\theta)$ is antisymmetric about $\hat{\theta}$, the process will oscillate back and forth between θ_0 and θ_1.

(2) θ_0 may be chosen near a local minimum and the process is caught in a cycle that does not converge. This is illustrated in Figure 2a.

(3) θ_0 may be chosen the wrong side of a local maximum and the process diverges. This is illustrated in Figure 3a.

(4) $s'(\hat{\theta})$ may be close to zero and cause difficulties; convergence may be slow to start with, and as $s'(\theta)$ approaches zero the newer estimates will be more poorly evaluated numerically, owing to the (relatively) larger rounding errors that occur. This is illustrated in Figure 4a.

Many other methods of straight-line fitting have been proposed, with a view to speeding up convergence to $\hat{\theta}$. This is done by fitting a line whose slope should be somewhere between $s'(\hat{\theta})$ and $s'(\theta_1)$. Although these methods will lead to much faster convergence in those cases where convergence does occur, they will all suffer from nonconvergence in the situations that have been depicted here.

1.4. PARABOLA FITTING

If we have any three points on the curve, say (s_0, θ_0), (s_1, θ_1) and (s_2, θ_2), we can fit a parabola through them to calculate (s_3, θ_3), the point at which this parabola crosses the θ-axis. Assuming a quadratic function of the form

$$s_j = a + b\theta_j + c\theta_j^2 \tag{4}$$

we naturally find two solutions for θ_3, and we can quite easily take that value that is closest to the initial values θ_0, θ_1 and θ_2. This method is due to Muller (1956), and the best computational formula is given by Traub (1964).

Instead of (4), we can assume s is an inverse quadratic function, i.e. the quadratic function is of the form

$$\theta_j = a + bs_j + cs_j^2 . \qquad (5)$$

There is then only one solution for θ_3, given by

$$\theta_3 = \frac{\theta_0 s_2 s_1 (s_2 - s_1) + \theta_1 s_2 s_0 (s_0 - s_2) + \theta_2 s_1 s_0 (s_1 - s_0)}{(s_2 - s_1)(s_2 - s_0)(s_1 - s_0)}$$

This can equivalently be written

$$\theta_3 = \theta_2 - \frac{s_2}{s(\theta_2, \theta_1)} + \frac{s_2 s_1}{s_2 - s_0} \left\{ \frac{1}{s(\theta_2, \theta_1)} - \frac{1}{s(\theta_1, \theta_0)} \right\} \qquad (6)$$

where, generalizing (2),

$$s(\theta_j, \theta_{j-1}) = (s_j - s_{j-1})/(\theta_j - \theta_{j-1}) \qquad (7)$$

The formulation given in (6) has computational advantages, as well as illustrating that it is an extension of (1), the first two terms above being in fact completely analogous to (1). Furthermore, it is clear how the derivatives s_1' and s_2', if available, can be used in place of $s(\theta_1, \theta_0)$ and $s(\theta_2, \theta_1)$.

Parabola fitting entails more computation at each step than straight-line fitting, but many fewer iterations are required to reach any given degree of precision. The process may or may not converge, but it is more likely to converge than straight-line fitting. Figure 1a shows a situation where parabola fitting will always converge. In the cases illustrated by Figures 2a, 3a and 4a parabola fitting may or may not lead to convergence, depending on which initial points are taken.

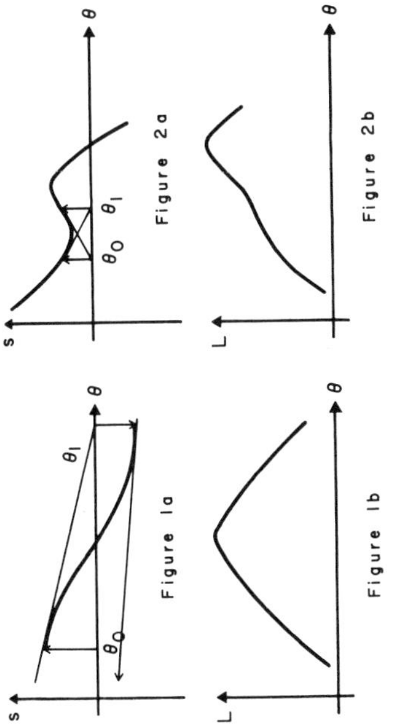

Figure 1a Figure 1b Figure 2a Figure 2b

1.5. FUNCTIONAL ITERATION

Write $s(\theta) = 0$ in the form $\theta = t(\theta)$. Now if

$$\left|\frac{dt}{d\theta}\right| < 1 \qquad (8)$$

for all θ, then $t(\theta_1)$ is always a better solution than θ_1, i.e.

$$|\hat{\theta} - t(\theta_1)| < |\hat{\theta} - \theta_1| .$$

In fact, to be sure of convergence, it is sufficient that (8) should hold in the interval $[\hat{\theta}, \theta_1]$ and in the interval $[\hat{\theta}, t(\theta_1)]$. (The proof of this is clear on considering a graph of $t(\theta)$; a formal proof of a similar theorem is given by Isaacson and Keller, 1966.)

If $\hat{\theta}$ satisfies $s(\theta) = 0$, then it also satisfies

$$\theta = t(\theta) \equiv \theta + ms(\theta) . \qquad (9)$$

We can therefore use the iterative scheme

$$\theta_j = \theta_{j-1} + ms(\theta_{j-1}) \qquad (10)$$

and be sure of convergence if a constant m can be chosen such that (8) holds. This is always possible if $s'(\theta)$ is always finite, non-zero and never changes sign. It is convenient, but not necessary, for m to be a constant; in fact, the Newton-Raphson method (3) can be considered a special case of (10) with $m = -1/s'(\theta)$. When m is a constant, convergence will usually be slow; but, if m is appropriately chosen, we have the advantage that there will in fact be convergence.

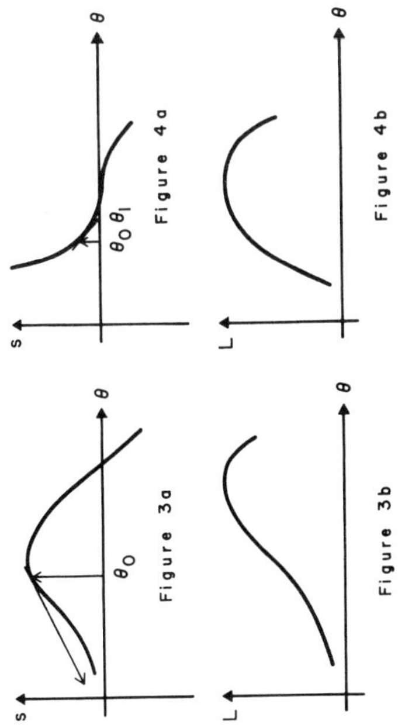

Figure 3a Figure 3b Figure 4a Figure 4b

2. MAXIMUM LIKELIHOOD ESTIMATION

The maximum likelihood estimate of a parameter θ is that value $\hat{\theta}$ that maximizes the likelihood, or, equivalently, that maximizes the logarithm of the likelihood. It is customary to use the logarithm of the likelihood, say $L(\theta)$, both to simplify the computations and to obtain an estimate of the variance of $\hat{\theta}$.

By definition the maximum likelihood estimate must satisfy the equation

$$L'(\theta) = 0, \tag{11}$$

so by taking $s(\theta) = L'(\theta)$ we can use any of the methods discussed in section 1 to obtain $\hat{\theta}$. The function $s(\theta)$ evaluated at a particular θ_j is known in this connection as a score.

2.1. CONVERGENCE TO THE RIGHT ESTIMATE $\hat{\theta}$

Equation (11) may have more than one solution, but only one is the value $\hat{\theta}$ that we desire. However, I do not think this is a major problem: we are unlikely to find a value $\hat{\theta}$ that satisfies (11) and think it to be the maximum likelihood estimate when in fact it is not. If $L(\theta)$ has a local minimum at $\hat{\theta}$, then the second derivative at that point, $L''(\hat{\theta})$, will be positive; and since the variance of $\hat{\theta}$ is asymptotically

$$V(\hat{\theta}) = -1/E(L''(\theta)) \doteq -1/L''(\theta), \tag{12}$$

we shall estimate that $\hat{\theta}$ has a negative variance and know immediately that something is wrong. It is not unknown for $L(\theta)$ to have a local minimum at a point which is outside the possible range for the true parameter θ (which is another indication that something is wrong). Intuitively, I find it difficult to believe that there can be many cases where $L(\theta)$ has a local minimum within the possible parameter range of θ. (I have recently been working on a genetic problem where θ was vector-valued and this occurred.) If $L(\theta)$ has a point of inflexion at $\hat{\theta}$ then $L''(\hat{\theta}) = 0$ and this would lead computationally to a variance of very large absolute magnitude; should this variance be positive, we should at least know that $\hat{\theta}$ is a poor estimate. I know of no examples in which $L(\theta)$ has such a point, but I do not think anyone has really examined cases to find one. The last possibility is that $L(\theta)$ should have more than one local maximum within the parameter range. However, this would imply the existence of a local minimum within the parameter range, which, as I have said, I consider to be unlikely. This discussion assumes, of course, that the likelihood is continuous.

It is for the reasons just stated that throughout section 1 I assumed the existence of only one value $\hat{\theta}$; and I think the assumption that this is so, at least within the possible range for the true parameter, will rarely cause trouble.

In section 1 I compared several methods for solving (11) with respect to their speed of convergence. Here again, however, I do not think this is a major problem. The speed of present day computers is such that we need be less and less worried about computer time. Many computer systems are now so arranged that a program is compiled each time it is executed, so any increase in program running time due to a more slowly convergent process may well be negligible. It is for this reason that I have not discussed the various processes, such as Aitken's ∂^2 method, designed specifically to speed up convergence. What we should be concerned with is finding a process that will converge, even though slowly. Furthermore we should like a process that will converge for a wide variety of different functions $L(\theta)$, so that one general computer program can be used for many different problems; clearly this process should also be available when θ is vector-valued, and I shall return to this problem in section 3. First, however, let us consider in more detail how the methods in section 1 are used in genetic applications.

2.2. GENETIC APPLICATIONS

Methods that can be considered to be variations of the graphical method and/or the bisection method have been advocated for specific genetic problems, e.g. the estimation of the frequency of a recessive condition with truncate selection (Lejeune, 1958). Parabola fitting was mentioned by Smith (1957) as an acceleration formula for estimating gene frequency by gene counting. Functional iteration was the method originally proposed by Haldane (1938) to estimate the frequency of a recessive condition with truncate selection. The most commonly used method, however, both in the field of genetics and in general, is straight-line fitting, especially by the Newton-Raphson method or modifications of it. Figures 1b, 2b, 3b and 4b show the general shapes of functions $L(\theta)$ that give rise to the corresponding $L'(\theta) = s(\theta)$ in Figures 1a, 2a, 3a and 4a. There is no reason to think that such $L(\theta)$ do not occur in genetic applications, and in fact they probably do account for many of the cases encountered in which the Newton-Raphson method does not converge.

Another possible reason for the failure of convergence is that a modification of the Newton-Raphson method is often used, instead of the Newton-Raphson method itself. Two modifications are commonly used, and they are virtually the same as the Newton-Raphson method if the sample size is large and the data fit the underlying genetic model. Assume, as so often is the case in genetic applications, that the observations x_i are multinomially distributed with parameters (p_i, n), the p_i being functions of the unknown parameter θ and

$$\sum_i x_i = n.$$

(What follows can be easily generalized to a product multinomial distribution). Then the logarithm of the likelihood is

$$L = \text{constant} + \sum_i x_i \ln p_i, \tag{13}$$

$$L' = s = \sum_i x_i \frac{p'_i}{p_i} \tag{14}$$

and

$$L'' = s' = \sum_i x_i \frac{(p_i p''_i - p'^2_i)}{p_i^2} \tag{15}$$

Thus the unmodified Newton-Raphson method uses (14) and (15), evaluated at a particular point θ_1, in (3).

One modification of this is to take the expectation of (15), instead of the actual second derivative observed; i.e., instead of (15), we substitute for s_1' in (3) the following expression evaluated at θ_1:

$$E(L'') = -E(L')^2 = -n\sum_i \frac{p'^2_i}{p_i} \tag{16}$$

This is a special result for the multinomial distribution, and has two advantages. Firstly, it enables us to obtain the variance of $\hat{\theta}$ as a byproduct of the iterative scheme (for a finite sample it is better to use the expected value of $L''(\theta)$, as indicated in (12), to obtain the variance of θ); secondly, it enables us to proceed without obtaining - either algebraically or numerically - the second derivatives p''_i. I do not consider either of these advantages of great importance if a computer is available that is capable of fast multiprecision arithmetic, thus allowing easy and accurate numerical differentiation. On the other hand, whenever (16) is a poor approximation to (15), there is a serious likelihood that this modification of the Newton-Raphson method may never converge.

The second modification that is commonly used makes the following approximation to (16), to be used instead of s':

$$E(L'') \doteq -\sum_i x_i \left(\frac{p'_i}{p_i}\right)^2 \tag{17}$$

I can see no advantage over (16) in this modification, and do not know why it is ever used. I should expect it to be an even poorer approximation to s*, and so even more likely to lead to non-convergence.

3. SEVERAL UNKNOWN PARAMETERS

Although I have spent much time on the case where only one parameter is to be estimated, in order to clarify the main issues involved, the case where θ is vector-valued is much more important in practice. Denote the vector-valued parameter variable by $\underline{\theta}$, and assume its k elements θ_u are functionally independent (it is always possible to arrange for this to be so in practice).

If the parameters all lie within a specified interval, such as is the case if gene frequencies are being estimated, the "graphical" method can always be used to advantage; it can be used directly to find the maximum of $L(\underline{\theta})$, without the need for any differentiation; as k increases, however, the amount of computation involved may become excessive. I do not see how the bisection method can be generalized to the case of several unknown parameters, and to my knowledge the method of parabola fitting has not yet been generalized; I think it may be possible to generalize this latter method, and an attempt to do so may lead to a useful result. Otherwise, of the methods discussed in section 1, we are left with straight-line fitting and functional iteration; I shall now consider these in more detail, restricting myself in the former case to the Newton-Raphson method and its modifications.

3.1. GENERAL NEWTON-RAPHSON METHOD

We now wish to find the maximum of $L(\underline{\theta})$, and so we solve the k equations

$$\underline{s}(\underline{\theta}) = \underline{0} \ , \tag{18}$$

where the element s_u of the vector $\underline{s}$ is the partial derivative of L with respect to θ_u $(u = 1, 2, \ldots, k)$. The generalization of (3) to solve (18) iteratively is

$$\underline{\theta}_2 = \underline{\theta}_1 + J_1^{-1}\underline{s}_1 \ , \tag{19}$$

where J is a k x k symmetric nonsingular matrix whose (u, v)-th element is

$$-\partial^2 L/\partial\theta_u\partial\theta_v,$$

and the subscript 1 denotes evaluation at $\underline{\theta}_1$. For the special case of an underlying multinomial distribution this element becomes, analogous to (15),

$$\sum_i x_i \frac{p_i(\partial^2 p_i/\partial\theta_u\partial\theta_v) - (\partial p_i/\partial_u)(\partial p_i/\partial_v)}{p_i^2}, \tag{20}$$

but this is rarely used in practice. Instead, the following approximations analogous to (16) and (17) are commonly used:

$$E(\partial^2 L/\partial\theta_u\partial\theta_v) =$$

$$-n\sum_i \frac{(\partial p_i/\partial\theta_u)(\partial p_i/\partial\theta_v)}{p_i} \doteq$$

$$-\sum_i x_i \frac{(\partial p_i/\partial\theta_u)(\partial p_i/\partial\theta_v)}{p_i^2} \tag{21}$$

As in the single-parameter case, I believe it would usually be better to use (20), rather than either of the expressions in (21); even though more computation may be involved, there is thereby less chance of having non-convergence. It should be noted that use of (20) does not necessarily involve more computation: it leads to more computation per iteration, but very frequently to fewer iterations being necessary (especially when use of (21) does not give convergence!). Sufficient conditions for the general unmodified Newton-Raphson method to converge have been given (see e.g., Isaacson and Keller, 1966, or Saaty and Bram, 1964).

3.2 GENERAL FUNCTIONAL ITERATION

We now write (18) in the form $\underline{\theta} = \underline{t}(\underline{\theta})$, the elements of $\underline{t}(\underline{\theta})$ being $t_u(\underline{\theta})$. If, corresponding to (8),

$$\left|\frac{\partial t_u}{\partial\theta_v}\right| < \frac{1}{k} \tag{22}$$

for all $\underline{\theta}$, u, v=1, 2, ..., k, then $\underline{t}(\underline{\theta}_1)$ is always a better solution than $\underline{\theta}_1$ and the iterative process will converge. (For a formal proof, see Isaacson and Keller, 1966.) We therefore try to solve

$$\underline{\theta} = \underline{t}(\underline{\theta}) \equiv \underline{\theta} + M\underline{s}(\underline{\theta}) \tag{23}$$

using the iterative scheme

$$\underline{\theta}_j = \underline{\theta}_{j-1} + M\underline{s}(\underline{\theta}_{j-1}) \ . \tag{24}$$

(23) and (24) are analogous to (9) and (10), m being replaced by a k x k matrix M. Just as before, if we take $M = J^{-1}$ we obtain the general Newton-Raphson method. It is difficult to choose M so that (22) is always satisfied, but Isaacson and Keller (1966) suggest that (24) will converge for a large variety of practical problems if M is chosen to be a diagonal matrix with elements

$$m_u = (1 - \partial t_u/\partial\theta_u)^{-1} = -(\partial s_u/\partial\theta_u)^{-1} = -(\partial^2 L/\partial\theta_u^2)^{-1} \ . \tag{25}$$

Using this value of M, (24) reduces to a process very similar to the Newton-Raphson method (19), the difference being that only the diagonal elements of J are computed - the off-diagonal elements are simply set equal to zero. I do not know if there are situations when this will lead to convergence while the Newton-Raphson method diverges. It is computationally a much simpler method than the Newton-Raphson method, and should perhaps be tried more often.

3.3. CONCLUDING REMARKS

It is clear from the foregoing that the problem of writing a general computer program to obtain maximum likelihood estimates is far from trivial, in view of the iteration problems that may arise. I have not mentioned all the techniques possible, and in particular I wonder whether the method of steepest ascent, as applied to finding the maximum of the likelihood directly, may not sometimes be of use. Also, as with the single-parameter case, acceleration techniques are available for the k-parameter case.

Edwards (1967) has described a program in FORTRAN which maps the likelihood surface on the assumption it is a hyperparabola, and so finds the maximum. This should have, as far as convergence is concerned, both the advantages and the disadvantages of the Newton-Raphson method.

Although I have discussed many problems, I have by no means touched upon all of them. I have assumed that the likelihood surface is continuous, whereas in fact it may not be - a particular example of this in genetic work was described by Murphy and Bolling (1967) Even if we are in a simple situation where our iterative process seems to converge, how do we know when we have reached the solution to the desired degree of precision? The usual test for this is to compare successive estimates and stop when they are close enough together. This test is probably good enough in practice if we are using the Newton-Raphson method, but it is theoretically possible, using this test, to come to a stop when the process really diverges. The use of acceleration techniques will certainly help counter this problem.

In conclusion, I would recommend that any general computer program for obtaining maximum likelihood estimates should incorporate more than one iterative scheme, automatically switching from one to another if there does not appear to be convergence within a specified number of iterations. In those situations where difficulties are encountered care should be taken in choosing the initial estimates, using other methods of estimation for this purpose. And the possibility of bypassing the problem altogether, by using other asymptotically efficient methods of estimation, should always be borne in mind. Such general methods have been proposed by Tweedie (see Smith, 1966); a special case was applied to a genetic problem by Elston (1966) necessitating simply the solution to a set of linear equations.

ACKNOWLEDGEMENT

This investigation was supported by a Public Health Service Career Development Award (No. 1-K3-GM-31, 732-03) from the National Institute of General Medical Sciences.

REFERENCES

Edwards, J. H. 1967. A generalized program for maximum likelihood estimation (Abstract 1283). Biometrics 23:175.

Elston, R. C. 1966. On testing whether one locus can account for the genetic difference in susceptibility between two homozygous lines. Genet. 54:89.

Haldane, J. B. S. 1938. The estimation of the frequencies of recessive conditions in man. Ann. Eugen. Lond. 8:255.

Isaacson, E. and Keller, H. B. 1966. Analysis of numerical methods. Wiley, New York.

Lejeune, J. 1958. Sur une solution "a priori" de la méthode "a posteriori" de Haldane. Biometrics 14:513.

Muller, D. E. 1956. A method for solving algebraic equations using an automatic computer. Math. Tables Aids Comput. 10:208.

Murphy, E. A. and Bolling, D. R. 1967. Testing of single locus hypotheses where there is incomplete separation of the phenotypes. Amer. J. Hum. Genet. 19:322.

Saaty, T. L. and Bram, J. 1964. Nonlinear mathematics. McGraw-Hill, New York.

Smith, C. A. B. 1966. Biomathematics, Vol. II. Hafner, New York.

Smith, C. A. B. 1957. Counting methods in genetical statistics. Ann. Hum. Genet., Lond. 21:254.

Traub, J. F. 1964. Iterative methods for the solution of equations. Prentice-Hall, Englewood Cliffs, N. J.

DISCUSSION ON ITERATION PROBLEMS

S. M. Robinson, N. E. Morton, and W. R. Harvey

ROBINSON: (This discussion expresses the personal views of its author, and should not be construed as reflecting the position of the Department of the Army or of any other agency of the United States Government.) The paper under discussion makes a laudable effort to present a broad survey of iterative methods to an audience whose primary interests are not in the field of numerical analysis. A great many methods are presented, and many valid comments are made concerning these. Some disputable remarks are also made, and it is principally to these that I will address this discussion.

Let us consider first the question of straight-line fitting. It should be pointed out that Equation (1) is much better numerically than is the form given just above it. Examples can be constructed in which the form

$$\theta_2 = (s_1\theta_0 - s_0\theta_1)/(s_1 - s_0)$$

will lead to cycling, because of rounding errors, whereas (1) will converge if written in the form

$$\Delta\theta_1 = -s_1\theta_{10}$$

where

$$\Delta\theta_j = \theta_{j+1} - \theta_j \quad \text{and} \quad \theta_{j,j-1} = \Delta\theta_{j-1}/(s_j - s_{j-1}).$$

With this formula, the values of $\Delta\theta_j$ may be stored to full precision. This technique avoids the loss of precision involved in taking differences of quantities (the θ_j) which are very close to a single non-zero value. Another point to notice in this section is the statement that the Newton-Raphson technique "is identical with the secant method if we differentiate s numerically..." The point here is that if we differentiate s numerically we are not using the Newton-Raphson method. This is not a trivial distinction: the convergence order of Newton's method is 2, whereas that of the secant method is about 1.62, so numerical differentiation makes quite a difference. Even so, the secant method very often saves time; Jeeves (1958) estimated that if computing the derivative s' required more than 42 per cent of the time needed to compute s, the secant method would be the more efficient of the two.

Some additional remarks should be made about the discussion of parabola-fitting. First, it is possible to use this method with only one point, if we have the first and second derivatives available.

Also, Equation (6) may be written as

$$\Delta\theta_2 = -s_2(\theta_{21} - s_1\theta_{210})$$

where

$$\theta_{210} = \Delta\theta_{10}/(s_2 - s_0) = (\theta_{21} - \theta_{10})/(s_2 - s_0).$$

Again, the use of the successive differences $\Delta\theta_j$ avoids the loss of precision caused by operating directly with the θ_j. Also, as an extension of the author's comments on the use of the first derivative, we may observe that

$$\theta_{j,j-1} = d\theta(\sigma_j)/ds = 1/s'(\xi_j),$$

where

$$\xi_j = \theta_{j-1} + \phi_j(\theta_j - \theta_{j-1}),\ 0 \le \phi_j \le 1$$
$$\sigma_j = s(\xi_j),$$

and

$$\theta_{210} = 1/2\, d^2\theta(\sigma_0)/ds^2 = -1/2\,[\, s'(\xi_0)]^{-3}\, s''(\xi_0),$$

where

$$\xi_0 = \theta_0 + \phi_0(\theta_2 - \theta_0), \qquad 0 \le \phi_0 \le 1$$

$$\sigma_0 = s(\xi_0),$$

so it is clear how both the first and second derivatives of s may be employed. If derivatives are used in place of divided differences, the convergence order changes (for the better). All of this requires, of course, that $s'(\hat{\theta}) \neq 0$, and that the function s be sufficiently differentiable. The above also illustrates the fact that Equations (1) and (6) are truncations of the Newtonian difference expansion of $\theta(0)$, and are thus analogous to the successive partial sums of the corresponding Maclaurin series:

$$\theta(0) = \sum_{n=0}^{\infty} \frac{s^n}{n!} (-1)^n \theta^{(n)}(s).$$

Formulas for any number of points, or for any number of derivatives at one point, may be readily developed from these expansions.

In the discussion of functional iteration, a serious error occurs: the conditions for convergence stated between Equations (8) and (9) are false, as illustrated by the following counterexample: let $s(\theta) = \theta - t(\theta)$, with

$$t(\theta) = \begin{cases} .99[\theta-(1+\frac{1}{2}\sqrt{2})] + .495(1+\sqrt{2}) + .01, & 1+\frac{1}{2}\sqrt{2} \le \theta < \infty \\ .99[\theta-\frac{1}{2}(1+\sqrt{2})]^2 + .495(\frac{1}{2}+\sqrt{2}) + .01, & 1 \le \theta < 1+\frac{1}{2}\sqrt{2} \\ 1 + .99(\sqrt{2}-1)(1-\theta)/[\,T(\theta-1)+1\,], & 1-1/T < \theta < 1 \end{cases}$$

where $T = .99(\sqrt{2} - 2) + (.74242575 - .4949505\sqrt{2})^{-1}$. The function $t(\theta)$ then has the following properties:

a. $t(\theta)$ is continuously differentiable for $1-1/T < \theta < \infty$.

b. $|t'(\theta)| \le .99$ for $1 \le \theta \le 1 + \frac{1}{2}\sqrt{2}$.

c. $t(\theta) = \theta$ for $\theta = 1$ and for no other θ.

Thus, if θ_1 and $t(\theta_1)$ are in the interval $[1, 1 + \frac{1}{2}\sqrt{2}]$, the convergence conditions are met.

Now let

$$\theta_1 = 1 + \frac{1}{2}\sqrt{2};$$

then

$$\theta_2 = .505 + .495\sqrt{2} \doteq 1.20504,$$

$$\theta_3 = .25757425 + .4949505\sqrt{2} \doteq .95754,$$

and

$$\theta_4 = \theta_1,$$

so that the sequence $\{\theta_j\}$ does not converge.

A valid convergence criterion would be

$$|t'(\theta)| \le p < 1 \quad \text{for} \quad |\hat{\theta} - \theta| < |\hat{\theta} - \theta_1|,$$

from which we can deduce

$$|\hat{\theta} - \theta_j| \le p^{j-1} |\hat{\theta} - \theta_1|,$$

so that $\{\theta_j\} \to \hat{\theta}$. This is not a unique criterion, as there are many others available.

Two comments might be made concerning the discussion between Equations (16) and (17); the first is a reiteration of the earlier caveat concerning numerical differentiation, and the second is that if s_i' in (3) is replaced by any expression which does not approach $s'(\hat{\theta})$ as $\theta \to \hat{\theta}$ then the convergence, if any, will at best be linear (of order 1). In other words, one loses the principal advantage of the Newton iteration, and is then in essence using the method of Equation (10).

Turning to the case of several variables, the bisection method has in fact been generalized (Steinberg, 1967), though it requires a very large amount of computational labor and thus appears to be impractical. The method of parabola-fitting (at one point) is also available for several variables, but has the disadvantage of requiring the computation of the second derivative of $\underline{s}(\underline{\theta})$, which is a k-linear operator containing k^3 elements. See Necepurenko (1954).

I entirely agree with the comments as to the undesirability of using the expressions in (21). However, there is available another class of methods, which generalize the secant method to the case of several variables. One such method is given in Robinson (1966), together with a list of references. These methods have superlinear convergence (the one mentioned above, for example, has a convergence order of about 1.62), yet do not require the computation of any derivatives. From the point of view of saving time for the human programmer, these may be the most efficient methods available.

In the discussion of generalized functional iteration, the convergence criterion is certainly valid, though very strong. Much less restrictive criteria are given in the book by Isaacson and Keller, cited by the author, and in other works. The specific criteria vary with the norm used in each case, but the important point is that the derivatives do not have to be uniformly small, as long as an appropriate combination of them is bounded by a quantity less than one.

The choice of M suggested in (25) was, I think, first given by Lieberstein (1958). In general, its convergence will be only linear.

Very little can be said in general about the method of steepest ascent, except that its convergence is generally linear. Its use, however, requires the computation of the first partial derivatives of $L(\underline{\theta})$, and if these are to be computed there seems to be no reason not to use the methods discussed in Robinson (1966) or its references, since these processes yield super-linear convergence.

As for determing when to stop the iteration, error tests are available if we can obtain bounds for the inverse of the derivative of the function being solved. For example, in the case of one variable, if L is twice continuously differentiable,

$$0 = dL(\hat{\theta})/d\theta = dL(\theta_j)/d\theta + [\, d^2L(\xi)/d\theta^2\,](\hat{\theta} - \theta_j)$$

with

$$\xi = \hat{\theta} + \phi(\theta_j - \hat{\theta}), \quad 0 \le \phi \le 1 .$$

If

$$|d^2L(\xi)/d\theta^2|^{-1} \le B$$

for all such ξ, then

$$|\hat{\theta} - \theta_j| \leq B\, |dL(\theta_j)/d\theta| .$$

For several variables, there are numerous criteria analogous to the one just mentioned; again, these vary with different norms. Most, however, are quite difficult to compute, except for special cases.

Jeeves, T. E. 1958. Secant modification of Newton's method. Comm. ACM 1, 8:9

Lieberstein, H. M. 1958. Technical Summary Report 80, Mathematics Research Center, U.S. Army, University of Wisconsin.

Necepurenko, M. I. 1954. On Cebysev's Method for Functional Equations. Uspehi Matem. Nauk 9:163-170 (Russian). Available in English as Technical Summary Report 648, Mathematics Research Center, U.S. Army, University of Wisconsin.

Robinson, Stephen M. 1966. Interpolative solution of systems of nonlinear equation.. SIAM J. Numer. Anal. 1:650-658.

Steinberg, J. 1967. Personal communication to the author.

of convergence. The binomial distribution provides an example of convergence of (16) in a single iteration for any trial value in the interval from 0 to 1, whereas Newton-Raphson iteration converges asymptotically .

MORTON: Dr. Elston asked why the multi-parameter analog of (17), the "observed" amount of information, has been used. In our case it was introduced in the early stage of the SEGRAN program because of limited machine memory. Each family was scored as it was read instead of being assembled into a summary table, and we therefore did not calculate the expected number. In many applications one is more concerned to have an accurate estimate of variances and covariances. In this case the multivariate analog of equation (16) using the "expected" amount of information has an advantage even if convergence is slower than for Newton-Raphson iteration (providing of course that the process does converge). For example, in the Brazilian data we were led to suspect heterogeneity of estimates of the inbreeding coefficient among systems, which disappeared when the expected amount of information was substituted for the observed amount. The expected amount of information included large contributions from rare classes, which being by chance missing in the data greatly underestimated the variance by equation (17) and correspondingly inflated χ^2. We have decided therefore to abandon use of the observed amount of information, using instead the multi-parameter analog of (16).

I am bothered by the assumption that if A and B are two iterative procedures, and A is an approximation to B, then A is less likely to converge than B. It seems to me that the conditions for convergence of equations 16 and 17 have not been investigated, although in the neighborhood of the root by large sample theory the convergence conditions must be the same as Newton-Raphson iteration. Farther from the root the convergence criteria might be more or less likely to be realized. I know of no theory or experience that would permit a clear choice among Newton-Raphson iteration and its various approximations on the basis of likelihood or speed

HARVEY: I must make it clear from the beginning of my discussion that I have had essentially no experience in solving nonlinear equations by iteration procedures. My experience with iteration methods has primarily been in the solving of linear equations and this experience has been very limited indeed, as indicated in the paper I presented earlier.

The only contribution which I might make to a discussion of iteration methods for the solving of nonlinear equations would be to describe a modification of the Newton-Raphson method that was developed recently by one of our graduate students (Mrs. Jean Hensel) with the assistance of Dr. Herman Chernoff of Stanford University. Although the procedure has not been used extensively enough as yet to know how useful it might be, it appears to have some very desirable features which should bring about convergence in many cases where other methods would not converge.

As described by Elston, the problem is to find the vector $\hat{\theta}$ which will maximize $L(\theta)$. The general Newton-Raphson method solves the equations

$$s(\theta)=0$$

iteratively by computing

$$\theta_{j+1} = \theta_j + J_j^{-1} s_j$$

where the symbols are the same as defined by Elston. Now for simplicity let

$$J_j^{-1} s_j = \Delta\theta_j .$$

The modification of the Newton-Raphson procedure developed by Mrs. Hensel makes use of the principle of steepest ascent. That is, the procedure to be described simultaneously maximizes the likelihood function while at the same time the roots of the likelihood equations are obtained. This is accomplished by multiplying $\Delta\theta_j$

by several constants during each iteration, adding each of these modified $\Delta\theta_j$ to θ_j, computing ln $L(\theta)$ for each set and locating the largest value.

The procedure is carried out in a stepwise manner as described below:

1. At the beginning of each iteration compute ln $L(\theta)$ at the following seven points:

   (a) $\theta_j - \frac{3}{2}\Delta\theta_j$ (d) θ_j

   (b) $\theta_j - \Delta\theta_j$ (e) $\theta_j + \frac{1}{2}\Delta\theta_j$

   (c) $\theta_j - \frac{1}{2}\Delta\theta_j$ (f) $\theta_j + \Delta\theta_j$

   (g) $\theta_j + \frac{3}{2}\Delta\theta_j$

2. If ln $L(\theta)$ is maximum at either (a), (b), (f) or (g) choose that point as θ_{j+1} and proceed with the next iteration.

3. If ln $L(\theta)$ is maximum at point (c) compute ln $L(\theta)$ at the following additional points:

   (h) $\theta_j - \frac{3}{4}\Delta\theta_j$

   (i) $\theta_j - \frac{1}{4}\Delta\theta_j$

   Now compare ln $L(\theta)$ at points (c), (h) and (i). If the maximum value is found at (c) or (h) choose that point as θ_{j+1} and proceed with the next iteration. If ln $L(\theta)$ is maximum at (i) compute ln $L(\theta)$ at the following additional points:

   (j) $\theta_j - \frac{3}{8}\Delta\theta_j$

   (k) $\theta_j - \frac{1}{8}\Delta\theta_j$

   Now locate the maximum ln $L(\theta)$ at points (i), (j) and (k) and choose that point as θ_{j+1} and proceed with the next iteration.

4. In comparing ln $L(\theta)$ at the points (a) through (g) in (1) above, if (d) is found to be maximum compute ln $L(\theta)$ at the following two additional points:

   (l) $\theta_j - \frac{1}{4}\Delta\theta_j$

   (m) $\theta_j + \frac{1}{4}\Delta\theta_j$

   Now compare ln $L(\theta)$ at the points (d), (l), and (m). If the maximum is at point (m) compute ln $L(\theta)$ at the following two additional points:

   (n) $\theta_j + \frac{1}{8}\Delta\theta_j$

   (o) $\theta_j + \frac{3}{8}\Delta\theta_j$

   Now locate the maximum ln $L(\theta)$ at points (m), (n) and (o) and choose that point as θ_{j+1} and proceed with the next iteration.

   In comparing ln $L(\theta)$ at points (d), (l) and (m), if the maximum is at point (l) compute the following two additional points:

   (p) $\theta_j - \frac{3}{8}\Delta\theta_j$

   (q) $\theta_j - \frac{1}{8}\Delta\theta_j$

   Now locate the maximum ln $L(\theta)$ at points (l), (p) and (q) and choose that point as θ_{j+1} and proceed with the next iteration.

   In comparing ln $L(\theta)$ at points (d), (l) and (m), if the maximum is at point (d) compute ln $L(\theta)$ at the following two additional points:

   (r) $\theta_j - \frac{1}{8}\Delta\theta_j$

   (s) $\theta_j + \frac{1}{8}\Delta\theta_j$

   Now locate the maximum of ln $L(\theta)$ at points (d), (r), and (s) and choose that point as θ_{j+1} and proceed with the next iteration. Of course, if ln $L(\theta)$ is still maximum at point (d) the solution has been obtained.

5. In step (1) if ln $L(\theta)$ is maximum at point (e) compute ln $L(\theta)$ at the following two additional points:

   (t) $\theta_j + \frac{1}{4}\Delta\theta_j$

   (u) $\theta_j + \frac{3}{4}\Delta\theta_j$

   Now locate the maximum of ln $L(\theta)$ at points (e), (t) and (u). If the maximum is at point (e) or point (u) choose that point as θ_{j+1} and proceed with the next iteration. If the maximum is at point (t) compute ln $L(\theta)$ at the following two additional points:

   (v) $\theta_j + \frac{1}{8}\Delta\theta_j$

   (w) $\theta_j + \frac{3}{8}\Delta\theta_j$

   Now locate the maximum of ln $L(\theta)$ at points (t), (v) and (w) and choose that point as θ_{j+1} and proceed with the next iteration.

I have with me a copy of the FORTRAN IV program used by Mrs. Hensel in a biomedical application of the above procedure with a six parameter model. Considerable work would be required to generalize such a program to many different models, especially if one included several additional iterative schemes, as suggested by Elston. However, if anyone is interested in securing a copy of this special application program, I can make arrangements for him to acquire a copy.

SIMULATION OF GENETIC SYSTEMS

B. R. Levin

Division of Biological and Medical Sciences
Brown University
Providence, Rhode Island

More than eleven years have now passed since the first simulation of a genetic system was programmed for a digital computer (Fraser, 1957a). During these years there have been tremendous advances in computer technology and increases in computer availability. There has also been a fair increase in the application of simulation techniques to theoretical problems of population genetics. In fact, the use of this technique has become sufficiently common that the significance of studies employing it has already been questioned (Robertson, 1967). In this discussion I am going to consider the general technique of simulation of genetic systems, some variations on it, and its assets and liabilities. I shall then consider a few of the problems to which it has been and is now being applied, and some of the results obtained.

THE TECHNIQUE OF GENETIC SIMULATION

In the majority of simulations of genetic systems, the mechanisms of segregation and recombination are mimicked through the use of Monte Carlo methods. This generally involves the use of a random number decision process. A random number x_j from a rectangular distribution with limits of zero and one is generated and a determination made of where x_j lies along the cumulative probability distribution. If, for example, an event E_i, from a mutually exclusive and exhaustive class of events $(E_1, E_2 \ldots E_i \ldots E_n)$, occurs with probability P_i, then event E_i is considered to have occurred if

$$\sum_{k=1}^{i-1} P_k \leq x_j \leq \sum_{k=1}^{i} P_k$$

In the genetic simulations these probabilities are those involved in segregation, recombination, and gametic and zygotic survival. On the basis of these comparisons it is possible make repeated decisions about the nature of the gametes produced and the zygotes surviving. The task of making these decisions is obviously facilitated by the high computation speed and large storage capacity of digital computers. In addition the binary arithmetic used in most machines has many properties which are well suited to the simulation of genetic systems (Fraser, 1957a, 1961).

Since each step in the process of offspring production is mimicked through the use of the random number decision process, the models produced are direct analogues of the genetic system including the operation of random or stochastic elements. There have been some variations on this basic technique. In a study of selection of homostyly in Primula (Bodmer, 1960) and in another of linkage and selection in finite populations (Latter, 1965), the random number decision process was used only to sample genotypes from a population of potential parents whose frequencies were defined algebraically. In a study of selection in systems of species competition (Levin, 1968) random perturbations in the values of the variables were obtained by adding the product of the expected standard deviation of the variables and a random number from a standard normal distribution ($u=0, \sigma=1$) to the computed values of the variables. (See appendix for a comparison of these approaches to the incorporation of stochastic elements into otherwise deterministic models.) In a series of studies of the interaction of linkage and selection (Lewontin, 1964a, b) the transformation of gametes from one generation to the next was accomplished through the use of a matrix technique ("genetic operators"), as opposed to the random number decision process.

Whether or not these variations are called simulations appears to depend primarily on the author of the study employing the technique. There doesn't seem to be a very clear distinction between that which is a simulation and that which is a succession of numerical solutions to a series of equations. In fact, if one wishes to stretch a point, any mathematical analogue of a system may be considered a simulation of that system. This is however just semantics and not of any fundamental biological significance. But to facilitate this discussion I would like to distinguish three types of simulations: (1) a 'direct stochastic simulation'---for that in which each step in the system is mimicked through the use of the random number decision process (e.g., the models used by Fraser); (2) a 'partial stochastic simulation'---for that in which a major portion of the system is defined (e.g., the models used by Bodmer, Latter, and Levin); and (3) a 'deterministic simulation'---for that in which the transition from generation to generation is through an algebraic mechanism and no random numbers are generated (e.g., the model used by Lewontin).

ASSETS AND LIABILITIES OF THE SIMULATION TECHNIQUE

The major asset of the simulation technique is that it sets very broad limits on the complexity of the models to be constructed. It allows for the consideration of large numbers of parameters and extremely complex interactions. This enables the construction of far more realistic analogues of the system than can be obtained through the more rigorous mathematical approaches to model building. Another asset is that it allows for the consideration of theoretical problems without resorting to abstract mathematics. The primary requisite for the construction of a simulation is an understanding of the biology of the system. The mathematical knowledge required is minimal and programming computers is becoming an increasingly simple task.

In spite of its assets, the simulation technique is far from a panacea. It has a number of very real liabilities. A simulation study is an experimental study; one attempts to draw inference about the nature of a system by manipulating some components of the system. This is done through the manipulation of controlling parameters, and as the complexity of the simulation increases the number of parameters to be manipulated also increases. Without some knowledge about the nature of the system beforehand, it is difficult to know which parameters and combinations of parameters are to be manipulated and to what degree. Without this a priori knowledge of the system allowing for the formulation of a rigorous series of hypotheses, it is possible to obtain from the simulation little more than a description of effects rather than an understanding of mechanisms. Finally, as the complexity of the simulation increases, the chance of making a logical error in programming or in manipulation of parameters also increases. Although these errors are generally gross and easily noticed, they may also be subtle and lead to interesting but erroneous conclusions.

I believe that to a great extent these liabilities can be overcome. The investigator using the simulation technique is not working in a vacuum; there are fifty years of theoretical and experimental population genetics literature behind him. He can obtain the a priori knowledge of the system to develop the rigorous series of hypotheses. With these hypotheses he can get at the desired cause and effect relationships and obtain a mechanistic understanding of the system. The simulation experimeter has all the opportunities of any other experimental worker. In addition he is working with an ideal experimental system -- one in which results are obtained very rapidly, and one that can be completely dissected. Finally, if the programs used for the simulation studies are available, any erroneous conclusions obtained through programming error will, at least in theory, be picked up by other investigators.

Whether the assets of the simulation technique outweigh its liabilities depends on the nature of the problem in which it is being applied. The more a system can be defined algebraically, the fewer the liabilities involved. This would imply that partial stochastic and deterministic simulations are somewhat safer. There are however problems to which even these cannot be applied. In these cases it seems far better to use a direct stochastic simulation than to avoid a problem because of its complexity or to simplify it until it lacks any semblance of reality.

MULTI-LOCUS SIMULATIONS

Most of our notions about changes in the genetic structure of populations through selection have been derived from considerations of single loci or from simple extensions of single-locus theory. These extensions have often avoided the complication of linkage primarily due to the difficulty of exact mathematical treatment when linkage is considered. The desire to obtain insight into the effects of linkage on selection for multiple locus characters motivated the first simulation studies (Fraser, 1957b) and has been the motivation for the majority of simulation studies that have followed (cf. Martin and Cockerham, 1960; Gill, 1965; Latter, 1966; and Young, 1967). Much of this work is to be considered in a forthcoming book by Dr. Alex Fraser, appropriately titled Computer Genetics, and will not be reviewed here.

Recently a great amount of attention has been given to the role of multi-locus relationships such as linkage and epistasis in the maintenance of genetic variability. Generally single-locus theory takes the view that loci associate at random and any non-random association caused by initial linkage relationships (linkage disequilibria) will be broken down by recombination. Accordingly if there were n heterotic loci each having two alleles and homozygote fitnesses of $(1\text{-}s)$, then, assuming a multiplicative fitness relationship, the mean fitness of the population at gene frequency equilibrium would be $(1\text{-}s/2)^n$. If, however, the loci were not associating at random but were held in complete disequilibrium such that there were only two chromosomes and three genotypes, then under the assumption of a multiplicative fitness relationship the two homozygotes would each have a fitness of $(1\text{-}s)^n$ and the heterozygote a fitness of 1.0. At gene frequency equilibrium, the mean fitness of the population would then be $1/2{+}1/2(1\text{-}s)^n$. To be sure, this model is extremely artificial, but it is able to illustrate a point. If, for example, $s{=}0.10$ and $n{=}100$, then with linkage equilibrium the mean fitness of the population would be only 0.005, but with complete disequilibrium it would still be greater than 1/2. Consequently, the maintenance of disequilibrium could allow for a higher mean fitness. Simulation techniques have played a role in determining the conditions under which such disequilibria could be maintained.

Using a two-locus deterministic model and a symmetric fitness relationship, Lewontin and Kojima (1960) demonstrated that permanent linkage disequilibrium would result from epistatic interaction of selection values. This result was confirmed and extended to non-symmetric cases in a study of a five-locus deterministic simulation, (Lewontin, 1964). This study demonstrated (1) an increase in mean fitness resulting from the disequilibrium, (2) that with strong epistasis between loci a disequilibrium could be maintained even for genes that are completely unlinked and (3) the disequilibrium effect was cumulative; i.e., somewhat distant loci could be held out of equilibrium by genes between them on the chromosome. The epistasis required could be obtained from the multiplicative fitness relations similar to that considered above; but for a significant amount of disequilibrium, selection had to be relatively intense, linkage relatively tight and the epistatic relationship rather strong. In addition, since the models used in this study were deterministic, no consideration was given to finite population size.

In a recent study employing a direct stochastic simulation of a 180-locus system, Sved (1968) has demonstrated that the approach to linkage equilibrium through recombination may be offset by random fluctuations in finite populations, yielding a net disequilibrium. This disequilibrium, although unpredictable in its magnitude and direction, occurs even in the absence of selection and has the same effect as that obtained through epistasis between selection values; i.e., loci tend to act as linkage units. With a series of linked heterotic loci, individual selective values reinforce each other, and thus afford greater stability for individual loci. In addition, the stability of neutral loci linked to the heterotic loci may also be enhanced.

To illustrate the increase in stability of neutral, selected loci ascribed to linkage relationships, a series of runs were made with a three-locus direct stochastic simulation developed by Mr. Timothy Deering. In these runs one locus was neutral and two were under rather intense selection ($s(A_iA_i){=}0.25$, $s(A_ia_i){=}0.00$, and $s(a_ia_i) = 1.00$). A multiplicative fitness relationship was assumed to hold for the total genotype (e.g. $W(A_1A_1A_2A_2A_3A_3) = (1 - .25)^2 =$

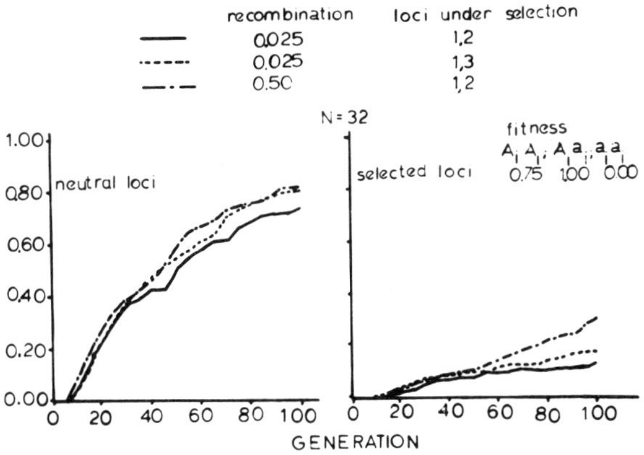

Figure 1: Frequency of fixation for three series of 100 runs each with two loci under intense selection and one locus neutral.

0.5625 and $W(A_1A_1a_2a_2A_3a_3) = (1 - 0.25)(1 - 1.00)(1.00) = 0.00$. Three series of 100 runs of 100 generations each were made with three different linkage relationships: (1) three independent loci (0.50 recombination); (2) fairly tight linkage between adjacent loci (0.025 recombination) with the central locus neutral; (3) relatively tight linkage between adjacent loci (0.025 recombination) with an outer locus neutral. All runs were initiated with complete linkage equilibrium and a frequency of A_i of 0.20 (considerably below the equilibrium level). The cumulative proportion of fixed runs are plotted at five generation intervals using the mean proportion for the two selected loci (see Figure 1). As indicated by the consistantly lower frequency of fixation for both neutral and selected loci, the tighter the linkage between the loci upon which selection is operating, the greater the stability of the polymorphisms. Since all runs were initiated with all loci in linkage equilibrium, the disequilibrium would progressively build up in time and would not have an important effect in the earlier generations. It should be noted that in these runs linkage disequilibrium resulted both from epistasis obtained through the multiplicative fitness relationship and from random fluctuations.

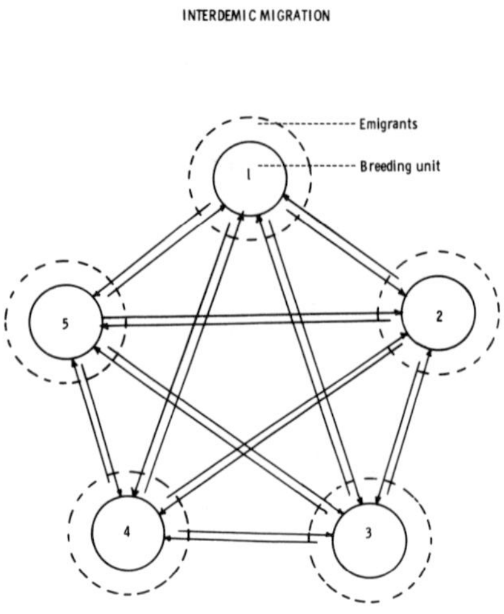

Figure 2: Inter-demic migration model.

SINGLE LOCUS SIMULATION STUDIES

In the multi-locus studies the simulation technique was employed to develop models which allow for increased complexity in internal genetic structure. Another application of simulation in genetic studies allows for the increase of complexity in the external structure of populations in space and in time. To be sure, the external structure of a population is dependent on the ecology of the species and as such involves a consideration of much more specific cases. But the simulation technique is perhaps best suited to deal with specific cases. An example of this application of simulation is in the studies of the t allele polymorphism of the house mouse Mus musculus.

Most house mouse populations studied are polymorphic for alleles at the brachury T locus. These allels are designated t^W and are generally recessive lethals or male steriles and appear to be maintained by an abnormal segregation ratio in heterozygous males (see Dunn and Suckling, 1956; Dunn 1957). On the average (for feral populations) 95 per cent of the effective sperm of a heterozygote bear the t allele. On the assumption of equal viability for the homozygous normal +/+ and heterozygote +/+ and lethal homozygous t/t, Bruck (1957) produced the following formulation for the equilibrium frequency of the normal allele:

$$p = 1/2 + \sqrt{m(1-m)/2m}$$

where m is the proportion of t bearing gametes in the effective sperm pool of the heterozygote (the transmission ratio). With m=0.95, an equilibrium frequency of 0.385 is expected for the t allele. The observed frequency in natural populations is generally half that value.

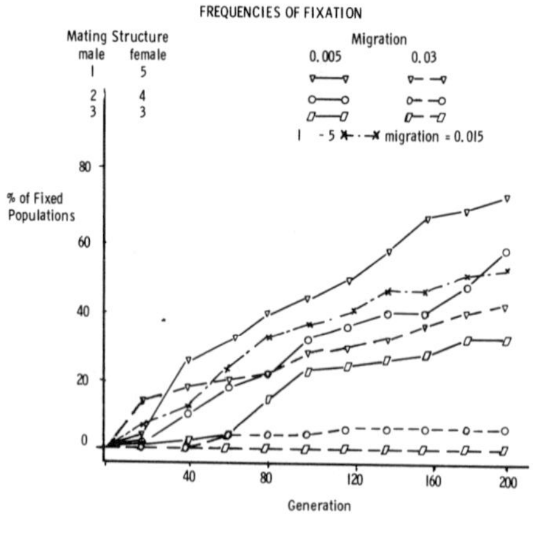

Figure 3: Frequency of fixation of t allele in a population composed of five demes each with six individuals. (50 runs for each set of input parameters.)

In an effort to explain the discrepancy between the observed and expected frequencies of the lethal t alleles, Lewontin and Dunn (1960) employed a direct stochastic simulation of this single locus system. In this study it was assumed that mice live in small family units, or demes, which are somewhat isolated from others, an assumption since borne out by some empirical findings (Anderson, 1964; Petras, 1967a). As a result of the small size of the breeding units, a high proportion of the simulated demes became fixed for the normal allele while others remained approximately at the expected frequency. They concluded that a geographic population of the house mouse is composed of both demes fixed for the allele and demes in which the t allele is segregating in high frequency In sampling such populations mice from both types of demes are obtained and the estimated frequency of the t allele is below that expected on the basis of the deterministic model.

In order to determine how isolated the demes had to be for this result to hold, a simulation similar to that used by Lewontin and Dunn, but allowing for migration between demes, was programmed by Levin, Petras and Rasmussen (unpublished). In this simulation after completion of reproduction and selection, each individual in a deme was subject to replacement by a migrant from another deme. The sex ratio of migrants was controlled by a separate set of input parameters and may or may not reflect that of the demes. A diagram of this model is shown in Figure 2.

In the runs made, five demes each containing six individuals and an even migrant sex ratio were considered, as were a number of sex ratios in the demes and a variety of migration rates. Fifty runs were made with each set of input parameters. Each run was initiated with all individuals heterozygous and allowed to go for 200 generations, or until fixation. The proportion of fixed runs is presented in Figure 3. As would be expected, the stability of the t allele increases with increasing effective size (approach to even sex ratios) and increasing migration rate. With a migration rate of 0.03 and equal sex ratio in the demes, fixation did not occur in any of the fifty runs. The gene frequency distribution for the t allele and its mean frequency for generations 20 to 200 are presented in Figures 4 and 5 for migration rates of 0.005 and 0.03, respectively. At the lower migration rate there was a great deal of instability and only in the series with equal sex ratio was the mean t allele frequency at the empirical level. In all runs made at the higher migration rate (Figure 5) the mean t allele frequency was above the level observed in natural populations.

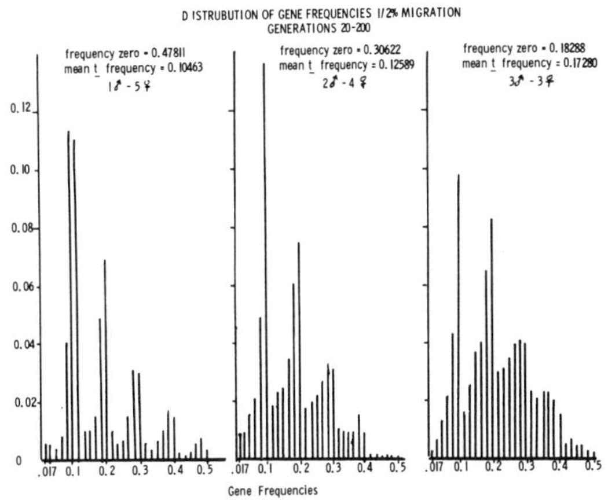

Figure 4: Distribution of t allele frequencies for generations 20-200 in a population composed of five demes each with six individuals and a migration rate of 0.005. (50 runs for each set of input parameters.)

These results indicate that the Lewontin and Dunn hypothesis for the low observed t frequency would only be valid if the migration rate were somewhat below 0.03, or less than one migrant every 5 1/2 generations. If, however, the migration rate were higher, an alternate hypothesis would be required to explain the low frequency of these alleles. Such an hypothesis was presented by Petras (1968) and suggests that the observed low frequency results from rather intense inbreeding. Using data obtained from an examination of the genotypic distribution of biochemical polymorphisms (Petras, 1967a), an estimate of the inbreeding coefficients was obtained and was of sufficient magnitude to account for the observed low t frequency. This too must be qualified. Since the inbreeding coefficient was estimated from genotypic frequencies, it may have two components, one from systematic inbreeding within demes and the other from the interdemic variance in gene frequency, or Wahlund effect (Wahlund, 1927). I believe that the Wahlund component would not affect t allele frequencies since it would not be altering genotypic frequencies within the deme, where selection is actually operating.

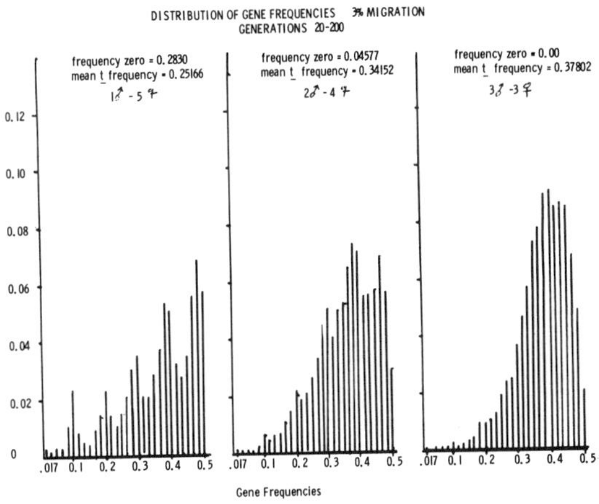

Figure 5: Distribution of t allele frequencies for generations 20-200 in a population composed of five demes each with six individuals and a migration rate of 0.03. (50 runs for each set of input parameters.)

SIMULATION OF SELECTION IN ECOLOGICAL SYSTEMS

Although there has been some consideration of variable selection (Clarke and O'Donald, 1964; Kojima and Yarbrough, 1967) and selection in a heterogeneous environment (Levene, 1953; Levins, 1962), most algebraic models of populations under selection assume fitness to be constant in intensity and direction. Consequently, models for selection in ecological systems have been extremely limited and most theories for such selection have been derived on intuitive grounds and presented through verbal arguments (see, for example, Pimentel, 1968). To be sure, the major limitation in developing models for selection in ecological systems is dealing with the large number of parameters to be considered for simultaneous changes in ecological and genetic structure. This may be overcome by use of simulation techniques; an example is in a study of selection in systems of species competition.

Loosely stated, the principle of competitive exclusion asserts that two species with the same ecology cannot coexist. This principle was derived without consideration of genetic variability. In a recent study (Pimentel et al., 1965) it was suggested that this genetic variability and the operation of selection on it may lead to coexistence of species sharing the same limiting resource. Reversals in numerical dominance were obtained in a housefly-blowfly competition system and attributed to what was termed a "genetic feedback" mechanism. That is, selection in the sparse species was presumed to be primarily for interspecific competitive performance and in the dominant species for intraspecific competitive performance. This would presumably lead to an improvement of the interspecific competitive performance of the sparse species without concommitant change in the dominant, and thus a reversal in the dominance relationship.

To determine conditions necessary for these reversals and whether or not they would stabilize the coexistence state between species sharing the same limiting resource, a partial stochastic simulation of selection in a system of species competition was constructed (Levin, 1968), based on the competition formulation of Gause (1934) but assuming the competition coefficients of these equations to be variables whose values were dependent upon the genetic structure of the competing populations. That is, the effect of an individual of species i in limiting the resources of species j depended upon the average inhibitory ability of the ith species on the jth and the average ability of the jth species to avoid competition from species i, this average being taken over all genotypes considered in the model. The fitness of a genotype was dependent upon its interspecific competitive performance and the number of individuals of the other species present, and its intraspecific competitive performance and the numbers of its own species present. The equations for the competition coefficients, changes in population size, fitness, and changes in the genetic composition were solved in a serial fashion and under the assumption of discrete time intervals. As indicated earlier, stochastic elements were incorporated by adding the product of the expected standard deviation of the variables (gene frequencies and population sizes) and a random number from a normal (0,1) distribution to the computed values of the variables.

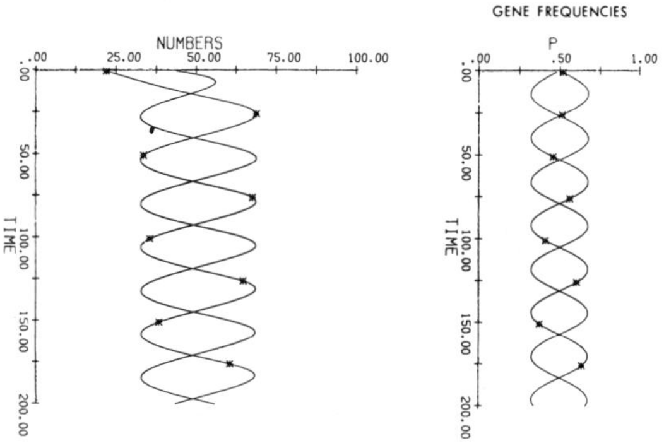

Figure 6: Single-locus deterministic model of selection for increased inhibitory ability in a system of species competition. Classical selection scheme with the homozygotes having opposite interspecific and intraspecific competitive properties. The two species differ only in their initial population size.

In the runs made, consideration was given to the two species case and to those situations where the range of genetic variability included possibilities for all four outcomes of competition considered by Gause and Witt (1935). A single-locus and a two-locus mode of inheritance of competitive performance and a variety of selection schemes were also considered.

Reversals in species dominance were obtained with selection operating both in the direction of increased inhibitory ability and increased avoidance ability. Examples of these runs are presented in Figures 6 and 7 for the one-locus and two-locus modes of inheritance, respectively. In these two runs selection was operating in the direction of increased inhibitory ability, and homozygotes had equal but opposite interspecific and intraspecific competitive properties. In the two-locus run competitive performance was additive and there was no linkage. From the runs made the following conditions appeared necessary for reversals to occur: (1) intraspecific and interspecific competitive performance be controlled by the same locus, (2) the selection scheme allow for continuous changes in gene frequency (i.e., does not form stable equilibria), and

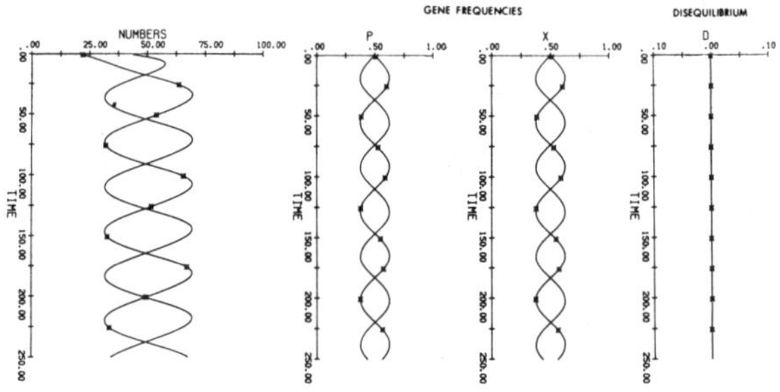

Figure 7: Two-locus deterministic model of selection for increased inhibitory ability in a system of species competition. Additive model with a classical selection scheme at both loci. Each locus controls both interspecific and intraspecific competitive performance with homozygotes having opposite interspecific and intraspecific competitive properties. There is no linkage nor an initial linkage disequilibrium. The two species differ only in their initial population size.

(3) genotypes at that locus have opposite interspecific and intraspecific competitive properties. These conditions had to be met for at least one locus controlling competitive performance, in at least one of the competing species.

The runs made with the stochastic model indicate that the stability of the coexistence state is not increased by the incorporation of genetic variability into the model. In those situations in which reversals in species dominance were obtained the effect of the random perturbations was to lead to oscillations of increasing amplitude (see Figure 8). These appear to be due to overcompensation by the regulatory mechanism.

It was concluded from the runs made with this model that a stable state of coexistence would result only if selection were operating towards increasing avoidance of competition, that is, towards ecological divergence.

SIMULATION OF THE GENETICS OF HUMAN POPULATIONS

Of all the species generally considered by population geneticists, the structure of human populations is in all probability the most complex and most diverse. Mating is rarely at random, generations are never discrete, and the distribution of populations in space is quite varied. Consequently human populations generally do not fit the assumptions made in the construction of most models used by population geneticists. The structures of some human populations have been extremely well studied, and rather accurate simulations of these populations can be constructed.

The simulation technique has been used in human population genetic studies in two general, but not necessarily mutually exclusive, ways: (1) to develop hypotheses about particular polymorphisms in somewhat generalized human populations (an approach similar to that taken in the t allele studies), and (2) to obtain some general notion of the effects of the demographic attributes of particular populations on their genetic structure. The work of Brues (1963) on the ABO polymorphism and that of Levin (1967) on the Rh polymorphism are examples of the first use. In the ABO study an attempt was made to determine the types of selection schemes and intensities of selection required to obtain the observed ABO blood group frequency distribution. In the Rh study consideration was given to the role of reproductive compensation in maintaining the Rh polymorphism in finite populations. Although these studies employed the direct stochastic simulation technique, no specific population was considered. The population sizes and fitness values chosen were to some degree arbitrary and not directly based on empirical estimates.

In the second use of simulation, the major demographic attributes of a particular population are precisely mimicked and the values of the controlling parameters are obtained whenever possible from estimates made in actual populations. The work of MacCluer (1967) on a model incorporating age and specific birth and death rates is an example of this latter use. In this simulation each individual in the population is accounted for and followed through each year of its life. Mating in the simulation was able to follow the pattern of the population upon which it is based and selection was able to operate at three levels: differential fertility, differential mortality and differential mating. This type of simulation may be used in both retrospective and prospective studies, that is, in ascertaining the effects of certain demographic attributes of the population in determining its current genetic structure and obtaining projections about the demographic and genetic future of the population.

Currently there is some concern about the genetic and demographic structure of contemporary human populations with primitive technology (see, for example Neel et al., 1964, 1967). One of the primary motivations behind this work is to obtain insight into the forces which shaped man's biological evolution. That is, through most of man's biological history, his breeding structure and technology was that of a hunter and gatherer and possibly quite similar to that of contemporary populations of that technological level. Although these studies are extensive in their detail, they are only able to deal with a very short period in the history of these populations. Therefore, much of the inference about the existing genetic composition and the evolutionary significance of various elements must be obtained through the use of models. The models developed through simulation techniques may play an important role in obtaining this inference.

A simulation of a primitive human population considering age structuring and variety of marriage practices common to some American Indian societies was developed (Schull and Levin, 1964). Currently, a much more extensive and much more precise simulation of a particular American Indian population (the Yamomama of southern Venezuela and northern Brazil) is under construction (MacCluer, personal communication). In this simulation each individual in the tribe is characterized by age, sex, lineage, village, genotype, marital status, reproductive history, and biological relationship to other individuals in the population. Mating is polygynous with provisions for periodic changes in alliances between lineages and frequent exchange of mates between sibships. Eventually this model is to consider a number of villages, exchange of mates between them, fusion of villages that have grown too small and fission of villages that have grown too large. This simulation is to be used in estimating expected levels of inbreeding and determining the evolutionary importance of various types of migration.

In a simulation of another primitive human population, the

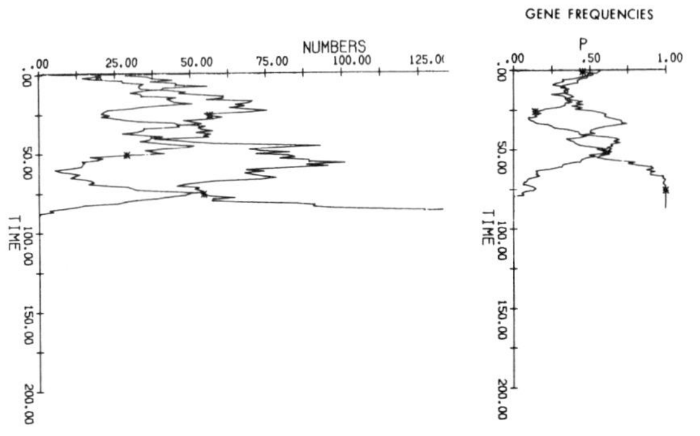

Figure 8: Single-locus stochastic model of selection for increased inhibitory ability in a system of species competition. Classical selection scheme with the homozygotes having opposite interspecific and intraspecific competitive properties. Single run with the loss of one of the species.

Birhor of India, a somewhat different population structure was considered (Williams, unpublished dissertation). In this model the members of the population lived in bands or demes, each of which was surrounded by six similar demes (see Figure 9). Female mates were chosen at random from surrounding demes and mating was monogamous with each pair producing one offspring of each sex which remain in the deme of the male parent. A single-locus mode of inheritance was considered and separate fitness values specified for each deme. This simulation was employed to obtain estimates of effective population size and to examine the effects of spatial heterogeneity on fitness values.

CONCLUSION

In the preceding discussion I have tried to give some notion of the types of problems to which the simulation technique has been, and is being, applied. The examples chosen have been taken from studies which I am most familiar with or interested in. They represent a very biased sample of, and only an abbreviated excursion into, the types of problems to which the simulation technique can be applied. I have mentioned only in passing the use of simulation to obtain estimates of genetic parameters and have not even considered the possible use of simulation as a teaching aid.

I believe that the simulation technique offers an opportunity to explore in a quantitative and precise manner problems which had previously been left only to conjecture based on intuition and often imprecise verbal arguments. Simulation is also an excellent tool for the examination of specific realistic cases, and if the future of theoretical population genetics does in fact lie in the examination of such cases (Lewontin 1967), then simulation will play an important role in that future.

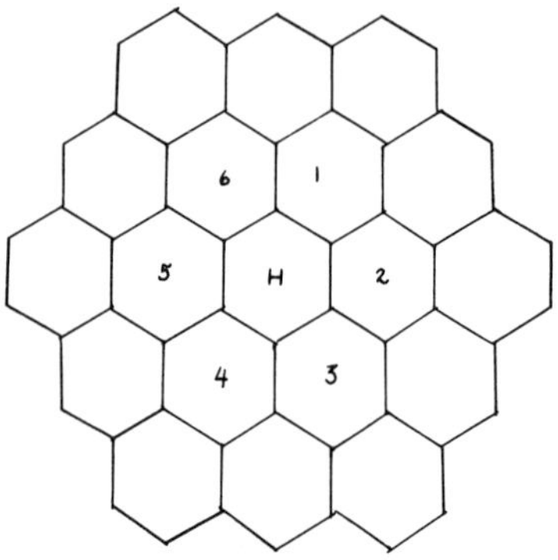

Figure 9: Relationship between bands in the simulation of a primitive human population (Williams, unpublished dissertation).

ACKNOWLEDGEMENTS

I would like to thank Drs. J. W. MacCluer and B. J. Williams for allowing me to borrow from their unpublished work and Dr. John Sved for allowing me to read the manuscript of his paper currently in press. I am especially grateful to Dr. Peter O'Donald and to Mr. Timothy Deering for their useful and often stimulating discussions.

Funds for computing time were provided in part by National Science Foundation Facilities Grant GP4825 to the Brown University Computing Laboratory.

APPENDIX

In a preliminary effort to compare the 'direct stochastic' and 'partial stochastic' simulation techniques, three models using different methods of simulation have been applied to the same problem. The first model is a direct stochastic simulation of a monoecious population with a single segregating locus and two alleles. In this simulation the random number decision process is used to select parents, choose the gamete produced by heterozygous parents and determine whether or not zygotes whose adaptive value is less than unity will survive. The other two models are partial stochastic simulations of the same system. In one of these the random number decision process is used to obtain a sample of parents from a population of potential parents whose frequencies are computed from the recursion equations for a single locus under selection. In the second partial stochastic simulation the gene frequencies are computed from the recursion equations and augmented each generation by the product of a random normal number (N(0, 1)) and the expected standard deviation for gene frequency change ($\sqrt{pq/2N}$). Copies of the programs used are available from the author.

The problem to be considered is the relative stability of selected and neutral polymorphisms in finite populations. One hundred runs were made with each program with the following sets of input parameters: (1) no selection, (2) selection favoring the heterozygote, with homozygotes having equal fitness and a mean population fitness of 0.95 at gene frequency equilibrium ($\hat{q}$ = 0.50), (3) selection favoring the heterozygote with the homozygotes having unequal fitness values and a mean population fitness of 0.95 at the gene frequency equilibrium ($\hat{q}$ = 0.20). A population size of 20 was considered and all runs initiated at the equilibrium gene frequency. Termination of a run would occur for one of two reasons: allelic fixation or completion of 100 generations.

The cumulative proportion of fixed runs plotted at five generation intervals is presented in Figure 10 for the systems with no selection, symmetric selection favoring the heterozygote, and for the asymmetric selection scheme favoring the heterozygote. Although the intensities of selection in the symmetric and asymmetric selection schemes are equal, the system with the symmetric selection values appears more stable. This is indicated by the lower proportion of fixed runs and the general form of the cumulative fixation curve. This effect of the level of the equilibrium in the determination of stability is in agreement with that considered in the more mathematical treatment of Robertson (1962).

The general form of the cumulative fixation curves generated by the different simulations appears quite similar for a given set of input parameters. Although these curves are quantitatively different, this difference does not appear to have any particular trend. For example, the simulation demonstrating the least stability in the absence of selection (the partial stochastic simulation selecting parents with the random number decision process) was most stable

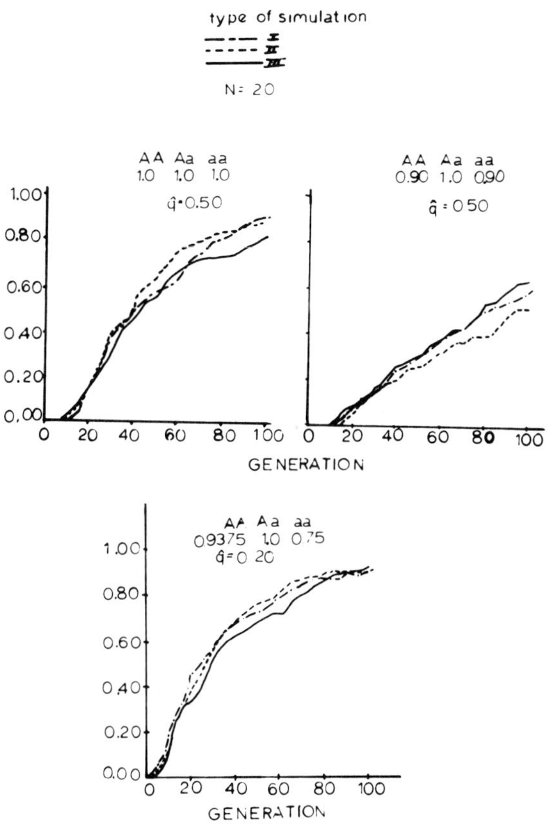

Figure 10: Comparison of simulation methods: Cumulative fixation curves for a system without selection, symmetric selection favoring the heterozygote and asymmetric selection favoring the heterozygote. Simulation: I. Partial stochastic simulation using random normal numbers. II. Partial stochastic simulation using the random number decision process to select parents. III. Direct stochastic simulation.

in the symmetric selection runs and intermediate in its stability in the asymmetric selection runs. From the data presented, it would be difficult to make any general statement about the difference in the curves generated by the different simulations.

In no way is it assumed that the three types of simulations examined would give quantitatively equivalent results. If a sufficient number of runs were made, systematic differences in frequencies of fixation would be expected. These are, in reality, different models. From the runs made, however, it seems that the differences between them are not too great.

The three simulations differ greatly in their use of computer time. The runs made in the direct simulation required a minimum of three times as much computer time as either partial simulation. This difference in required computer time would be greater with larger population sizes and more intense selection. In the direct simulation a minimum of two and as many as five random numbers may be required for each zygote generated, and in those runs where selection is operating, not all zygotes generated survive. In the simulation using the random number decision process to obtain a sample of parents, one random number is required for each individual in the population, independently of the intensity of selection. In the simulation using random normal numbers, one such number is required for each generation, independently of the size of the population or the intensity of selection. The random normal number generator currently used, however, requires twelve rectangularly distributed random numbers for each random normal number.

From this brief and preliminary investigation of the various simulation techniques I would conclude that, at least with regard to questions of relative stability, the partial stochastic simulation techniques offer adequate results. In addition to the fact that they are more economical in their use of computer time, they also include a complete algebraic definition of the system, an advantage considered in more detail in the main body of the paper.

REFERENCES

Anderson, P. K. 1964. Lethal alleles in Mus musculus. Local distribution and evidence for isolation of demes. Science 145:177-178.

Bodmer, W. F. 1960. The genetics of homostyly in populations of Primula vulgaris. Phil. Trans. Royal Soc. London series B 242:517-549.

Bruck, D. 1957. Male segregation ratio as a factor in maintaining lethal alleles in wild populations of house mice. Proc. Nat. Acad. Sci. U.S. 43:152-158.

Brues, A. M. 1963. Stochastic tests of selection in the ABO blood groups. Am. J. Phys. Anthr. 21:287-299.

Clarke, B. and O'Donald, P. 1964. Frequency-dependent selection. Heredity 19:201-206.

Dunn, L. C. 1957. Studies of genetic variability in wild populations of house mice, II. Analysis of eight additional alleles at locus T. Genetics 42:299-311.

Dunn, L. C. and Suckling, J. 1956. Studies of genetic variability in populations of wild house mice, I. Analysis of seven alleles at the locus T. Genetics 41:344-352.

Fraser, A. S. 1957a. Simulation of genetic systems by automatic digital computers. I. Introduction. Aust. J. Biol. Sci. 19:484-491.

Fraser, A. S. 1957b. Simulation of genetic systems by automatic digital computers. I. Effects of linkage on rates of advance under selection. Aust. J. Biol. Sci. 10:484-491.

Fraser, A. S. 1961. Simulation of genetic systems by automatic digital computers. 5-linkage, dominance and epistasis. In Biometrical Genetics (ed. O. Kempthorne), New York, Pergamon Press, 70-83.

Gause, G. F. 1934. The Struggle for Existence (reprinted 1954), New York, Hafner.

Gause, G. F. and Witt, A. A. 1935. Behavior in mixed populations and the problems of natural selection. Am. Nat. 69:596-609.

Gill, J. L. 1965. Selection and linkage in simulated genetic populations. Aust. J. Biol. Sci. 10:1171-1187.

Kojima, K. and Yarbrough, K. M. 1967. Frequency-dependent selection at the esterase 6 locus in Drosophila melanogaster. Proc. Natl. Acad. Sci. U.S. 57:645-649.

Latter, B. D. H. 1965. The response to artificial selection due to autosomal genes of large effect. I. The effects of linkage on limits to selection in finite populations. Aust. J. Biol. Sci. 18:1009-23.

Levene, H. 1953. Genetic equilibrium when more than one ecological niche is available. Amer. Natur. 87:311-333.

Levin, B. R. 1967. The effect of reproductive compensation on the long term maintenance of the Rh polymorphism: The Rh crossroad revisited. Am. J. of Hum. Genet. 19:288-302.

Levin, B. R. 1969. A model for selection in systems of species competition (in press). Concepts and models of Biomathematics: Simulation Techniques and Methods (eds. F. Heinmets and L. D. Cady). M. Dekker, New York.

Levins, R. 1962. Theory of fitness in a heterogeneous environment. Amer. Natur. 96:361-373.

Lewontin, R. C. 1964a. Interaction of selection and linkage. I. General considerations: Heterotic models. Genetics 49:49-67.

Lewontin, R. C. 1964b. Interaction of selection and linkage. II. Optimum models. Genetics 50:757-782.

Lewontin, R. C. 1967. Population Genetics. In Annual Review of Genetics, I:37-70 (ed. H. L. Roman). Academic Press, New York.

Lewontin, R. C. and Dunn, L. C. 1960. The evolutionary dynamics of a polymorphism in the house mouse. Genetics 45:705-722.

Lewontin, R. C. and Kojima, K. 1960. The evolutionary dynamics of complex polymorphisms. Evolution 14:458-472.

MacCluer, J. W. 1967. Monte Carlo methods in human population genetics: A computer model incorporating age-specific birth and death rates. Am. J. Hum. Genet. 19:303-312.

Martin, F. G. and Cockerham, C. C. 1960. High speed simulation studies. In Biometrical Genetics, pp. 35-45 (ed. O. Kempthorne). Pergamon Press, New York.

Neel, J. V., Salzano, F. M., Junqueira, P. C., Keiter, F. and Maybury-Lewis, D. 1964. Studies on the Xavante Indians of the Brazilian Mato Grosso. Am. J. Hum. Gen. 16:52-140.

Neel, J. V. and Salzano, F. M. 1967. Further studies on the Xavante Indians. X. Some hypotheses-generalizations resulting from these studies. Am. J. Hum. Gen. 19:554-574.

Petras, M. L. 1967. Studies of actual populations of Mus. II. Polymorphism at the T locus. Evol. 21:466-478.

Petras, M. L. 1967. Studies of natural populations of Mus. I. Biochemical polymorphisms and their bearing on breeding structure. Evol. 21:259-274.

Pimentel, D. 1968. Population regulation and the genetic feedback. Science 159:1432-37.

Pimentel, D., Feinberg, E. H., Wood, P. W., and Hayes, J. T. 1965. Selection, spatial distribution and the coexistence of competing fly species. Am. Nat. 99:97-108.

Robertson, A. 1962. Selection for heterozygotes in small populations. Genetics 47:1291-1300.

Robertson, A. 1967. Animal breeding. In Annual Review of Genetics, I:295-312 (ed. H. L. Roman), Academic Press, New York.

Schull, W. J. and Levin, B. R. 1967. Monte Carlo simulation: some uses in the genetic study of primitive man. In Proceedings of the Symposium on Stochastic Models in Medicine and Biology, pp. 179-198 (ed. J. Gurland), Univ. of Wisconsin Press, Madison.

Sved, J. A. 1968. The stability of linked loci with a small population size. Genetics 60:543-563.

Wahlund, S. 1928. Zusammensetzung von Populationen und Korrelationserscheinungen von Standpunkt der Vererbungslehre aus betrachtet. Hereditas 11:65-106.

Young, S. S. Y. 1966. Computer simulation of directional selection in large populations. I. The programme, the additive and dominance models. Genetics 53:189-205.

DISCUSSION ON MONTE CARLO SIMULATION

W. J. Schull and W. F. Bodmer

SCHULL: Three seemingly straightforward issues confront the investigator motivated to use Monte Carlo simulation, namely, what might profitably be simulated; under what circumstances, that is, when; and how should the simulation proceed. Though readily enumerated, in practice no one of these issues is easily resolved. Presumably when one turns to Monte Carlo simulation one does so because the problem of interest has proven intractable to more conventional methods of mathematical or numerical analysis. Patently, I am assuming that the problem at hand, when properly formulated, is complex. This complexity may be genetic in the sense that many loci are envisaged or involved interactions occur among loci or non-genetic, or both, of course. Obviously complexity for its own sake is to be eschewed, a pitfall not always avoided for it is often tempting to presume that complexity and reality are synonymous. The degree of complexity of a simulation is ultimately limited only by the storage capacity of the computer and the data available on the input parameters.

Successful simulation requires one to ask meaningful questions, to have access to information from which the required input can be obtained, and to formulate a model which is appropriate for investigating these questions. Dr. Levin has enumerated a number of circumstances under which simulation has occurred, and has described briefly the nature of that simulation. Let us therefore emphasize some of the difficulties and pitfalls.

Possibly the first such that the investigator must face is the level of programming to be used. Few investigators today have a sufficient command of machine language or the motivation to acquire that command to entertain seriously machine language simulation. It is obvious, however, that the greatest machine efficiency is to be achieved only if the simulation is formulated at this level, and even under the best of circumstances Monte Carlo techniques are expensive. One has to equate the time of the investigator to machine efficiency, of course, and cost considerations may often favor assembly or source language programming. It must also be borne in mind that machine language programming is the least flexible in that it is useful only for a specific model computer frequently. Source language programming is often the quickest means to build a complex and functional program. Its sole economy, however, is to be reckoned in terms of hours of the investigator's time. The machine inefficiency which follows from source language programming can be tolerated if the program is viewed as a preliminary screening to isolate problems which warrant more thorough consideration. Assembly language is often a satisfactory means out of this apparent dilemma. If interest in the simulation of genetic problems continues to mount, much is to be said for the development of a specific language more uniquely suited to simulation than the languages now in common use.

Once an operational program is achieved and experiments begin, a new issue arises. How are the data which are generated to be subsumed into meaningful units? Clearly, a single "experiment" or "run" has quite limited utility since it is merely the record of one chance combination of events. To my knowledge, there is at the moment no convention for the presentation of these data. As a consequence, one may see or present the mean of a number of "runs" and the more extreme ones, or a hodge-podge of all the runs, etc. It is difficult to say the least to integrate visually such results. Further thought on how to present information of this kind is to be encouraged.

As most of this audience is aware, the random numbers commonly generated for use in Monte Carlo simulation are not true random numbers, and are more properly termed pseudo-random numbers since they are generated in a completely deterministic way. The congruence or power residue method is probably the most widely used procedure for producing pseudo-random numbers. Of principal interest here, however, is the randomness of these pseudo-random numbers since patently this is central to an approach which presumes random events. Unfortunately, the tests of randomness which are generally applied to the numbers generated are often inadequate. Frequently they consist of nothing more than testing to see whether the distribution of the digits, 0 to 9, is rectangular. For many uses, this may be adequate but this would hardly seem the case in genetic simulations. If further tests are made, they often consist of chi-square contingency tables to see whether successive digits are random, or "poker" tests in which sets of five numbers are viewed as poker hands and the randomness of the latter studied. Unfortunately, weak serial correlations may escape attention, but these correlations may still be sufficiently large to perturb the simulation. We have observed serial correlations as great as 0.10 on occasions. Obviously some procedures are more sensitive to such correlations than are others. If, for example, one evaluated the inbreeding coefficient of an individual through a random selection of ancestral lines as suggested many years ago by Wright and McPhee, the results could be erroneous because of the correlation in the lines selected. In general, it seems best to minimize difficulties of this nature taking decisions transversely rather than perpendicularly. That is to say, it is more desirable to make all of the decisions of a particular kind--to view the particular decision as the unit of operation--and then of a second kind, etc. rather than to use the individual as the unit and complete all of the decisions with respect to that individual and then proceeding to the next person.

Monte Carlo simulation is or at least should be viewed as essentially an experimental technique, and like all such techniques must be refined and evaluated. Doubtlessly, like other experimental procedures, some investigators will prove more adroit in the use of this approach than others. This should, however, be viewed as a challenge to the investigator rather than a reflection on the techniques.

BODMER: As Dr. Levin has emphasized, there are a variety of approaches to Monte Carlo simulation, and he has given a valuable review of a number of different applications. I should especially like to reemphasize the use of a mixed analytical and numerical approach. Most simulation models, either direct, or mixed analytical and numerical, involve a relatively large number of parameters. Each simulation is really an experiment based on a relatively small sample of the total parameter space, from which one wants to make extrapolations to the whole or at least a major part of this space. Analytical results can clearly be a great help in directing the sampling of the parameter space and help in avoiding reaching incorrect general conclusions based on a relatively limited numerical survey.

As a further example of simulation of the human population taking into account demographic parameters, I should like to mention a study done by Dr. A. Jacquard while he was visiting us at Stanford. A general program was written by him to simulate the human reproductive process. It is based, in general terms, on the fitness flow sheet discussed by Bodmer and Cavalli-Sforza (1965), which is an attempt to interrelate all the factors contributing to reproductivity, such as mortality, distribution of age of marriage, monthly probability of conception, etc., all as functions of age. Similar models have been discussed by Ridley and Sheps (1966) and MacCluer (1967). Jacquard (1967) included in his model a contraceptive regime coupled with a desired family size. The efficiency of contraception is specified by a parameter E multiplied by the monthly probability of conception. Contraception is only practiced during a given period of a few months after marriage and after each childbirth until the desired family size is achieved. The full reproductive history of one female at a time is constructed following the dictates of the flow diagram, until she either dies or passes out of the reproductive period. Demographic variables such as the mean and variance of the number of offspring are computed from the combined histories of 100 women. As expected, for a given desired family size the net reproductive rate and the variance in the number of offspring decreases with increasing efficiency of contraception.

The model has been used further to determine the relative contributions of variation in the age of marriage and mortality and in the probabilities of divorce and widowhood on the variance of the offspring distribution (Jacquard and Bodmer, 1968). If, for example, the age of marriage is held constant, while all other distributions remain unaltered, the resulting change in the variance of the number of offspring is a measure of the contribution of varying age of marriage to this variance. In the central, normal population, based mainly on French demographic data, the mean and variance of the offspring distribution in a non-contracepting cohort were 8.6 and 11.4, respectively. If there were no mortality, widowhood, or divorce the mean and variance were 9.0 and 10.0, while if in addition all marriages occurred at age 21 they were 10.7 and 1.5, respectively. This shows clearly that a large fraction of the variation of the offspring distribution in this population is due to variation in the age of marriage and that mortality and divorce contribute relatively little to the variance. A simulation of the Hutterite population, based on the data of Eaton and Maier, showed that variation in the natural fecundity or monthly probability of conception is most probably a major contributor to the observed variance of the offspring distribution in this population. These simulations show how useful results on the effects of variation in a number of demographic parameters on the fertility distribution can be obtained even with very complex and quite realistic models.

Bodmer, W. F. and Cavalli-Sforza, L. L., 1965. Perspectives in Genetic Demography, Proc. Second World Population Conference, 455-459.

Jacquard, A. 1967. La Reproduction Humaine en regime Malthusien. Population 5:897-920.

Jacquard, A. and Bodmer, W. F. 1968. Etude de la variance de la taille des familles. Population (in Press).

Ridley, J. and Sheps, M. 1966. An Analytic Simulation Model of Human Reproduction with Demographic and Biological Components. Population Studies, XIX, 3:297-310.

MODELS OF QUANTITATIVE GENETIC VARIATION AND COMPUTER SIMULATION OF SELECTION RESPONSE

B. D. H. Latter
Iowa State University
Ames, Iowa

(On leave from Division of Plant Industry, C. S. I. R. O., Canberra, A. C. T., Australia.)

An understanding of the nature of quantitative genetic variation among individuals in a population is basic to theoretical and experimental studies of population structure, evolutionary change, and directed improvement in domestic species. Evolutionary progress in a population must be effected primarily by selection for quantitative differences in structure, behavior and performance, and the maintenance of polymorphic variation in human and animal populations (Neel and Schull, 1968) is almost certainly to be explained in terms of the effects of individual loci on quantitative traits of this sort.

Exploratory studies with Drosophila have clearly demonstrated the inherent complexity of quantitative genetic variation, and of the response patterns shown by small populations under directional selection. In populations with as many as 40 breeding individuals per generation, chance deviations between replicates due to genetic sampling have been detected after only five generations of directional selection (Clayton et al., 1957). A decline in reproductive capacity can be anticipated as a probable correlated response to selection for traits of apparently trivial adaptive significance under laboratory conditions (Mather and Harrison, 1949; Latter and Robertson, 1962). Analyses of populations produced by inbreeding and selection have shown quantitative gene action to involve both dominance and interaction between non-homologous chromosomes (Robertson, 1955; Breese and Mather, 1960), and continued selection is known in many instances to lead to the preservation of genetic variation rather than to uniformity (Falconer, 1960). Component traits may differ qualitatively in their contribution to responses to selection, inbreeding and environmental stress, as demonstrated by Robertson (1959a, b) for cell size and number as determinants of body size in Drosophila. More recent analyses of quantitative variation in sternopleural hair number in this species (Thoday, 1967) have in addition underlined the possible importance of genes of large and identifiable effects.

During the past decade, computer simulation techniques have been available which in principle allow any specified genetic model to be studied in great detail, though the exercise may be expensive in terms of computer time (Fraser, 1957). Yet little progress has been made towards the development of a sufficient genetic model for any experimental trait, even in Drosophila. We have, for example, few indications of the number of loci contributing to variation in a quantitative character, of the gene effects and gene frequency distributions, or of the distribution of dominance effects over the loci concerned. Non-additive genetic effects generated by directional selection are in all probability to be explained by the interactions of component developmental variables, but few attempts have been made to explore models of this sort, despite the availability of computer techniques. There is an urgent need for a combined program of experimentation and computer simulation to develop models which can be accepted as realistic, at least in the sense of being sufficient to explain the most conspicuous features of response to selection in finite populations. The present discussion deals with the initial steps in such a project and the philosophy and methods of approach which have so far been developed.

BASIC QUANTITATIVE GENETIC MODELS

Some aspects of the mechanics of selection response in finite populations have been studied in considerable detail for simplified genetic models involving one or two genes, assumed to be segregating against a constant background of genetic and environmental variation (Kimura, 1964; Latter, 1966b; Hill and Robertson, 1966). The objectives of these investigations have been the elucidation of the joint effects of population size, gene frequency, linkage, gene effects and the mode of gene action on parameters such as total selection response and the time scale of the process (Robertson, 1960). Computer simulation techniques have been used in this work simply to extend the existing mathematical framework to somewhat more complex models.

A closely related though virtually unexplored approach is the construction of simple yet complete genetic models of continuous variation, with a view to computer simulation tests of the goodness-of-fit of alternative models to observed response patterns in laboratory experimentation. Provided the computer studies involved are programmed so that a complete analysis of the selection process is possible, this sort of approach can be expected to complement the study of simple genetic systems, and in addition to aid directly in the design of selection experiments whose prime objective is the discrimination amongst possible genetic models. The aim initially should then be to make use of a number of possible models which can be completely specified by only one or two unknown parameters, but which are readily extended to include any desired element of genetic complexity.

The choice of models and decisions as to the strategy of experimentation and simulation are quite obviously going to be interdependent in such an approach, given the single objective of providing a progressively more accurate picture of the nature of quantitative genetic variation. To some extent the specification of exactly what constitutes simplicity in a genetic model must be subjective, depending on the ideas one has about the minimum requirements for a simple yet realistic model. My own view, for example, is that provision for multiallelic variation in a model does not constitute undue complexity, provided the total system can be defined by a small number of parameters. Another worker may have concluded that two alleles per locus is an acceptable restriction, provided a satisfactory fit to the available data can be obtained without introducing multiple alleles. Decisions of this sort are of course dependent on the nature of the experimental system which one has in mind.

In my initial attempts at model-building, I have introduced the following general restrictions:

(i) Gene action is assumed to be additive, no dominance or epistasis being involved in the basic models. In future work, non-additive gene effects are to be introduced by considering component variables, some of which are subject to threshold behavior.

(ii) The basic models have symmetrical properties, so that selection for decreasing performance is precisely the mirror-image of selection in the positive direction. Observed asymmetry of response is therefore assumed to be of a type which can be eliminated by a suitable scale transformation (Falconer, 1960).

(iii) Linkage is introduced when required by a random scattering of the specified number of loci throughout the linkage map of the organism concerned.

(iv) The base population is assumed to be in equilibrium under random mating, despite the fact that the genetic variation in some models is supposed to have been generated by a combination of natural selection for an intermediate optimum and recurrent mutation.

Working within the framework set by these restrictions, a set of five basic genetic models has been used (Table 1). Future work will undoubtedly require the examination of other symmetric gene models, and genetic asymmetry will also be introduced, but the five basic models are sufficient for our present purpose. These are numbered below for cross-reference in this and other papers.

Models 1, 2

In both models 1 and 2 it is assumed that the additive genetic variance in the base population is due to segregation at n loci, each with two alleles at initial frequencies of 0.5. The contribution of each locus to the total variation is specified in terms of its proportionate effect (Falconer, 1960), i.e., the difference in mean value between sub-populations carrying the two alternative homozygotes at the locus, denoted by a, relative to the initial phenotypic standard deviation, σ. A typical locus may therefore be represented as follows:

Genotypes	A_1A_1	A_1A_2	A_2A_2
Initial frequencies	0.25	0.50	0.25
Effects	$-\frac{1}{2}a$	0	$+\frac{1}{2}a$

The heritability of the character is then initially equal to

$$h^2 = \sum_{i=1}^{n} g_i^2 = \frac{1}{8}\sum_{i=1}^{n}\left(\frac{a_i}{\sigma}\right)^2 = \frac{n}{8}E\left(\frac{a_i}{\sigma}\right)^2 \qquad (1)$$

and the total possible response to unidirectional selection, measured in phenotypic standard deviations, is

$$\max\left(\frac{r}{\sigma}\right) = \frac{n}{2}E\left(\frac{a}{\sigma}\right). \qquad (2)$$

In model 1 it is supposed that the segregating loci do not contribute equally to the observed variation in the character, the proportionate effects being a random sample from an exponential distribution, specified by a single parameter λ. An exponential distribution of gene effects gives one of the simplest representations of a population in which there are many loci segregating with comparatively small effects on the character under selection, together with a few genes of large individual effects. The density function is

$$f(x) = \lambda e^{-\lambda x}; \quad 0 < x < \infty \qquad (3)$$

with $E(x) = \lambda^{-1}$ and $E(x^2) = 2\lambda^{-2}$, so that under model 1 the value of λ is related to the initial heritability by the expression

$$\lambda_1 = \left[\frac{n}{4h^2}\right]^{\frac{1}{2}} \qquad (4)$$

and the total possible response is therefore

$$\max\left(\frac{r}{\sigma}\right)_1 = \frac{n}{2\lambda_1} = (nh^2)^{\frac{1}{2}}. \qquad (5)$$

The parameter λ of the exponential distribution can be given an alternative interpretation which indicates the magnitude of effect of the major genes involved in the model. The area under the distribution curve corresponding to values of $\frac{a}{\sigma}$ greater than x_0, i.e.,

$$\int_{x_0}^{+\infty} \lambda e^{-\lambda x}\,dx = e^{-\lambda x_0} \qquad (6)$$

is equal to 0.05 for $\lambda x_0 = 3$. Five percent of segregating loci are therefore expected to have effects greater than $\frac{3}{\lambda}$ phenotypic standard deviations. A value of $\lambda = 3$ then corresponds to a model in which only five percent of genes have effects greater than one phenotypic standard deviation. Table 2 gives a more detailed breakdown of frequencies in an exponential distribution of gene effects.

In model 2 it is assumed that all loci make an equal contribution to the total variance displayed in the base population. Since in this model we are specifying initial gene frequencies of 0.5, $\frac{a}{\sigma}$ is the same for all loci and we have from (1) and (2)

$$\max\left(\frac{r}{\sigma}\right)_2 = \frac{na}{2\sigma} = (2nh^2)^{\frac{1}{2}}. \qquad (7)$$

Comparison of (5) and (7) shows that in model 1 with an exponential distribution of gene effects, the "effective" number of loci is equal to one-half the actual number (Wright, 1952).

Models 3, 4

In genetic models 3 and 4 it is assumed that the initial genetic variance in the trait under selection is due to a set of n multiallelic loci with additive effects. A typical locus is supposed to carry a very large number of alleles A_i, each of which is present at low frequency in the base population. The array of alleles can be considered to have arisen by a long-continued process involving (i) the accumulation of plus and minus mutations of individually small effect, and (ii) the action of natural selection favoring an intermediate optimum for the metric trait concerned. Kimura (1965) has shown the equilibrium array of allelic effects to be normally distributed, so that the alleles concerned can be taken to represent a set of isoalleles - all qualitatively of "normal" activity but differing quantitatively, e.g., in respect of the level of enzyme activity promoted. Mutations with gross phenotypic effects are presumed to have been eliminated from the population by natural selection.

TABLE 1

The properties of five symmetrical genetic models.

Model number	Initial gene frequencies*	Alleles per locus	Gene effect distribution	Unknown parameters†
1	0.5	2	exponential	n
2	0.5	2	equal effects	n
3	$\frac{1}{2N}$	2N	exponential	n
4	$\frac{1}{2N}$	2N	equal effects	n
5	$\phi(\frac{a}{\sigma})$	2	exponential	n, λ

*N denotes the effective breeding population size; $\frac{a}{\sigma}$ represents the proportionate effect of the locus on the character under selection.

†n denotes the number of loci contributing genetic variance to the character under selection; λ is the parameter specifying the exponential distribution of gene effects.

The behavior of such loci under unidirectional selection in finite populations has been extensively studied by Latter and Novitski (1969), with particular reference to the expected total response to selection and variation among replicate populations at the limit. It is of course obvious that in dealing with models of this sort, the concept of the maximum possible response to selection (cf. equations 2, 5, 7) is inappropriate, in contrast to the two-allele models previously discussed. At a multiallelic locus carrying alleles with normally distributed effects, it is always possible that a more extensive sampling of the base population will produce an allele of more extreme effect, so that the total possible response is infinite if the base population is considered to be unlimited in size.

Since every allele is supposed to be rare in the base population, the number of distinct alleles represented in the initial set of N selected parents is taken to be equal to 2N, each with frequency $\frac{1}{2N}$ (Table 1). The genetic model is therefore completely specified if the contribution of each locus, $g_i^2 \sigma^2$, to the total phenotypic variance in the base population is given. In model 3 it is assumed that the loci make unequal contributions, the value of g_i being exponentially distributed over loci (equation 3). The value of λ in the density function is related to the initial heritability of the character by the expression

$$\lambda_3 = [\frac{2n}{h^2}]^{\frac{1}{2}} \qquad (8)$$

where

$$h^2 = nE(g_i^2) \quad . \qquad (9)$$

In model 4 each of the set of n multiallelic loci makes the same contribution $g^2 \sigma^2$ to the total phenotypic variance in the base population.

TABLE 2

Characteristics of an exponential distribution of gene effects*.

Value of λx_0	Proportion of loci with $\frac{a}{\sigma} > x_0$
0.1	0.905
0.5	0.606
1.0	0.368
1.5	0.223
2.0	0.135
2.5	0.082
3.0	0.050
3.5	0.030
4.0	0.018

*Proportionate effects ($\frac{a}{\sigma}$) distributed according to equation 3.

It is important to stress that the interpretation of these two multiallelic models can be somewhat more general than has been indicated above. An individual "locus" can be taken to represent a specific chromosome or chromosome segment, within which are located a number of genes contributing to genetic variation in the character under selection, where recombination among the loci is effectively suppressed. The "allelic" effects then correspond to aggregate effects of the array of possible gametic configurations in the chromosome segment (Latter and Novitski, 1969).

Note that in models 1, 2, 3, and 4, a total of four parameters is necessary to specify both the genetic model and the regime of selection based on individual performance; viz. N, the effective breeding population size; $\bar{i}$, the standardized selection differential; h^2, the heritability of the trait in the base population; and n, the number of segregating loci. However in practice the values of N, $\bar{i}$, and h^2 are known, so that n is the only variable to be determined (Table 1). It is of course assumed here that the number of chromosomes and their map lengths are known for the species concerned, so that linkage can be simulated by a random allocation of loci to the chromosome set.

Model 5

Genetic model 5 is closely related to models 1 and 2, but involves a single additional parameter to enable relaxation of the assumption that all gene frequencies are initially equal to 0.5. The additive genetic variation in the base population is supposed to be due to segregation at n loci, each with two alleles. The proportionate effects of the loci are assumed to be a random sample from an exponential distribution with parameter λ (equation 3), as in model 1. Those loci with effects less than a particular value x_0 have initial gene frequencies of 0.5, such loci representing a proportion π of the total number n, where

$$\pi = 1 - e^{-\lambda x_0} \quad . \qquad (10)$$

The remaining fraction, $1-\pi$, of loci with proportionate effects greater than or equal to x_0, all contribute equally to the initial phenotypic variance. The genetic variance due to an individual locus in the base population is therefore

$$g_i^2 \sigma^2 = \begin{cases} \frac{1}{8} a_i^2 , & a_i < x_0 \sigma \\ \frac{1}{8} x_0^2 \sigma^2 , & a_i \geq x_0 \sigma \end{cases} \qquad (11)$$

and the initial gene frequencies are

$$q_i = \begin{cases} \frac{1}{2} , & a_i < x_0 \sigma \\ \frac{1}{2} [1 \pm \sqrt{1 - k_i^2}] , & a_i \geq x_0 \sigma \end{cases} \qquad (12)$$

where

$$k_i = x_0 \frac{\sigma}{a_i} \quad .$$

It can readily be shown that the parameters h^2, n, and λ uniquely determine x_0, and hence the value of π and the q_i for individual loci. The expression relating the various constants is

$$h^2 = nE(g_i^2)$$

$$= \frac{n}{4\lambda^2 \sigma^2} [1-e^{-\lambda x_0} (1+x_0 \lambda)] \quad . \qquad (13)$$

The genetic situation envisaged in this model is one in which a small number of loci of comparatively large effect are segregating in the population, with "plus" alleles initially at extreme frequencies, i.e., close to zero or unity. Loci of smaller effect are more common, with intermediate gene frequencies converging to 0.5 for the loci of smallest effect. A simple model of natural selection for an intermediate optimum value of the quantitative character, together with uniform mutation rates involving only two alleles per locus, leads to an equilibrium configuration of the sort represented by this model (Latter, 1960). Model 5 therefore provides the counterpart of model 3, with only two possible allelic states per locus.

In the present discussion the gene frequency distribution is taken to be symmetrical, though a further parameter can readily be introduced to specify the relative frequency of loci with the plus allele at low frequency ($q_i < 0.5$) vs. those at high frequency. An excess of loci with plus alleles at low frequency, for example, might be expected if the prior regime of natural selection had been initially for reduced expression of the trait concerned, followed by stabilizing selection for an optimal mean level of expression. As far as I am aware, the precise characteristics of such a gene frequency distribution have not been investigated.

For the symmetrical distribution of gene frequency, the maximum possible response under model 5 is given by

$$\max \left(\frac{r}{\sigma}\right)_5 = \frac{n}{2\lambda_5} \qquad (14)$$

Since n and λ can be varied independently, a decrease in the value of λ corresponds to an increase in the maximum possible response (cf. model 1).

THE STRATEGY OF COMPUTER SIMULATION

Comparisons of the results of laboratory or field experimentation with expectations based on particular genetic models may lead to the eventual rejection of some models as incompatible with the available data and to the progressive modification of those models which appear to be most appropriate. In any discussion of the strategy to be adopted in making such comparisons the following three questions arise immediately:

1. Which parameters can be expected to be of greatest value in discriminating among possible genetic models?
2. What experimental design is likely to give maximum sensitivity in comparisons of alternative models, for equal expenditure of time and effort in estimating the specified parameters?
3. What step-wise procedure of computer simulation should be adopted in fitting a given model to the available experimental data?

The choice among possible parameters is severely limited by the nature of the usual patterns of response to inbreeding and selection. Changes in the overall population mean as a function of the degree of inbreeding in the absence of selection provide the most direct test for the importance of directional dominance. The study of chromosome substitution lines and hybrids between selected populations gives further direct evidence as to the importance of non-additive genetic effects in the genetic system concerned. In the present discussion it will be assumed that genetic variation in the base population has been shown by such tests to be essentially additive, so that the basic genetic models described above are potentially appropriate.

The response of the population to directional selection, and the resulting variability among replicate breeding lines, are known to depend on the characteristics of the genetic system underlying quantitative variation in the initial population. This is, in addition, the phenomenon of most immediate practical importance, and the following discussion will deal specifically with this phase of study. The initial average rate of response to selection is simply a function of the heritability of the character, which can be directly estimated from observations of the base population itself: all models will therefore give rise to similar response patterns in the short term, unless genes of large effect are involved (Latter, 1965a). Measurements of long-term response are therefore required for tests of goodness-of-fit of possible genetic models, and the simplest descriptive parameters are the following (Robertson, 1960):

(i) The total response (r) at the limit, i.e., $r = \lim_{t \to \infty} x(t)$, where x(t) denotes the mean of the trait averaged over all replicate populations in generation t, coded so that x(t) is zero for $t = 0$.

(ii) The component of variance (σ_L^2) among replicate population means at the limit.

(iii) The 95 percent life of the selection process (L_{95}), or some equivalent measure of the time scale of response to selection; the L_{95} is defined to be that value of t for which $x(t) = 0.95r$.

It should be emphasized that little is known of the sensitivity of these parameters to important differences in the initial genetic system.

We have as yet virtually no answer to the question concerning experimental design, and one of the objectives of the present project was to determine the most useful sequence of experimental manipulations of breeding population size and selection intensity. Some of the results derived from an examination of the five basic models described above are to be given in the present section. With regard to the steps involved in the matching of experimental observations by computer simulation, the prime consideration must be the total computer time, in view of the considerable expense of techniques of direct simulation. The available procedures are discussed below in order of increasing complexity and hence increasing expense per unit simulation.

The use of prediction formulae

The expected total response to truncation selection can be predicted for independently segregating loci each with two alleles, by use of the formula provided by Kimura (1957) for the probability of fixation, u(q), of an allele with selective advantage $\bar{i}\frac{a}{\sigma}$ and initial frequency q. In the case of additive genes (models 1, 2, 5) the solution is

$$u(q_i) = [1 - \exp(-2v_i q_i)]/[1 - \exp(-2v_i)] \qquad (15)$$

where

$$v_i = N\bar{i}\frac{a_i}{\sigma},$$

so that

$$E\left(\frac{r}{\sigma}\right) = \sum_{i=1}^{n} [u(q_i) - q_i]\left(\frac{a_i}{\sigma}\right). \qquad (16)$$

The variance component among means of replicate populations at the selection limit is predicted to be

$$\frac{\sigma_L^2}{\sigma^2} = \sum_{i=1}^{n} [u(q_i)][1 - u(q_i)]\left(\frac{a_i}{\sigma}\right)^2. \qquad (17)$$

No corresponding formula has yet been derived for predictions of the time scale of the process, so that more elaborate methods of simulation are necessary if the L_{95} is to be used for comparison of observation and theory.

Since the joint distribution of the a_i and q_i are specified in models 1, 2, and 5, the task of evaluating the expectations (16) and (17) for independently segregating genes presents no difficulty in principle. In the case of model 2, since all loci are identical the calculations are elementary. For models 1 and 5 involving exponentially distributed values of $\frac{a_i}{\sigma}$, the total response and variance among lines at the limit can be estimated by repeated sampling from an exponential distribution of gene effects, or by standard techniques of numerical integration.

Comparable algebraic solutions have not yet been derived for independently segregating multiallelic loci of the sort involved in models 3 and 4. A satisfactory empirical prediction equation has however been provided by Latter and Novitski (1969) for the total expected response as a function of the parameter combination $z_i = N\bar{i}g_i$, viz.

$$E(\frac{r}{\sigma}) = 3.68 \sum_{i=1}^{n} \left[\frac{Z_i - 1}{Z_i + 1} \right] g_i \qquad (18)$$

where

$$Z_i = (1 + z_i)^{1.09} \quad .$$

This expression has been shown to be satisfactory for a wide range of parameter combinations in the range $0 \leq z_i \leq 16$, with $5 \leq N \leq 50$, $\bar{i} \leq 1.8$, and $g_i^2 \leq 0.10$. A corresponding prediction of the expected variance among lines at the selection limit is given by

$$\frac{\sigma_L^2}{\sigma^2} = 2 \sum_{i=1}^{n} \frac{g_i^2}{(1 + cz_i)^c} \qquad (19)$$

where

$$c = 0.685 \ .$$

It must be emphasized that expressions (18) and 19) have been given simply as an aid to numerical evaluation within the prescribed limits for z_i, N, $\bar{i}$, and g_i. Their form does not in any way indicate the nature of the true underlying mathematical relationships.

For any given combination of the four parameters N, $\bar{i}$, h^2 and n, the expected total response and ultimate drift variance can then readily be calculated for models 1 - 4, assuming free recombination among the loci concerned. If λ is independently specified, similar predictions can be made for model 5. Such calculations give a valuable indication of the expected behavior of these particular quantitative genetic models under directional selection, despite the fact that they (i) ignore changes in the total phenotypic variance (σ^2) with time; (ii) make use of selective values which may be inadequate for genes of large effect (Latter, 1965a); and (iii) make no allowance for linkage among the loci concerned. Because of these limitations, the formulae are most useful for quantitative traits of low heritability with relatively few loci per chromosome.

In practical applications of these procedures, the unknowns are n, the number of independently segregating loci, and λ specifying the scale of genetic effects in the case of model 5. The heritability h^2 is of course fixed by the choice of trait and base population, and the remaining parameters N and $\bar{i}$ can be manipulated by the experimentalist. In Figure 1 are graphed the predictions of total selection response for models 1 - 4, with $h^2 = 0.33$, $\bar{i} = 1.40$. The chosen value of heritability is that of scutellar bristle number in the Canberra population (Latter, 1964), and the selection differential corresponds to a frequently used intensity of truncation selection in *Drosophila* experiments. Comparable graphs for model 5 are shown in Figure 2 with values of λ ranging from 0.5 to 3.0.

It is instructive to make comparisons among the five genetic models on the basis of the graphs in Figures 1 and 2, together with the corresponding predictions of drift variance at the limit given by equations (17) and (19). At low values of effective population size, e.g., N = 5, it is clear that the total response realized is not conspicuously different for any one of the five models: the curves relating total response to the number of loci for N = 5, show considerable overlap in the range of values of $\frac{r}{\sigma}$ predicted. In the absence of prior information as to the probable value of n, observations of $\frac{r}{\sigma}$ at low values of N cannot therefore be expected

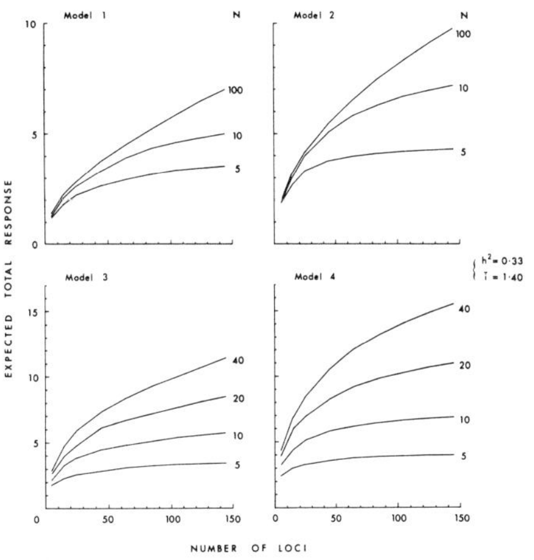

Figure 1. The expected total response at the limit ($\frac{r}{\sigma}$) for models 1-4, predicted by equations (16) and (18) as a function of the number of independently segregating loci (n), and the effective breeding population size (N).

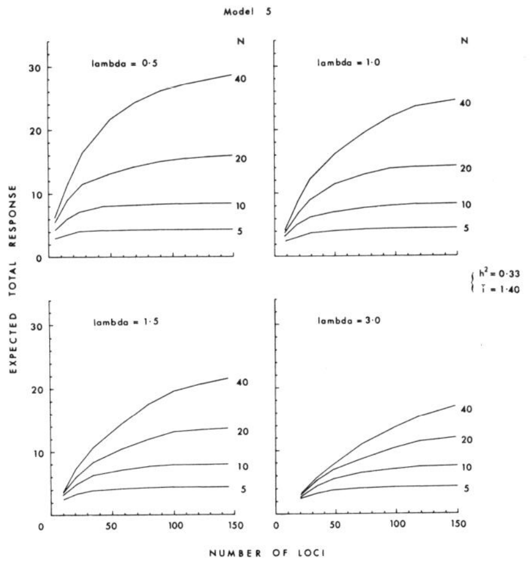

Figure 2. The expected total response at the limit ($\frac{r}{\sigma}$) for model 5, predicted by equation (16) for a range of values of λ, specifying the scale of genetic effects [equations (3), (13)] .

to aid in discrimination among these models.

However, the ratio of the drift variance expected at the limit, $\frac{\sigma_L^2}{\sigma^2}$, to the expected total response, $\frac{r}{\sigma}$, may have considerable diagnostic value even at very low levels of effective population size. The values of this ratio predicted by equations (16) and (17) for model 5 are conspicuously greater than for the other four models, unless λ is large and n small. The characteristic feature of model 5 is that response in small populations is subject to pronounced chance variation, being due to some extent to genes of large effect initially at low frequency. However, as λ increases with n held constant, model 5 becomes identical with model 1 so that no contrast can be expected under these limiting conditions. Models 3 and 4 show values of the ratio of drift variance to total response which are intermediate between those of models 1 and 2 on the one hand, and model 5 on the other.

It can also be seen from Figures 1 and 2 that marked differences among the models are to be expected in the total response realized at different levels of breeding population size. For purposes of illustration, Table 3 gives the predicted responses at the limit for N = 100, as a multiple of the corresponding response expected for N = 20, calculated for each of the models as a function of the number of independently segregating loci. Here again models 1 and 2 behave very differently from model 5, unless λ is large and n small in the two-parameter model. Models 3 and 4 are intermediate as before.

Despite the obvious limitations of the formulae involved in making these calculations, they are of value in leading to quite definite expectations which can be tested using more detailed techniques of simulation. In particular, the foregoing suggests that (i) the most informative experiment per unit of time and effort may be an extensively replicated set of breeding lines subjected to a regime of selection emphasizing chance effects in the early generations; (ii) comparisons of response in large populations vs. that in smaller breeding lines may have considerable diagnostic value, though here the linkage effects will have to be taken into account by more direct simulation; (iii) comparisons of models involving equal vs. exponentially distributed gene effects (models 1 and 2, 3 and 4) indicate that they do not differ appreciably in either response or variability between lines at low values of N in the absence of linkage. This is also true of the sensitivity of total response to experimental manipulation of effective population size (Table 3).

The effect of linkage

The use of prediction formulae for independently segregating loci may be thought to give no information regarding the effect of linkage on the parameters concerned. However, use can be made of equations (16), (18) and (17), (19) to assess the maximum possible effect of linkage on the total realized response and the variability among lines at the limit, i.e., the effect of a complete suppression of crossing-over within particular chromosomal units (Latter and Novitski, 1969). Provided the numbers of loci located in each chromosome segment are reasonably large, and gene frequencies at the individual loci are not too extreme, the distribution of aggregate effects for the segments will be approximately normally distributed for all models. Predictions for models 3 and 4 will then give a rough guide to the response and variability between lines due to chromosomal genetic variation. Table 4 sets out a number of comparisons of response expected with free recombination versus that given with no recombination within linkage blocks. Model 4 simulates linkage blocks of equal size, i.e., making equal contributions to the initial genetic variance; model 3 corresponds to a genome with either (i) a marked tendency to "clumping" of loci

TABLE 3

Predicted total response ($\frac{r}{\sigma}$)* at an effective breeding population size of N = 100, as a multiple of that expected at N = 20 [h^2 = 0.33, $\bar{i}$ = 1.40 as in Figures 1 and 2].

Model	Number of independent loci (n) 10	20	50	100
1	1.00	1.01	1.03	1.10
2	1.00	1.00	1.00	1.02
3	1.24	1.32	1.40	1.58
4	1.25	1.34	1.54	1.73
5; λ = 3.0	1.00	1.02	1.15	1.50
λ = 1.5	1.02	1.12	1.58	2.16
λ = 1.0	1.17	1.32	1.92	2.53
λ = 0.5	1.43	1.78	2.53	3.11

*Based on equations (16) and (18).

TABLE 4

Expected total response under free recombination, expressed as a multiple of that predicted with complete suppression of crossing-over. The ratio has been estimated as $\frac{r_i}{r_j}$, where subscripts denote the basic genetic models involved; i = 1-5, j = 3,4.

Parameters* N	$\frac{n}{c}$	z	Linkage blocks of equal size (j=4) Genetic model (i) 1	2	3	4	5;λ=3	Unequal linkage blocks† (j=3)
5	50 10	1.8	1.15	1.59	1.19	1.50	1.58	1.38
	100 10	1.3	1.14	1.47	1.16	1.33	1.47	1.36
	100 20	1.8	1.35	1.74	1.38	1.58	1.75	1.38
	200 10	0.9	1.12	1.37	1.16	1.27	1.39	1.31
	200 20	1.3	1.26	1.53	1.30	1.42	1.56	1.36
10	50 10	3.6	1.09	1.64	1.43	1.81	1.77	1.45
	100 10	2.5	1.14	1.65	1.31	1.64	1.80	1.38
	100 20	3.6	1.41	2.05	1.63	2.04	2.24	1.45
	200 10	1.8	1.11	1.56	1.19	1.48	1.67	1.33
	200 20	2.5	1.35	1.90	1.44	1.80	2.02	1.38
20	50 10	7.2	1.01	1.46	1.62	2.18	1.85	1.45
	100 10	5.1	1.03	1.57	1.50	2.00	2.08	1.49
	100 20	7.2	1.35	2.06	1.96	2.62	2.72	1.45
	200 10	3.6	1.03	1.66	1.36	1.83	2.11	1.42
	200 20	5.1	1.31	2.11	1.72	2.32	2.67	1.49

*N represents effective population size; n the total number of loci contributing to response under free recombination; c the number of linkage blocks; and z is the expected value of N $\bar{i}$g for the units of segregation in model j.

†Expected ratios for unequal linkage blocks are given by multiplying each value in the row by this factor.

concerned with the quantitative trait, or (ii) an irregular pattern of chiasma localization, with an essentially random allocation of loci along the physical length of the chromosome.

A number of conclusions can be drawn from the results presented in Table 4, which help in the specification of numerical values for parameters used in direct simulations involving linkage:

(i) The effect of linkage in reducing total response is expected to be minimal in very small populations where chance fixation of alleles plays the dominant role, and also at high levels of effective population size where the most extreme chromosomal configuration is unlikely to be lost by chance in the early generations of selection. The calculations presented in Table 4 show the maximum effect of linkage to be reached at values of $N \leq 20$ for models 1 and 2 with 100 segregating loci or less, provided the number of linkage blocks is greater than five. The maximum effect for the remaining three models involving either multiallelic loci, or loci with extreme rather than intermediate gene frequencies, occurs at appreciably higher values of N. For the multiallelic loci of models 3 and 4, the maximum appears to be at a value of N considerably greater than 50, and certainly beyond the range of validity of the approximate calculations we are making here.

(ii) At values of population size in the vicinity of $N = 5$, the reduction in response due to linkage may be a very minor effect of the order of 10 percent or less. However, with as few as five linkage blocks, the reduction is potentially as great as 60 percent if 100 or more loci are involved.

(iii) Comparison of models 1 and 2, and of models 3 and 4, shows up the consistently greater proportional reduction in response due to linkage among loci of equal effect, when contrasted with a model involving variable gene effects.

(iv) The numerical calculations based on model 5 indicate that linkage effects are most pronounced when response is due predominantly to loci of large effect, with the favored allele at low frequency in the base population.

(v) The breakdown of the total genome into linkage blocks of very different size ($j = 3$ in Table 4) can be seen to lead to an additional reduction in total response, by comparison with that expected for linkage blocks making equal contributions to the initial genetic variance. The effect can obviously be quite appreciable when a small number of gene blocks account for a major fraction of the variation.

It is tempting to extend the range of parameter combinations beyond those encompassed by Table 4. However, it must be stressed that many approximations are involved in these predictions, and more elaborate methods of simulation are required to set the limits beyond which they may be misleading. We have been deliberately conservative in drawing up the table, since its purpose is primarily to illustrate the most conspicuous effects of linkage on total response.

Techniques of partial and direct simulation

Partial simulation techniques may very effectively be used in the exploration of almost all models of quantitative genetic variation. Direct simulation can then be used specifically to check out the conclusions suggested by the use of the simpler techniques, and to introduce linkage and epistatic interactions in all their complexity as a final step in the process of simulation. In essence, the use of partial simulation procedures involves an examination of the behavior of a subset of loci, assuming the remainder of the genotype to be segregating independently; the background variance contributed by loci not contained in the subset is assumed to remain unchanged throughout the selection process.

Changes in gametic or zygotic frequencies from one generation to the next are frequently simulated by algebraic transformations, followed by a random sampling of the required number of gametes or parents from the population. It may not be necessary to generate a finite group of offspring of the selected parents, nor need the existence of separate sexes be introduced unless sex-linkage or sex-limited effects are involved in the model.

The use of partial simulation has been explored in particular by Latter (1965a, b; 1966a, b), Hill and Robertson (1966) and Latter and Novitski (1969). The techniques have been used to study the effects of linkage between neighboring loci, to simulate multiple alleles, and to estimate parameters such as the L_{95} (95 percent life) for which algebraic prediction formulae are not available. Furthermore, selection regimes may involve values of N and $\bar{i}$ which change during the process of selection, as for example in the use of a "bottleneck" in population size in the early generations (Robertson, 1960). Here again these simple techniques of simulation may be used, to provide the same information as that given by prediction formulae for the regimes of constant populations size and intensity of selection.

SIMULATION OF RESPONSE TO SELECTION IN DROSOPHILA

The comparisons of the previous section have suggested that a selection experiment in which the outcome is influenced both by continuous selection pressure and marked sampling fluctuations in gene frequency, may be valuable for two reasons: (i) the experiment is expected to be of short duration and can ultimately be extended by intercrossing of the plateaued populations, and (ii) observations of the extent of replicate variability should enable useful tests to be made of the goodness-of-fit of the simple genetic models previously discussed in this paper. In this section we shall present the results of such an experiment involving scutellar bristle number in *Drosophila*, and give a brief account of computer simulation of the response patterns observed. The only comparable study reported in the literature for any trait is that of Fraser and Hansche (1963).

Scutellar bristle number in the Canberra population is almost invariably four in both males and females, but the incidence of females with extra bristles is such that selection for increased bristle number in this sex can readily be practiced. The heritability of the trait on the underlying probit scale is $h^2 = 0.33 \pm 0.01$, the non-genetic variance being due primarily to chance effects during development. True environmental variation among individuals accounts for less than 10 percent of the total phenotypic variance in normal cultures (Latter, 1964). Comparison of the means of random inbred lines and their F_1's shows inbreeding to lead to an increase in the mean on the probit scale corresponding to roughly $0.6\ \sigma$, whereas selection may produce an accumulated change approaching $10\ \sigma$. In the region of the scale concerned in the present experiment, the response to selection is linear in large breeding populations (Latter, 1966c), and both intercrosses among lines and chromosome substitution studies suggest that the gene effects concerned are predominantly additive.

We shall consider a set of 15 replicate breeding lines which were derived from the Canberra population, each being initiated by a single pair of parents randomly chosen from the population. The following selection regime was followed in each of the 15 populations:

in the first generation two females with extra-bristles and randomly chosen males were mated in two single-pair cultures; four extra-bristle females and four random males were mass-mated in a single culture in the second generation; and thereafter the ten females of highest bristle score were mass-mated with ten random males in each generation of selection. The objective of the scheme was to pass each of the populations through a "bottleneck" as far as the sampling of genes is concerned, but to increase the breeding population size in subsequent generations to prevent loss of lines due to the accumulated effects of inbreeding.

The effective standardized selection differential, calculated in the manner described by Latter (1964) for a univariate model, averaged 1.10, 0.85, 1.00, and 0.92 in the first four generations, thereafter approximating 0.70 for the duration of the experiment. The actual values of $\bar{i}$ were not precisely the same for each of the replicate populations in the early generations. Due to the threshold nature of the character, populations with lower than average mean values on the underlying scale were subjected to more intense selection pressure to secure the required number of extra-bristle females. There would therefore have been a tendency to reduce the variability among replicates, but the effect is likely to have been of a minor order in view of the very small breeding population sizes involved. No attempt has been made to take this factor into consideration in computer simulation of the observed results.

The response pattern is shown in Figure 3 for the mean of all fifteen populations, and the two most extreme replicates in each direction from the mean, chosen on the basis of their average level over the last five generations of selection. The observed total response over the period of 22 generations was x(22) = 1.78 ± .17 on the underlying probit scale, the phenotypic standard deviation

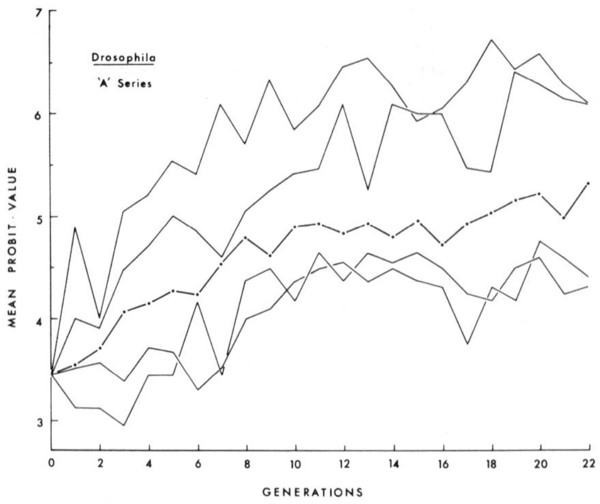

Figure 3. Observed response to selection for increased scutellar bristle number in Drosophila melanogaster. The broken line shows the average value of fifteen replicates, and the two most extreme replicates in each direction are plotted individually. Each replicate was initiated as a random full-sib group from the Canberra population.

in the base population being unity (Falconer, 1960). The variance among population means, averaged over the last five generations of selection, was $\sigma_L^2 = 0.43 \pm .15$, i.e., roughly 1.3 times the additive genetic variance in the base population. Over the first ten generations, the linear response corresponded to a realized heritability of 0.18 ± .02, compared with $h^2 = 0.33 \pm .01$ in the base population.

Results of partial simulation

Attempts have been made to mimic the observed response pattern with each of the five basic genetic models, by means of a fast computer program based on partial simulation techniques. The program traces the fate of each locus independently of the remainder of the genotype, with the background variance held constant, assuming exclusively autosomal transmission and monoecious reproduction. Since selection in the Drosophila experiment was confined to females, the neglect of sex-linked inheritance in simulation does not appreciably influence the expected change in gene frequency per generation due to selection, but the effects of genetic sampling in contributing to between-replicate variability are to some extent underestimated. However, it was felt that this constituted a minor deficiency in the simulation procedure, by comparison with our lack of precise knowledge of effective population size in Drosophila populations. In the experiment described above, the parents of generations 0 and 1 were mated as single pairs in separate vials, and the effective population size for simulation purposes is therefore known exactly. Thereafter the selected parents were mass-mated in a single culture, and the effective population size has been taken as 0.6N to allow for variation in progeny group size, due to reproductive differences among the N parental individuals (Crow, 1954).

Simulation of artificial selection in the program is achieved

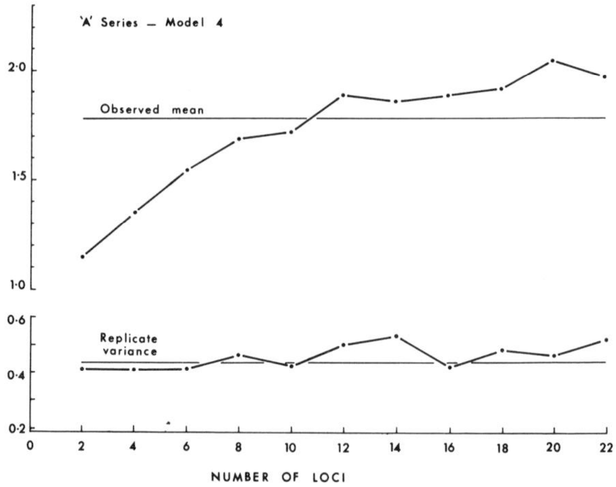

Figure 4. Estimation of the number of independently segregating loci for genetic model 4, from a partial simulation study of response to selection in Drosophila. The predicted mean and component of variance among replicates at generation 22 are plotted as a function of n, the number of loci, for comparison with the experimental observations.

by the use of the relative selective values described by Latter (1965a). In general, an approximation is used for the relative probability of a normally distributed population contributing individuals of a specified genotype to the region of the scale beyond the point of truncation. However, simulation based on genetic model 5 makes use of a subroutine which calculates the required probability accurately, since genes of large effect may be involved (Latter, 1965a). The value of $\bar{i}$ used in runs with this program was one-half that imposed experimentally on females, since no selection was practiced among males. Sampling of gene effects from an exponential distribution for models 1, 3 and 5 was achieved by transforming a random number θ_i in the range 0, 1 by the relationship

$$\frac{a_i}{\sigma} = -\log_e(\theta_i)/\lambda \quad . \tag{20}$$

Table 5 summarizes the results of these preliminary attempts at computer simulation. For models 1 - 4, the estimated numbers of loci are those giving best fit to the observed mean response at the conclusion of the experiment. The corresponding drift variance between replicates is tabulated for each of the four models, and Figure 4 illustrates the procedure of estimation of the number of loci for genetic model 4. In the case of model 5, the values of the two variables n and λ in Table 5 are those giving the best fit to both the response and drift variance observed, though the range of values of λ was not extensively sampled.

In drawing inferences from the data of Table 5 it is obvious that the deficiencies are on the side of the experimental estimates, particularly that of the drift variance component based on only 15 populations. However, the experiment concerned is only one of a series of selection studies with this character, and no firm conclusions regarding the suitability of any particular model can be arrived at until all the experiments have been considered. For the present we shall therefore take the statistics in Table 5 at face value. They suggest (i) that a modest number of loci, of the order of 10-20 or less, can account for the experimental observations via models 4 or 5 if linkage is ignored; (ii) that model 3 is compatible with the data, but a large number of loci would have to be postulated if allowance is to be made for the effects of linkage in the model; and (iii) that models 1 and 2 appear to be inadequate, in that the drift variance expected with either model is considerably less than that observed.

The computer results derived from model 5 proved particularly instructive. One of the reasons for including this model in the study was that it allows for the possibility of genes of major effect, initially at low frequency in the population and therefore contributing only a small proportion of the observed genetic variation. Since model 5 has two independent parameters to be specified, viz. n and λ, and we are attempting to fit only two experimental estimates, viz. total response and drift variance, it is to be expected that a satisfactory fit should be possible. Nevertheless, the partial simulation studies show quite clearly that values of λ less than unity are not compatible with the data, since they lead to predicted values of total response far less than that observed, and a drift variance considerably greater than the experimental estimate. Of the range of values of λ tested (Table 5), λ = 3.0 proved to be of the right order of magnitude, corresponding to a model with the genes of largest effect approximately equal to one phenotypic standard deviation. Figure 5 shows the expected gene frequency distribution in the base population for model 5 with n = 20, λ = 3.0.

Direct simulation

A computer program described by Latter and Novitski (1969)

TABLE 5

Goodness-of-fit of five basic genetic models to the observed response in Drosophila, determined by partial-simulation techniques.

Genetic Model	Number of loci* (n)	Total response $x(22)/\sigma$	Drift variance σ_L^2/σ^2
1	40	1.8	0.33
2	10	1.7	0.27
3	50	1.8	0.48
4	10	1.7	0.42
5	20 (λ = 3.0)†	1.7	0.46
Experiment		1.78 ± .17	0.43 ± .15

*The value of n for models 1 - 4 is that giving best agreement between the observed and simulated value of x(22).
†The values of λ tested were 0.5, 1.0, 1.5, 3.0, and 6.0.

TABLE 6

The regime involved in direct computer simulation of selection response in Drosophila

Generation	Selection intensity* Females	Males	$\bar{i}$
Base	1/1	1/1	0.00
0	2/60	2/2	1.06
1	3/33	3/3	0.86
2	6/100	6/6	0.97
3	6/75	6/6	0.91
4-21	6/30	6/6	0.68

*N/T denotes selection of the N individuals of highest phenotypic value from among T scored.

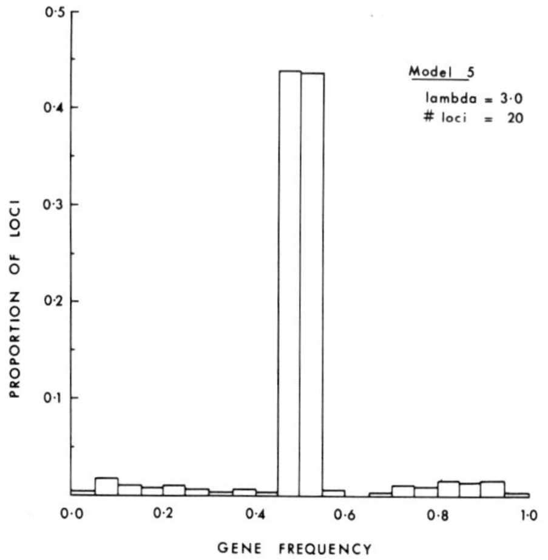

Figure 5. The initial gene frequency distribution implied by model 5 with h^2 = 0.33 and n = 20, λ = 3.0; the distribution was derived by the use of 1000 randomly chosen values from an exponential distribution of gene effects.

has been used as a direct check on some of the predictions flowing from the partial simulation approach. The program is one in which binary digits are used to represent the alleles at each of a number of loci, and the operations of logical algebra are adapted to simulate genetic processes of recombination and gamete formation (Fraser, 1957). A population of individuals of each sex is actually generated in the computer, and then scored for the metric trait, ranked, culled, and mated each generation as in laboratory experimentation. A genetic system of any degree of complexity can be specified, and the mating scheme can be modified to simulate any desired pattern of reproduction.

For each replicate simulation run, a random male and a random female were drawn from the base population and mated to give a full-sib group of offspring in generation 0. Thereafter the selection intensities and breeding population sizes listed in Table 6 were applied, using truncation selection in females and random choice of males. Production of offspring in mass-mating cultures was simulated by the following procedure for each individual: (i) a random male parent and a random female parent were sampled with replacement; (ii) a random gamete was derived from each, and combined to give the required offspring individual. Sex-linked inheritance was not introduced into the simulation procedure for reasons discussed in the previous section.

The partial simulation study indicated that the observed response pattern could be represented by the behavior of 10 independently segregating multiallelic loci, or 10 linkage blocks within which recombination is negligible, and which behave independently under selection (Table 5). The first use of direct simulation was to check on this prediction, using 10 freely recombining units each represented by 4 bits of the computer word. Recombination values between successive bits of the unit were set to zero, and the configurations 0000, 0001, 0010,...,1111 used to represent five alleles with a symmetrical binomial distribution of effects on the character under selection; or alternatively to represent the 16 possible gametic types produced by a linkage block with four loci of equal effect, and initial frequencies of 0.5. The initial state of each bit in individuals sampled from the base population was decided by a separate random event, the alternative configurations 0 or 1 having equal probability.

This direct simulation based on model 4 can be seen from Table 7 to be compatible with the experimental observations of both total response over the 22-generation period, and the component of variance among lines, within the limits of statistical accuracy of the estimates. The observed response pattern of the 15 computer replicates is depicted in Figure 6 for comparison with the experimental pattern of Figure 3. The realized heritability of the simulated response over the first 10-generation period was calculated to be $0.20 \pm .02$, compared with the experimental value of $0.18 \pm .02$.

Direct simulation results for models 1, 2, and 5, both with and without linkage, are also summarized in Table 7. The maximum number of loci available in the computer program was 46, and this number was used in each of the sets of runs to be discussed. It was anticipated from the partial simulation results that such a value of n, with the introduction of linkage, would be too few to give the required total response under model 1, but that the number should be roughly appropriate for models 2 and 5. The random arrangement of loci used for the simulation of linkage is depicted in Figure 7; the chromosome map lengths indicated are one-half those normally shown for the chromosomes of *Drosophila,* to allow for the absence of crossing-over in the male.

The conclusions from the results of direct simulation are in

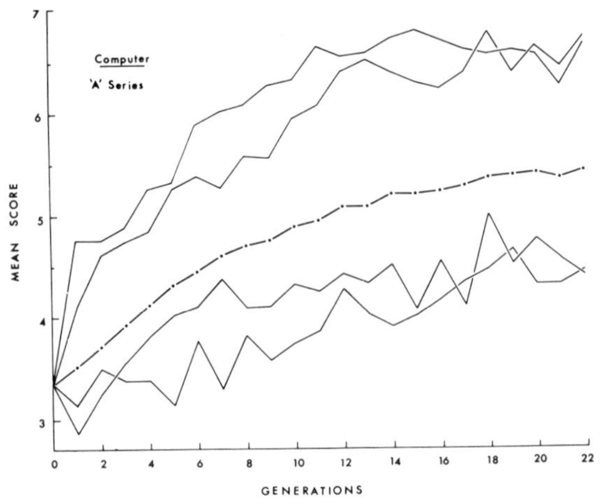

Figure 6. Simulated response to selection for increased scutellar bristle number, based on model 4 with ten independently segregating multiallelic loci of equal effect. As in Figure 3 the mean of 15 replicates is plotted, together with the two most extreme replicates in each direction.

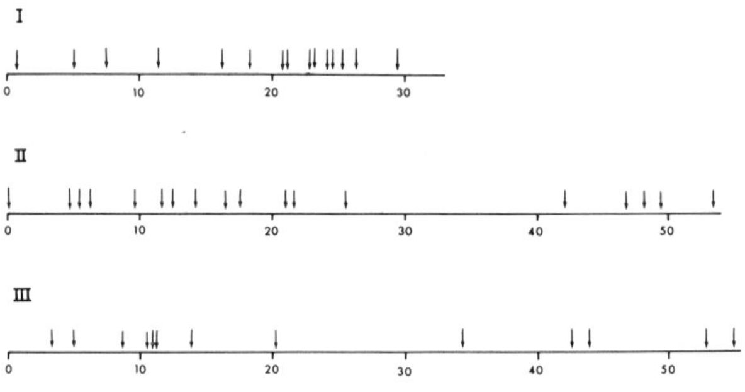

Figure 7. Random arrangement of 46 loci throughout the linkage map of *Drosophila,* used in direct simulation studies of response to selection under models 1, 2, and 5 (Table 7). Map distances allow for the absence of crossing-over in the male.

general agreement with expectations based on the partial simulation study: (i) models 1 and 2 are inappropriate in giving appreciably less drift variance among replicates than that observed; (ii) model 5 with 40-50 loci randomly scattered through the three major chromosomes, can readily be seen from Table 7 to give an adequate representation of the experimental observations; and (iii) model 4 with approximately 10 unlinked multiallelic loci has also been shown to provide a sufficient genetic basis for the simple selection experiment under discussion.

CONCLUSIONS

In reviewing progress since 1957 in this area of computer simulation of genetic processes, it is fair to say that simulation techniques have to date been of very limited value in the interpretation of laboratory selection experiments, and have only recently been used to advantage in studying the behavior of genetic systems under selection. I believe there are two main reasons why these procedures have not been of greater practical application.

First, quantitative genetics is an experimental science, and few replicated experiments capable of quantitative interpretation are available: in the absence of fully documented observations of population behavior under selection and inbreeding, the computer examination of genetic models can be of value only in the design of future experimentation. Second, a systematic approach to model building using computer simulation techniques has not so far been followed, though the work of Kimura (1957) and Robertson (1960) has provided the necessary framework for such an approach.

The theoretical discussion and experimental results given in this paper indicate that computer simulation can be remarkably successful in discriminating among simple contrasting genetic models and in specifying the range of acceptable values of the parameters involved in models compatible with the data. The five basic genetic models include loci of equal or variable effects on the quantitative character under selection, with intermediate or extreme gene frequencies; the loci may in addition be represented by only two alleles, or by many alleles each at low frequency in the initial population. Both additive gene action and genetic symmetry have been assumed for each of the models in the simulation studies reported in this paper.

It has been demonstrated that two simple models are compatible with the detailed pattern of response shown by scutellar bristle number in Drosophila, in an experiment making use of a "bottleneck" in the early generations to emphasize the effects of genetic sampling. Somewhat more significant is the observation that a range of values of one parameter (λ in model 5) are totally inappropriate, thereby excluding an hypothesis that response in this population is due to a number of genes at very low frequency in the base population, with major effects on the character. However, it must be stressed that the models are intended to provide no more than a sufficient representation of the genetic system involved, with reference to one particular set of data. Without further modification, the two models are not capable of explaining other response patterns which have been observed in experiments of quite different design, though the complications to be introduced may be few. It is possible that allowance for the following may prove sufficient: (i) an asymmetrical gene frequency distribution; (ii) antagonistic effects due to natural selection at some loci; and (iii) threshold behavior of some individual components of total scutellar bristle number.

Of particular importance for the present symposium is the strategy of computer simulation which has been outlined in this paper, and its potential use in the design and interpretation of experiments in quantitative genetics. What is also needed, of course, is an extensive series of experiments designed with this end in view. I am quite convinced that it is only via joint exploration by experimentation and computer simulation, that we can hope to build up an accurate picture of the nature of quantitative genetic variation in natural populations.

TABLE 7

Goodness-of-fit of four basic genetic models to the observed response in Drosophila, determined by direct simulation

Recombination between loci	Genetic model	Number of loci (n)	Total response $x(22)/\sigma$	Drift variance σ_L^2/σ^2
Free	1	46	1.78±.14	0.25±.10
	2	46	2.62±.17	0.34±.13
	4	10	2.05±.17	0.48±.17
	5 (λ =3.5)	46	2.10±.20	0.51±.19
Random*	1	46	1.43±.11	0.18±.06
	2	46	1.78±.14	0.27±.10
	5 (λ =3.5)	46	1.76±.20	0.49±.19
	Experiment		1.78±.17	0.43±.15

*Loci randomly scattered throughout Drosophila linkage map as in Figure 7.

Journal paper No. J6128, Iowa Agriculture and Home Economics Experiment Station, Ames, supported by National Institute of Health Grant No. GM 13827.

REFERENCES

Breese, E. L. and Mather, K. 1960. The organization of polygenic activity within a chromosome in Drosophila. II. Viability. Heredity 14: 375-400.

Clayton, G. A., Morris, J. A. and Robertson, A. 1957. An experimental check on quantitative genetical theory. I. Short term responses to selection. J. Genet. 55: 131-151.

Crow, J. F. 1954. Breeding structure of populations. II. Effective population number. In Statistics in Mathematics and Biology (Ed. O. Kempthorne) Hafner, New York. 543-556.

Falconer, D. S. 1960. Introduction to quantitative genetics. Oliver and Boyd, Edinburgh.

Fraser, A. S. 1957. Simulation of genetic systems by automatic digital computers. I. Introduction. Aust. J. Biol. Sci. 10: 484-491.

Fraser, A. S. and Hansche, P. E. 1963. Simulation of genetic systems: major and minor loci. In Genetics Today. (Ed. S. J. Geerts). Pergamon Press, London. 507-516.

Hill, W. G. and Robertson, A. 1966. The effect of linkage on limits to artificial selection. Genet. Res. 8: 269-294.

Kimura, M. 1957. Some problems of stochastic processes in genetics. Ann. Math. Statist. 28: 882-901.

Kimura, M. 1964. Diffusion Models in Population Genetics. Methuen, London.

Kimura, M. 1965. A stochastic model concerning the maintenance of genetic variability in quantitative characters. Proc. Nat. Acad. Sci. 54: 731-736.

Latter, B. D. H. 1960. Natural selection for an intermediate optimum. Aust. J. Biol. Sci. 13: 30-35.

Latter, B. D. H. 1964. Selection for a threshold character in Drosophila. I. An analysis of the phenotypic variance on the underlying scale. Genet. Res. 5: 198-210.

Latter, B. D. H. 1965a. The response to artificial selection due to autosomal genes of large effect. I. Changes in gene frequency at an additive locus. Aust. J. Biol. Sci. 18: 585-598.

Latter, B. D. H. 1965b. The response to artificial selection due to autosomal genes of large effect. II. The effects of linkage on limits to selection in finite populations. Aust. J. Biol. Sci. 18: 1009-1023.

Latter, B. D. H. 1966a. The response to artificial selection due to autosomal genes of large effect. III. The effects of linkage on the rate of advance and approach to fixation in finite populations. Aust. J. Biol. Sci. 19: 131-146.

Latter, B. D. H. 1966b. The interaction between effective population size and linkage intensity under artificial selection. Genet. Res. 7: 313-323.

Latter, B. D. H. 1966c. Selection for a threshold character in Drosophila. II. Homeostatic behavior on relaxation of selection. Genet. Res. 8: 205-218.

Latter, B. D. H. and Novitski, C. E. 1969. Selection in finite populations with multiple alleles. I. Limits to directional selection. Genetics. (In press).

Latter, B. D. H. and Robertson, A. 1962. The effects of inbreeding and artificial selection on reproductive fitness. Genet. Res. 3: 110-138.

Mather, K. and Harrison, B. J. 1949. The manifold effect of selection. Heredity 3: 1-52.

Neel, J. V. and Schull, W. J. 1968. On some trends in understanding the genetics of man. Perspect. Biol. Med. 11: 565-602.

Robertson, A. 1960. A theory of limits in artificial selection. Proc. Roy. Soc., B. 153: 234-249.

Robertson, F. W. 1955. Selection response and the properties of genetic variation. Cold Spr. Harb. Symp. Quant. Biol. 20: 166-177.

Robertson, F. W. 1959a. Studies in quantitative inheritance. XII. Cell size and number in relation to genetic and environmental variation of body size in Drosophila. Genetics. 44: 869-896.

Robertson, F. W. 1959b. Studies in quantitative inheritance. XIII. Interrelations between genetic behavior and development in the cellular constitution of the Drosophila wing. Genetics 44: 1113-1130.

Thoday, J. M. 1967. Selection and genetic heterogeneity. In Genetic Diversity and Human Behavior, (Ed. J. N. Spuhler). Aldine Publishing Company, Chicago.

Wright, S. 1952. The genetics of quantitative variability. In Quantitative Inheritance, (Eds. E. C. R. Reeve and C. H. Waddington). H. M. S. O., London. 5-41.

POPULATION STRUCTURE

Newton E. Morton
Population Genetics Laboratory
University of Hawaii

In 1962, at the beginning of my interest in this area, I wrote: "The structure of human populations comprises all those factors which determine mating preference. As defined, it poses a number of questions to the historian, sociologist, demographer, and other students of human behavior. Only two aspects of the problem are relevant to genetics: first, the effect of population structure on gene frequencies; second, the relation between gene frequencies and genotype frequencies." All subsequent experience has justified this abstraction of genetic parameters from social ones, which have led other investigators away from real populations to Monte Carlo simulation.

My particular interest has been to use biological data (including phenotypes, surnames, and even metrics) to measure consanguinity too remote to be accurately determined from pedigrees. The immediate objective is descriptive: to determine for any pair of individuals an appropriate coefficient of kinship, thereby giving the relation of gene frequencies to phenotype and mating type frequencies. A long-range objective, equally but no more interesting to me than the immediate goal, is analytical: to account for differences among observed population structures in terms of systematic pressures and random genetic drift. Some of the questions which I hope to answer are the following:

1. Is the mean coefficient of inbreeding in primitive populations large enough to dominate random homozygosity and heterozygous effects of rare recessive genes?
2. What is the effect of preferential consanguineous mating (first-cousin or uncle-niece) on inbreeding levels?
3. What ecologies and social systems lead to the greatest differentiation of local communities, and is there much variation among systems? For example, are inhabitants of coral atolls more inbred than New Guinea highlanders?
4. What are the coefficients of kinship among races and regions for a particular class of genes?
5. Is the divergence measured in this way due to disruptive selection or random genetic drift?
6. What are the systematic pressures acting on various genetic systems, and to what extent are they due to selection, mutation, and long-range migration?

Until recently there were few serious attempts to apply mathematical theories of population structure to real (or simulated) populations. The choice of a model is therefore still partly a matter of taste and convenience, but I believe that evidence now available indicates that approaches through microtaxonomy, migration matrices, estimates of effective population size, and descriptions of kinship systems offer far less than the theory of Malecot which depends on the concept of distance. Intuitively one expects that the mean coefficient of kinship for a pair of individuals will decline monotonically with the distance between their origins, and the asymptote for large distances will be zero in the absence of a systematic gradient (cline) in gene frequencies, but negative in the presence of a cline. These two situations are shown in Figure 1 together with assortative pairing (preferential consanguineous marriage) which has been observed for inbreeding from pedigrees in northeastern Brazil (Azevedo et al., 1969).

Malecot considered only random pairing in the absence of a cline. He concluded that as distance (d) increases, the mean coefficient of kinship is approximately

{1} $$\varphi(d) = ae^{-bd}d^{-c},$$

where c = 0, 1/2, and 1 for isotropic migration in 1, 2, 3 dimensions. This result was first obtained by Malecot (1948, 1959) for one and two dimensions, and by Kimura and Weiss (1964) in three dimensions. Equation {1} holds as d increases (in the absence of a cline) for any istropic migration, but for small d the form of $\varphi(d)$ depends on the migration law in time and space. Malecot (1962) showed that exponential migration in one dimension yields equation {1} exactly, with c=0. For any value of c, Malecot found that

{2} $$b = \frac{\sqrt{8m}}{\sigma_d},$$

where m is the linearized systematic pressure, defined as

{3} $$m = -\partial\Delta/\partial q \,|\, q^*.$$

Here Δ is the rate of systematic change of gene frequency per generation from a value q toward an equilibrium value q^*, and σ_d^2 is the variance of short-range migration measured as the distance between birthplaces of mates.

At this early stage in population structure analysis it is tempting to suppose that Malecot's result holds approximately in real populations, despite clines and varying patterns of migration. The advantage of Malecot's clear hypothesis over the models that have been proposed for Monte Carlo simulation is that equation {2} leads to testable genetic consequences. I shall be delighted if equation {2} is generally valid, since it provides answers to question 6 above. Even if equation {2} is not validated, equation {1} is justified both intuitively and empirically and is sufficient to answer the first five questions.

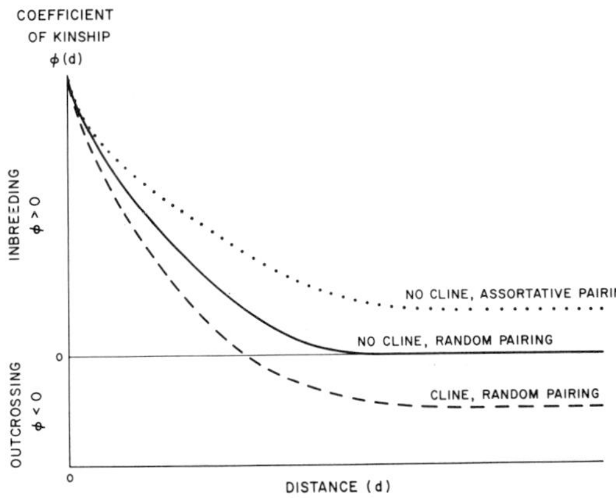

Figure 1. Three models for isolation by distance.

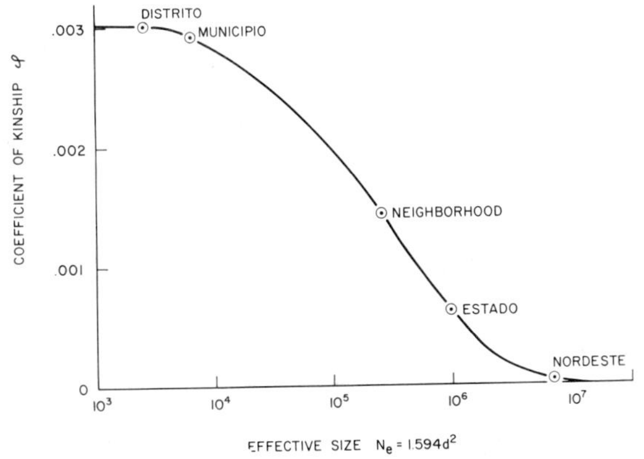

Figure 2. Coefficients of kinship of subregions relative to the total

SUBDIVISION OF A SINGLE POPULATION

The geneticist commonly studies a large population whose gene frequencies can be estimated with precision, subdivided into local communities whose gene frequencies usually cannot be estimated precisely. Michigan, Switzerland, Hutterite colonies, northeastern Brazil, Micronesia, and western New Guinea are examples of such populations, in which the relation between gene and disomic genotype frequencies may be taken as

{4} $$P(G^iG^i) = q_i^2 + q_i(1-q_i)\,F$$
$$P(G^iG^j) = 2q_iq_j(1-F) \qquad (i \neq j)$$

where F is the coefficient of inbreeding. If individuals are paired at random, F is to be replaced by φ, the coefficient of kinship. Either parameter may be subscripted if the coefficient varies among alleles (Yasuda, 1968). These parameters are expected to vary among loci subject to different systematic pressures.

Consider two individuals drawn either by marriage or at random from the same community or two different communities, supposing until a later section that these are homogeneous in size and mating pattern. Since the number of communities will typically be large, and their size small, it is hardly feasible to estimate F for a particular community or pair of communities, and we almost never know the pedigree of the whole population well enough to specify exactly the coefficient of kinship for every pair of individuals. The only practical solution for real populations seems to be to express the coefficients of kinship and inbreeding as functions of the distance d between communities of birth or origin of the two individuals concerned, say $\varphi(d)$ for kinship and f(d) for inbreeding. If we are interested only in the means of these quantities for a given value of d, all relationship may be included. If we are interested in a particular pair of individuals, and if close kinship (say, $\varphi \geq 1/32$) can be ascertained directly, then a more accurate estimate is

{5} $$\varphi_T = \varphi_C + \varphi(d)_R$$

where T, C, and R denote total, close, and remote kinship, respectively, and $\varphi(d)_R$ is computed excluding close relatives. The corresponding equation for the coefficient of inbreeding is

{6} $$F_T = F_C + f(d)_R$$

If consanguineous marriage is preferential because of age or other social factors, the substitution of $\varphi(d)$ for $f(d)_R$ gives an approximation which slightly overestimates F_T (Azevedo et al., 1969). The practical problems in applying these methods therefore reduce to measuring d, $\varphi(d)$, and f(d).

Before turning to these questions, we may consider briefly two alternative methods to study population structure. One, proposed by Sewall Wright (1951), converts distance into relative population size by the transform

{7} $$N = Cd^2,$$

where C is a positive constant, and arranges communities into a hierarchy. In northeastern Brazil we distinguished 2,725 distritos grouped into 1,083 municipios, which in turn were successively grouped into 27 neighborhoods and 7 estados, the whole constituting the single region nordeste (Azevedo et al., 1969). In Switzerland we recognize communes within cantons, and in primitive populations clans within villages within tribes. Individuals paired at random within a subregion have a distance distribution whose "typical" value is roughly

{8} $$d_n = \sqrt{\sigma^2/n}$$

where

{9} $$\sigma^2 = E(d^2)$$

is the variance of distances between individuals in the whole region and n is the number of subregions at a given hierarchical level. For example, measuring distance as we did in northeastern Brazil, $\sigma^2 = 4.4 \times 10^6$, so that the "typical" distance associated with two individuals drawn at random from the same estado (n=7) is $d_n = \sqrt{4.4 \times 10^6/7} = 790$.

This hierarchical description of population structure introduces the further approximation that the mean coefficient of kinship in a subregion is $\varphi(d_n)$ rather than the exact but not readily calculable expression $\Sigma g(d)\varphi(d)$, where g(d) is the distribution of distance within the subpopulation. Furthermore, the hierarchical model clearly loses much of the information in $\varphi(d)$, leading instead to statements of the following type: "Two randomly chosen communities from the same subregion (or from randomly chosen subregions) have a certain coefficient of kinship φ", without otherwise distinguishing whether the communities are close or distant. Figure 2 shows the transformation of $\varphi(d)$ into a hierarchical partition for northeastern Brazil.

A second approach to population structure uses a migration matrix, which in practice requires that the number of communities be small and the frequency of migration between each pair of communities be accurately known and be random with respect to genotype and kinship. Systematic pressure is confounded with migration into the region, which is assumed to come from an equilibrium population, rather than (as is usually the case) from neighboring communities. In Monte Carlo simulation random trials give a sample of possible outcomes under the model, and these results are compared with $\varphi(d)$ or some other observation on the population.

The assumptions required for extrapolation from a migration matrix are restrictive. Real populations continuous in space cannot be precisely represented by a finite matrix, even of a size too large to be manageable. Migration is often not random with respect to kinship. Systematic pressures may not be negligible, and certainly should not be confounded with migration into the region, which is not equivalent to long-range migration. Deterministic simulation appears more applicable than Monte Carlo trials, but even a striking fit to data could be due to compensatory errors in estimating systematic pressure and migration. Simmons et al. (1965) have stressed the indeterminacy of population structure from the migration pattern in a single generation: "The vicissitudes of history (viz. typhoon, tidal wave, volcanic eruption, canoe sinkings, accidental canoe voyages, plagues, pestilence, famine, and the ravages of warfare and vendetta, together with transient and faddish changes in sexual practices) can, in such small groups, result in very few of the members giving rise to the succeeding generation, and these few represent a fortuitous, rarely any 'mean or average' sampling from the gene pool of the community. Thus, only a detailed long-term genealogical history of each community would offer adequate explanation for the gene pattern currently found in it." If we admit the force of this argument, we must question the utility of migration matrices for the study of real populations, however useful they may be to explore a mathematical model. Only the abstraction and power of Malecot's concept of distance seem to provide a simple description of population structure, and without such a description no test of hypotheses is possible.

THE "SAME PLACE" PROBLEM

A single community is often not geographically structured: a man may be no more likely to marry the girl next door than one on the other side of the village. Our first problem in applying the concept of distance is therefore to assign a distance r to pairs of individuals from the same community. Yasuda (1966) assumed that the appropriate value is r=0, and therefore concluded that the distribution of marital distance has a mode at zero and can be well represented by a modified beta function, which is mathematically intractable because the second moment is infinite. Recently Yasuda (personal communication) found that the mode is non-zero when marital distance is measured very precisely.

I have preferred to take r as the radius of an average community,

{10} $$r = \sqrt{A/n\pi} \ ,$$

where A is the area of a population containing n communities within which an accurate determination of distance is impractical or irrelevant, uninhabited areas being omitted. Consequently the density for d=0 is null, and we may reasonably entertain the hypothesis that the probability of a distance d between random individuals is the augmented gamma distribution

{11} $\mu(r)=h$ for individuals from the same place and
$\mu(d)=(1-h)d^g z^{g+1} e^{-dz}/g!$ otherwise.

I also assume that the distance distribution for mates m(d) has the same form, with different values of h, g, and z. The theory of Malecot (1968) applies to equation {11}, which has been found to hold to a reasonable approximation in various populations (Morton et al., 1968b; Azevedo et al., 1969).

As a consequence of the assumption that $r>0$, the intercept a in equation {1} is not to be interpreted as the mean coefficient of kinship of actual communities $\varphi(r)$, but only (for $c \geq 0$) as an upper limit.

VARIABILITY OF THE SAMPLE SPACE

Malecot's theory neglects at least six kinds of population heterogeneity, which we may call distributional, migrational, reproductive, temporal, clinal, and preferential:

Distributional heterogeneity. Birthplaces are neither distributed uniformly over a region nor clustered into communities of equal size arrayed in a regular lattice. Instead, population density may be highly variable, with a non-uniform mixture of uninhabited areas, isolated houses, villages, and cities.

Migrational heterogeneity. Partly as a consequence of the above, and partly because of mountains, deserts, and other natural obstacles, as well as for social and historical reasons, linear geographic distance may be only a rough measure of genetic isolation, especially over short periods of time.

Reproductive heterogeneity. Random and selective factors make numbers of descendants highly variable, especially after several generations. This is not explicitly recognized in current theories.

Temporal heterogeneity. In successive generations there may be abrupt changes in migration and reproductive patterns. A substantial increase in the frequency of consanguineous marriage occurred during the last century in Europe, apparently because of increased and variable family size (Moroni, 1964). Again this is not explicitly admitted in current theories, although Malecot (1968) has recently extended his results to include temporal changes in migration, when these are known, and increased fertility has much the same effect as reduced migration.

Clinal heterogeneity. Malecot assumed that gene frequency does not vary systematically over the region. This assumption will not be true if selection differs among environments or if there is a transient gene frequency gradient due to migration between different gene pools. All cases of systematic gene frequency variation over moderate or large distances may be called clines. Their effect is to make pairs of individuals born at large distances less related than random pairs from the region, and therefore to have a negative coefficient of kinship.

Preferential heterogeneity. Mating at a given distance may not take place at random. Self-fertilization is excluded for many organisms, incest may be prohibited, and close consanguineous marriage may either be favored or avoided. Immigrants from large distances are not infrequently the descendants of former emigrants, returning preferentially to their place of origin. Mating may be assortative with respect to race, religion, class, or other non-geographic barrier.

The conjunction of an attractive theory with nasty complications generates several responses, depending on the personality and interests of the investigator. He may be discouraged by real populations, preferring instead to work with mathematical models and to regard actual data as illustrative but not critical, thus ignoring advances made by testing simple hypotheses in other branches of science. I have no patience with this approach, since the first five of the six questions asked in the introduction can be answered by distance analysis even if equation {2} is not valid. At this stage a judicious mixture of two other actions seems appropriate: to neglect certain complications until a more complex theory is developed, and to select or adjust the data to conform more closely to current theory.

DISTRIBUTIONAL HETEROGENEITY

If variable density is ignored, pairs from the same place will be representative of larger rather than smaller communities, since a sample of n individuals from a community generates n(n-1)/2 pairs. This may be no disadvantage if we merely wish to estimate mean coefficients of inbreeding for the whole population. If we are interested in studying communities of different sizes we may stratify the sample, for example by abstracting alpine villages from Switzerland (Morton et al., 1968b). This has the disadvantages that grouping may be rather coarse and arbitrary, and information between groups may be wasted. Simmons et al. (1965) sampled disproportionately, taking larger samples from the smaller communities in the Caroline Islands, so that their material is more representative of the smaller islands.

Alternatively, we may define a quantity C called density for each community. In some applications C may be the population per unit area, in which case the transform

{12} $$d' = d\sqrt{C}$$

gives a measure of distance such that a small circle of radius d′ includes the same number of individuals anywhere in the region, independent of density.

To apply this to real populations we need to define the mean density corresponding to a pair of communities with densities C_i and C_j. Our DISTAN program takes

{13} $$C = \frac{C_i C_j}{C_i + C_j}$$

by analogy with effective population size. To maintain comparability with crude distance, it is convenient to multiply density by

{14} $$K = (\bar{d}/\bar{d}')^2$$

where $\bar{d}$, $\bar{d}'$ are the mean distances before and after applying equation {12}. Then adjusted distance has the same mean as crude distance and may be interpreted approximately in the same units, although small communities are located closer together and large communities farther apart on the transformed scale. We found in Switzerland that the two measures of distance give closely similar values of b, but the intercept a is larger for adjusted distance. This was expected, since a for adjusted distance is the upper limit for pairs of individuals from very small communities, rather than from typical communities. It is therefore of interest to study both crude and adjusted distance.

In some applications it is desirable to define density as

community size, or more abstractly, for example as an estimate of $1/\varphi_i$, where φ_i is the mean coefficient of the ith community. Then the intercept a becomes the upper limit to the mean coefficient of kinship for highly inbred communities. We propose to estimate φ_i in Switzerland by regressing the inbreeding coefficient of individuals on demographic variables of their commune of origin, so that each commune can be scored for $1/\varphi_i$ even from a relatively small sample of individuals.

MIGRATIONAL HETEROGENEITY

Mates may be asked to estimate the distance between their birthplaces as they normally would travel, but this is highly inaccurate. It has been suggested that distance in a mountainous country like Switzerland should not be measured linearly but as people actually travel. We considered this possibility but rejected it for both practical and theoretical reasons. It is not feasible to specify how people migrate between all possible pairs of more than 3,000 communes. Any attempt to approximate this, for example by the shortest road distance, is arbitrary and imprecise, and exposes the measurement of distance to erratic changes with the activities of the highway department. There is certainly no reason to think that the probability of migration from one point to another, depending on economic and personal factors, is related in any simple, predictable way to the shortest road distance, or that all types of roads from footpaths to highways are equally traveled.

Finally, the theory of isolation by distance has treated migration as an n-dimensional vector, absolute distance being taken only in the generation under study. A theoretical treatment of absolute distance (however measured), compounded over many generations, has seemed impractical to competent mathematicians like Malecot and Kimura. The burden of proof is on the reader who is attracted by nonlinear distance to derive valid rules for its measurement, and a precise mathematical analysis of its consequences under locally anisotropic migration, and to show that actual data such as we have presented agree more closely with this prediction than with Malecot's. We leave this detour to those hopeful of its success. Meanwhile, we are impressed with how closely our results fit Malecot's theory. Linear distance (crude or adjusted) may be only a rough measure of isolation, but no better measure has been proposed.

REPRODUCTIVE HETEROGENEITY

The distribution of numbers of descendants, while not entering explicitly into Malecot's theory, determines the effective community size and therefore the expected value of a, which depends also on the dimensionality parameter c and on how distance is adjusted. We have attempted no theoretical analysis of a, preferring to regard it only as an upper limit to the mean coefficient of kinship of communities (for crude distance) or of small communities (under a given definition of density).

TEMPORAL HETEROGENEITY

Temporal changes in migration can in principle be handled (Malecot, 1968), but the information required is demanding, and the results lose the simplicity of equation {2}. Temporal changes in reproductive patterns lie outside current theory. Both complications may be reduced by omitting close relatives, so that the measured kinship is due to remote generations when the migration and reproductive patterns may have been stable. Omission of close relatives is usually difficult for random individuals, and therefore we prefer to regard equation {2} as an hypothesis to be tested, as in the last section below.

CLINAL HETEROGENEITY

Gene frequency clines have important effects as the size of a region increases, distances larger than some critical value d* corresponding to outcrosses, with a negative coefficient of kinship. The critical value is the only point at which panmixia holds: viz., $\varphi(d^*)=0$.

Cockerham (personal communication) suggested that negative values of $\varphi(d)$ be eliminated by expressing kinship relative to random pairs at the largest distances in the population. Let $L<0$ be their coefficient of kinship. Then

{15} $$\varphi'(d)=\frac{\varphi(d)-L}{1-L}$$

is the mean coefficient of kinship relative to large distances, corresponding to a greater region from which these distances would be typical. On the transformed scale the largest distances now have an asymptote of zero. This function {15} can therefore be fitted by equation {1}; equation {2} remains an hypothesis to be tested. Having fitted $\varphi'(d)$ to equation {1}, we may easily return to deviations from regional panmixia by inverting equation {15} to give

{16} $$\varphi(d) = L + (1-L)\varphi'(d)$$

Thus L measures differentiation of adjacent regions, and $\varphi(d)$ measures deviations from panmixia within a region.

The coefficient of kinship within a random subregion is obtained by replacing L in equation {15} with the corresponding mean for a subregion, which is approximated by $\varphi(d_n)$, where d_n is given by equation {8}.

PREFERENTIAL HETEROGENEITY

Nonrandom mating means $f(d) \neq \varphi(d)$, where $f(d)$, $\varphi(d)$ are the mean coefficients of kinship at distance d for mates and random individuals, respectively. In northeastern Brazil, the only population so far examined from this point of view, $f(d) > \varphi(d)$, most conspicuously at large distances, confirming other evidence for preferential consanguineous mating (Azevedo et al., 1969). Clearly f(d) is the relevant function for phenotype and mating type frequencies but confounds random kinship with preferential mating, while $\varphi(d)$ is expected to conform much more closely to Malecot's theory and to give a pure measure of genetic differentiation with distance.

DISTAN

A General Program For Analysis Of Population Structure

This program was written by Shirley Yee under my direction. There are three modes of input:

- Mates (M mode) correspond to pairs of individuals, occupying one record which includes their marital distance d and may include their two surnames, phenotypes, metrics, and coefficient of kinship. An assembly program PAIRFILE is available to generate M mode records of all n_1n_2 pairs of individuals between two files of size n_1, n_2.
- Individual records (I mode) to be paired with all others in the file have two fields for the rectangular coordinates and may have the surname, phenotype, metric, and coefficient of inbreeding of the individual.
- Sample records (S mode) bear a count N of the number of individuals in a sample, the rectangular coordinates of their community, and may have the phenotype vector, gene counts, and mean metric of the sample.

There are 8 classes of data input, which with the 3 modes generate 24 possible categories, 17 of which are included in DISTAN. Of these, 10 lead to an FD table, 3 to an MD table, and 4 to an RD table defined as follows:

An FD table contains estimates of the coefficient of kinship for mates or pairs of random individuals to be fitted by equation {1}.

An MD table contains frequencies of mates or pairs of random individuals to be fitted by equation {11}.

An RD table contains estimates of the coefficient of kinship from carriers of rare genes, either mates or random pairs, to be fitted by the posterior probability

{17} $$r(d) = \frac{\{Q + \varphi(d)\}\ \mu(d)}{Q + \alpha}$$

Input is governed by 12 types of control cards. The DC card gives the distance r to be assigned to pairs "from the same place": i.e., mates recorded at zero distance and individuals or samples with the same coordinates. This is followed by a string of up to 59 numbers, in increasing order, around which distance intervals are formed by an algorithm given in the write-up. Under control of the DC card all distances within an interval are represented by the midpoint.

For I and S modes a transformation is available to compute distance in kilometers from the rectangular coordinates X, Y expressed in degrees, using the formulae

{18} $$x = (X - \bar{X})\ (111.4175\cos\beta\ - .0940\cos3\beta\ + .0002\cos5\beta)$$
$$y = (Y - \bar{Y})\ (111.1363\ - .5623\cos2\beta\ + .0011\cos4\beta)$$

where $\bar{X}$, $\bar{Y}$, are the mean coordinates, and the middle latitude is $B = \frac{Y + \bar{Y}}{2}$. The coefficients in equation (18) are for the international (Hayford) spheroid. Distance between two individuals from places i and j is computed as

{19} $$d_{ij} = \sqrt{(X_i - X_j)^2 + (Y_i - Y_j)^2} \quad (i \neq j)$$
$$= r \quad (i = j)$$

This use of plane trigonometry is justified for the small distances with which population structure theory is concerned, large distances corresponding to the asymptote L.

The pairing logic is as follows. In M mode the pairs are preassigned. In I and S mode all possible pairs of objects within a defined group are formed, except that an object is never paired with itself. In I mode the object is always an individual. In S mode the object is an individual for phenotype and migration, and a sample for gene counts and metrics. S mode therefore loses all information about the within-sample coefficient of kinship for gene counts and metrics. Two or more samples with the same coordinates are treated as different samples at distance r.

Analysis is governed by 7 types of control cards which identify and select particular data tables, provide trial values of the parameters a, b, c, h, g, z, Q, and L, and iterate on any nonsingular set of the first 7 parameters by maximum likelihood scores (MD and RD tables) and pseudo-ML least squares (FD tables), taking the residual variance among distance classes as error (Morton et al., 1968a). Iteration may extend over two or more tables, which may either be kept separate or combined (if of the same class, say MD) to produce a summary table retained in memory. Large tables may be condensed for printing by an edit card, leaving the original table in memory, or distance classes may be irreversibly combined by a pool card.

The iteration logic fits all nuisance parameters such as gene frequencies, metrical means, and frequencies of common surnames to each distance class separately. This reduces noise due to a cline but makes distance classes containing small numbers of observations undesirable, since iteration may not converge. The relevant part of ALLTYPE (Yasuda, 1968) for simultaneous estimation of gene frequencies and the coefficient of kinship from pairs of phenotypes is incorporated into DISTAN. Each distance class is weighted by the number of pairs, and the information matrix is divided by the empirical variance among distance classes to give by inversion the covariance matrix of the estimates.

Maximum likelihood iteration uses an information matrix based on the expected class frequencies; i.e., if in class k the expected number in that cycle of iteration is m_k and each observation in k has scores u_{ik} and u_{jk} for parameters i and j, then the corresponding element of the information matrix is $K_{ij} = \sum_k m_k u_{ik} u_{jk}$.

ISONYMY (M1 AND I1)

A pair of identical surnames is said to be isonymous. Let n_d be the number of isonymous pairs among N_d total pairs in a distance interval m_d assigned class value d. Then the isonymy is

{20} $$I_d = n_d/N_d$$

which is related to the coefficient of kinship for surnames as

{21} $$\varphi(d) = \frac{I_d - \sum_i q_{id}^2}{4(1 - \sum_i q_{id}^2)} \qquad \text{(Crow and Mange, 1965)}$$

where q_{id} is the frequency of the ith surname in the given distance class. DISTAN computes q_{id} for up to 20 common surnames, the aggregate of rare surnames being assumed to make the same contribution $\sum_d r_d / \sum_d N_d$ to $\sum_i q_{id}^2$, where r_d is the number of pairs isonymous for rare surnames. Individuals may be grouped by sex to avoid spurious isonymy if wives do not retain their married names. Deviations from random consanguineous marriage are a source of error in M mode, but not in I mode where pairs are formed randomly.

This analysis has so far been carried out only for northeastern Brazil, where it gives higher estimates of the coefficient of kinship than other methods, corresponding roughly to the coefficient of inbreeding. This may reflect either the isolating effect of patrimony on surnames in rural areas, or the polyphyletic and spatially contagious origin of such names as Jesus and Conceicao, which are uncommon in other Portuguese populations but must often have been acquired at baptism by children of Indian and Negro women (Azevedo et al., 1969). The same contagion must have operated to some degree with less common surnames.

GENE FREQUENCY (S2)

Malecot's theory is expressed in terms of correlation of gene frequency, but we have found that direct estimation of this correlation is unsatisfactory. Dominance in finite populations of relatives creates unsolved problems in calculating the sampling variance of community gene frequencies, so that the observed variance of such estimates gives no reliable basis for computing the genic variance (cp. Nei and Imaizumi, 1966). Local gene frequency estimates raise troublesome questions. If a gene is not demonstrably present in a small sample, should we estimate its local frequency as zero? More generally, how do we estimate gene frequencies in a small sample when maximum likelihood iteration fails to converge? In view of these problems we have made little use of the relevant formula,

{22} $$\varphi(d) = \frac{q_i q_j' - \bar{q}\,\bar{q}'}{1 - \bar{q}\,\bar{q}'}$$

where q_i, q_j, and $\bar{q}$ are the vectors of gene frequency estimates for communities i and j and for the total sample, respectively.

PHENOTYPES (M3, I3, AND S3)

The theory of Yasuda (1968) for phenotype pairs does not suffer from the above estimation problems, is much more efficient, and permits calculation of the mean coefficient of kinship within communities. A distance class contains all phenotype pairs formed at that distance over the whole sample space. Unless the number of observations is very small, in which case adjacent distance classes may be pooled, there is no difficulty in simultaneous estimation of the gene frequency vector q_d and the coefficient of kinship $\varphi(d)$, genes with no stationary point in the interval 0, 1 being deleted. The sampling variance of $\varphi(d)$ when all possible pairs are formed is of course larger than for a multinomial distribution, but the empirical variance among distance classes gives a measure of error. Phenotype pairs therefore provide the best of all bioassays of kinship.

METRICS (M4, I4, AND S4)

Let X be an additively inherited metrical character in standard deviation units (i. e., with a mean of zero and unit variance), whose value is X_i, X_j for two random samples (or individuals) at distance d. Then

{23} $$E(X_iX_j) = 2h^2\varphi(d)\{1+h^2\alpha\},$$

where h^2 is the heritability and α is the mean inbreeding coefficient (Malecot, 1948). Since these quantities are often not known, the value of a in equation 1 cannot be established with precision, but b and c are unaffected if we take

{24} $$\varphi(d) = X_iX_j,$$

weighting by $w = n_in_j$. Clines due to nongenetic factors may be a source of error.

Estimates of b in northeastern Brazil are similar for metrics and phenotypes, suggesting that polymorphs and the genes of quantitative genetics are subject to similar systematic pressures, and may well be identical.

INBREEDING FROM PEDIGREES (M5)

Close consanguinity determined from pedigrees is a significant fraction of all inbreeding, and the part for which deviations from panmixia are most apparent. Even under favorable conditions the ascertainment of consanguinity must always be incomplete for remote relationship.

In northeastern Brazil the mean coefficient of inbreeding estimated in this way is consistently higher than the coefficient of kinship and has b=0, as expected since pedigrees ignore selection, mutation, and migration. About 70 percent of the total coefficient of inbreeding was ascertained through pedigrees.

This part of the program accepts data on any FD table, since each estimate of $\varphi(d)$ may be associated with a weight N_d corresponding usually to the number of observations. Computations of the inbreeding coefficient from a pedigree must be done prior to entering DISTAN.

MIGRATION (M6, I6, AND S6)

In M mode migration may include marital distance (between birthplaces of mates), parental distance (between birthplaces of parent and offspring), and grandparental distance (between birthplaces of grandparent and grandchild). In I and S modes migration may include random distance (between two individuals chosen at random) and central distance (between birthplace and the mean coordinates of the region). All these distributions are defined for unselected phenotypes. Place of origin or residence may be substituted for birthplace.

Approximate agreement with an augmented gamma distribution (equation 11) is generally found.

CARRIERS OF RARE GENES
(Augmented Gamma Migration, M7 and I7)

The posterior probability that two individuals hemizygous or heterozygous for an idiomorphic allele with frequency q have distance d is given by equation (17). If $\mu(d)$ is assumed to be an augmented gamma distribution as in the preceding section, the analysis is designated M7, I7 for mates and individuals, respectively, $\mu(d)$ and Q being defined accordingly. For a single locus,

{25} $$Q = \frac{q}{1-q} \quad \text{for hemizygotes}$$
$$= \frac{q}{1-.5q} \quad \text{for heterozygotes}$$

With a rare trait dependent on more than one locus, perhaps with admixture of sporadic cases,

{26} $$Q \doteq A/B \quad \text{for mates with one or more affected children}$$
$$\doteq B \quad \text{for random pairs of individuals with one or more affected children,}$$

where A and B are components of the genetic load (Morton et al., 1968a). This analysis, which appears to be powerful (Morton and Yasuda, 1962), is being applied to retinitis pigmentosa, myotonic dystrophy, and hemophilia in Switzerland.

CARRIERS OF RARE GENES
(Empirical Migration, M8 and I8)

This differs from the above in that $\mu(d)$ is taken from an MD table instead of being approximated by an augmented gamma distribution. Usually the MD and RD tables are controlled by the same DC card, so that the distance intervals are the same. Limited experience suggests that the results are essentially the same as for the preceding section, but the analysis is freed from parametric assumptions about the migration distribution.

THE MEAN INBREEDING COEFFICIENT (α):
Answers to Questions 1 and 2

We have considered four distance functions:

- $\varphi(d)$; the mean coefficient of kinship for random pairs
- f (d), the mean coefficient of kinship for mates (the inbreeding coefficient)
- $\mu(d)$, the distribution of distance between random pairs
- m(d), the distribution of distance between mates

The mean coefficient of inbreeding in the population is

{27} $$\alpha = \sum_d f(d)m(d)$$

which may be estimated either directly from a random sample of mates, ignoring distance, or from the functions f(d) and m(d), which give a more complete description of the population. DISTAN uses both equation {11} and the empirical migration function to estimate α.

We must distinguish between α and the mean coefficient of kinship,

{28} $$\varphi = \Sigma\varphi(d)\mu(d)$$

which are related by the equation

{29} $$F_T = \alpha + (1-\alpha)\varphi$$

where F_T is the total inbreeding coefficient that could be measured from complete genealogies. With populational gene frequencies α is the relevant measure of deviation from panmixia in equation {4}, and φ measures differentiation among populations.

No bioassay of the inbreeding coefficient in primitive populations has yet been made, and incomplete genealogies provide only unreliable estimates of F_T, which at best confounds local deviations from panmixia with genetic drift. It is suggestive that Ramah Navaho (Allen, 1965) and S-leut Hutterites (Crow and Mange, 1965) are virtually panmictic, so that inbreeding may probably be neglected except in populations distributed over areas large relative to their marital migration. Even such a population in northeastern Brazil had an inbreeding coefficient of only .008, of which 70 percent was ascertained through pedigrees (Azevedo et al., 1969). It remains to be seen whether primitive populations or societies with preferential consanguineous marriage have substantially larger coefficients of inbreeding, which would then be a more important force in determining the inbred load and frequencies of rare recessive genes than random homozygosity and heterozygous effects.

DIFFERENTIATION OF LOCAL COMMUNITIES (b): Answer to Question 3

Since a in equation {1} depends on dimensionality and how distance is measured, it is natural to use b to assay differentiation among local communities. Expressed in kilometers, we found values of .006, .023, and .158 for polymorphisms in northeastern Brazil, Switzerland, and Alpine villages, respectively, in agreement with other evidence which shows that marital migration decreases in the same order. Thus, distance is only about one-fourth as strong an isolating factor in sparsely populated northeastern Brazil as in Switzerland, where it is 7 times as important in Alpine villages as in the country as a whole. It remains to compare these results with populations under different cultural and ecological conditions.

REGIONAL COEFFICIENTS OF KINSHIP (φ): Answer to Questions 4 and 5

Equation {28} for a single region has a generalization for two or more regions, called Wahlund's principle:

{30} $$\varphi = \sigma_q^2/\bar{q}(1-\bar{q}),$$

where $\bar{q}$, σ_q^2 are the true mean and variance of gene frequencies among regions. Estimates of φ free from sampling errors in estimating gene frequencies may be obtained by forming pairs of phenotypes between regions, using either the PAIRFILE assembly program for I mode or an appropriate distance scale for I and S mode. Thus if coordinates are coded $0, 2^i$ for the ith region, each pair of origins can be distinguished in DISTAN output from a DC card coded 2^i (i=0, ...10). It is also possible to convert an F_1 outcrossing effect H for a trait with inbred load B into an equivalent coefficient of kinship between the parental populations,

{31} $$\varphi = -H/2B$$

By equation {28} we found .0004 for the ABO groups in Switzerland and .0010 for polymorphisms in northeastern Brazil, in good agreement with estimates by equation {31} of .0004 and .0009 for crosses of minor and major races in Hawaii (Morton et al., 1967). These values seem small enough to be attributed to genetic drift.

Much larger estimates, however, are obtained by equation {30} for polymorphisms in minor and major races. For example, among the Portuguese, Amerindian, and Negro ancestors of our Brazilian population, φ=.50 for Gm^3 and .27 for R^o. Genes with frequencies near 0 or 1 show much less variation, and the regression is

{32} $$\varphi = (.68 \pm .19)\bar{q}(1-\bar{q})$$

Drift would not lead to any systematic relation between gene frequency and φ, and the high values of φ for intermediate gene frequencies can only mean that polymorphisms are subject to variable selection among geographic regions. This polytypy is a strong stabilizing force, capable of balancing polymorphisms without heterosis. Rare idiomorphs and common monomorphs are much less variable in frequency, and must be subject to rather consistent selection over wide areas.

Measurement of genetic differentiation by φ has a clear biological meaning in terms of genotype frequencies by equation {4}. The measures of "biological" distance proposed by various authors have no simple interpretation. It is surprising that these less informative indices have not yet been abandoned.

DISTANCE IN CLINICAL GENETICS

Since distance provides a measure of kinship, it can be used to test genetic hypotheses, particularly for rare genes. For example, estimates of α by equation {27} and of Q by equation {17} give a basis for calculating genetic loads which does not depend on pedigree information.

Consider also two diagnostic groups, say retinitis pigmentosa with deaf mutism and the typical form, both of which are largely due to recessive genes. We wish to know whether some of these genes are alleles, in which case we assert that the two groups are in part genetically identical. The components A and B of the genetic load give an estimate of the numbers of loci involved in each group and when they are pooled, but distance may be a more powerful discriminant. Let r(d), r′(d), and r″(d) be posterior distance distributions, defined as follows:

r(d) for mates who have produced at least one affected child

r′(d) for random pairs of such mates within sex and diagnostic group

r″(d) for random pairs of such mates between groups

Then if the parameter Q estimated from r″(d) is significantly greater than for r′(d), we have evidence that the two groups are in part genetically different, whereas if g estimated from r″(d) is significantly less than for random pairs, μ(d), this is evidence that the diagnostic groups are in part genetically the same. Similarly, if Q for r′(d) is significantly greater than for r(d), we have evidence for either multiple loci or sporadic cases, the frequencies of which can be estimated from the values of Q and segregation analysis, independently of inbreeding from pedigrees.

ESTIMATION OF SYSTEMATIC PRESSURES: Answer to Question 6

So far we have not assumed the validity of equation {2}, which must be tested if we are to make inferences about systematic pressures. We may write the linearized systematic pressure in equation {3} as

{33} $$m = u + k,$$

where u is the frequency of migration from an equilibrium gene pool and k is due to selection.

Migration into a region has complicated effects on the parameter b. Migration over small distances within a region of homogeneous gene frequencies contributes to σ_d, but not to u. Migration over very large distances between regions of different gene frequency contributes to u but not σ_d. In the intervening range, migration into a region contributes both to u (to the extent that immigrants are representative of the species) and to σ_d (to the extent that immigrants are representative of the region). This complexity makes it impractical to estimate σ_d directly, either in the past or even in this generation. Instead, we must consider two or more rare genes, or groups of genes, the ith with known selective pressure k_i and distance parameter b_i. Then we have a set of linear relations from equation {2},

$$b_i^2 \sigma_d^2 - 8u - 8k_i = 0 \qquad \{34\}$$

which may be solved for σ_d^2 and $\underline{u}$. The test of Malecot's theory reduces to the goodness of fit of equation {34} for $i = 1, 2, .., n$, with n-2 degrees of freedom. Selective pressure for the jth polymorphism may be estimated from the values of σ_d^2 and $\underline{u}$ for the population and b_j for the polymorphism.

Simulation may be useful to test the validity of Malecot's theory under a variety of migration patterns, with different values of σ_d^2 and $\underline{u}$. The concept of distance gives simulation a raison d'etre, providing both parameters to estimate and an hypothesis to test, without which simulation methods are sterile.

SUMMARY

The logic is given of the DISTAN computer program to measure isolation by distance in real or simulated populations.

ACKNOWLEDGMENTS

PGL Paper No. 25.

This work was supported by U.S. Public Health Service grants GM 10424 and GM 15421 from the National Institutes of Health.

REFERENCES

Allen, G. 1965. Random and nonrandom inbreeding. Eugen. Quart. 12:181-198.

Azevedo, E., Morton, N. E., Miki, C., and Yee, S. 1969. Distance and kinship in northeastern Brazil. Amer. J. Human Genet. 21:1-22.

Crow, J. F. and Mange, A. P. 1965. Measurements of inbreeding from the frequency of marriages between persons of the same surname. Eugen. Quart. 12:199-203.

Kimura, M. and Weiss, G. H. 1964. The stepping stone model of population structure and the decrease of genetic correlation with distance. Genetics. 49:561-576.

Malécot, G. 1948. Les mathématiques de l'hérédité. Paris: Masson.

________. 1959. Les modèles stochastiques en génétique de population. Publ. Inst. Stat. (Univ. Paris) 8:173-210.

________. 1967. Identical loci and relationship. Vth Berkeley symp. on probability and statistics. 317-332

Malécot, G. 1962. Migration et parenté génétique moyenne. In J. Sutter (ed.). Les déplacements humains. Entret. Monaco Sc. Hum. I. Hachette.

Malécot, G. 1966. Probabilités et hérédité. Inst. Nat. Etudes Demogr. 47, 356 pp. Presses Univ. France.

Malécot, G. 1968. Conséquences statistiques de la parenté. Int. Stat. Inst., Proc. 35th Sess., Sydney, Australia. (In press)

Morton, N. E., and Yasuda, N. 1962. The genetical structure of human populations. In J. Sutter (ed.), Les déplacements humains. Entret. Monaco Sc. Hum. I. Hachette.

Morton, N. E., Miki, C., and Yee, S. 1968a. Bioassay of population structure under isolation by distance. Amer. J. Human Genet. 20:411-419

Morton, N. E., Yasuda, N., Miki, C., and Yee, S. 1968b. Population structure of the ABO blood groups in Switzerland. Amer. J. Human Genet. 20:420-429

Moroni, A. 1964. Evoluzione della frequenza dei matrimoni consanguinei in Italia negli ultimi cinquant'anni. Atti Ass. Genet. It., Pavia, 9:207-223.

Nei, M. and Imaizumi, Y. 1966. Genetic structure of human populations. Heredity 21:9-36.

Simmons, R. T., Graydon, J. J., Gajdusek, D. C., and Brown, P. 1965. Blood group genetic variations in natives of the Caroline Islands and in other parts of Micronesia. Oceania 35:132-170.

Wright, S. 1951. The genetic structure of populations. Ann. Eugen. (London) 15:323-354.

Yasuda, N. 1968. An extension of Wahlund's principle to evaluate mating type frequency. Amer. J. Hum. Genet. 20:1-23.

DISCUSSION ON POPULATION STRUCTURE

L. Cavalli-Sforza, W. F. Bodmer, I. Barrai, N. Yasuda

CAVALLI-SFORZA: The study of population structure in man has presented some challenges both at the level of theory, and of its application to real data. Wright, Kimura and Malecot have provided general theories of isolation which are elegant and concise. Unfortunately, their application to real data is usually not straightforward, and may leave considerable doubt as to the validity or the exact meaning of the estimates of the parameters thus obtained.

Ideally, one would like to be able to separate the effects of drift, migration, stabilizing selection and disruptive selection. But the mode of analysis must differ with the type of data available, which varies greatly, and it is difficult to believe that a single mode of treatment can cover all situations.

From the point of view of the populations being examined, isolated groups and smaller populations may offer more insight into the effects of drift, but estimates derived from them will inevitably be subject to large errors because of their small size. We are painfully learning that large numbers of individuals must be collected in order to make statements on the importance of any evolutionary factor, including selection. This fact may direct us to large populations, which however have other disadvantages, such as heterogeneity of various sorts, e.g. in population distribution, and in selective conditions. Moreover, every population has a particular history, which has inevitable consequences for population structure. The larger the population the more complicated is likely to be its history. In addition, the last few centuries have been times of drastic change for most human groups. The necessity of assuming that the steady state has been obtained, in order to apply present theories of isolation by distance, may therefore be highly misleading.

One problem that faces the user of these theories is the great irregularity of population size, distribution, and migration habits in the human species. If one wishes to segregate the effects of drift from those of selection one has, inevitably, to measure as exactly as possible migration, so as to gauge its effects in buffering drift. Migration data are conveniently collected, for genetic purposes, by obtaining birthplaces of individuals and their ascendants or descendants. For some populations birthplace may be less meaningful than other modes of indicating the origin of individuals; e.g., for a New Guinea population, clan and sub-clan to which an individual belongs may seem more informative than the birthplace, as exogamy rules are followed strictly. Any mode of indicating origin can, without serious loss of information, give rise to the formation of classes, and in fact the tendency of human populations to cluster in groups favors this procedure. When this is done (whether the classes are villages, areas, tribes, clans of birth or origin--we can call them "colonies") the natural way to represent data is a migration matrix in which rows are, e.g., birth colonies of offspring and columns are birth colonies of their parents. As long as these colonies have some degree of permanence, this method of representing migration is useful.

A migration matrix, like any two dimensional frequency distribution, is a rather complete, but of course expensive, way of storing or representing the information. One may want to condense it into a distribution of, say, migration as a function of distance between colonies. This can be done with partial success, but difficulties are encountered in practice in estimating the parameters of these distributions. This is in part the result of the fact that they tend to have almost infinite second order moments, so that the value of the second moment about zero, which is used in the standard theories of isolation by distance, is subject to a very high estimation error. Its value is greatly affected by those few individuals who are born at a very long distance from their parents (long-range migration). It has been suggested that short-range and long-range migration should be distinguished and that one should absorb long-range migrants in another class, which includes all other stabilizing factors in the parameter used to measure "systematic pressure". This seems somewhat arbitrary, as the distribution of migration distances is continuous and usually uni-modal. Moreover long-range migration, whether included in "systematic pressure" or not, has important effects in the theoretical models and is the quantity which is most likely to have changed in recent times.

It seemed, therefore, interesting to develop methods that would use the migration matrix directly for the prediction of drift. [Bodmer and Cavalli-Sforza, 1968, Genetics 59:565; see also the following contribution to the discussion by Bodmer]. These methods can be used even if migration is not constant in time, as long as the migration matrix of any one generation is known. This rarely happens, but there is one additional possibility, which allows one to test whether the migration matrix obtained, say, from the living population, can be considered as valid over a longer period of time. One can in fact, by use of the latent roots of the matrix, predict the distribution frequency of population sizes to be expected after equilibrium for migration has been obtained, and compare it with the observed distribution. This supplies a criterion for testing whether the migration pattern has been changing recently.

Using the migration matrix approach, computations have to be carried out automatically because they are numerically heavy, so much so that it is difficult to handle migration matrices larger than 40 x 40. Insofar as this is an analytical approach, however, it must be distinguished from Monte Carlo methods. We have been using also this last approach, feeding into the model demographic data from a real population and obtaining from the Monte Carlo trials expected consanguinity frequencies, as well as variations between village for gene frequencies due to drift. The comparison of values obtained in the real population with those expected from Monte Carlo trials was very interesting and satisfactory (Cavalli-Sforza and Zei, Proceedings of the 3rd International Congress of Human Genetics). The Monte Carlo approach can provide a closer approximation to the real population under many aspects, but gave essentially the same results as the migration matrix approach as far as migration and drift is concerned, and is more expensive in computer time.

Models of isolation by distance have the advantage of offering a more synthetic view. In them one has to rely, however, on distance as an ancillary variable to express migration. Among the assumptions hidden in this approach, one is that the migration distributions are compatible with the theoretical model. The Gaussian and some other distributions have been shown to be eligible, but not all are. Moreover, the fit of observed migration distributions to any theoretical one is usually poor, mostly because of the heterogeneity of population distribution. One can only hope that the loss of information incurred when summarizing the migration matrix with the migration distribution is not serious. When much care is taken to improve the approximation, and distances are used, I have a personal preference for computing them along roads rather than from geographic coordinates alone. Distances as the crow flies might be better for crows, rather than for human genetics, and may be unsatisfactory even for this sort of animals.

In general, however, differences between the various approaches should not be overstated. Every approach has special merits, and one approach or other may be best for application to a specified real situation. The parameters in the various approaches can, however, be translated one into the other. In any case we are bound to simplify reality greatly. When in doubt on the validity of some approximation, the Monte Carlo method, being more flexible than any analytical approach, can usually provide us with some answer. It should not be viewed as an evasion from the collection of real data, which are obviously the basic source of information, but as an aid to theory.

On the other hand, it seems important to keep in mind the differences between various types of data. The different values of relationship obtained from single alleles, or from pedigrees (whether directly, or from consanguinous matings or isonymy) have all different expectations, which may be sometimes difficult to predict, with sufficient accuracy. The choice of variances or of co-variances for comparison of expected and observed values does not seem a primary issue, and depends more on the nature of the data than of the model chosen. All models can supply expectations for either mode of estimation.

BODMER: I should like to take this opportunity to review briefly some of the key features of the migration matrix approach to the study of population structure, as developed in collaboration with Dr. Cavalli-Sforza (Bodmer and Cavalli-Sforza, 1968). The population is assumed to be distributed in discrete colonies within which there is random mating. The migration matrix M_{ij} gives the proportion of individuals in colony i who came from colony j. Assuming that each colony is segregating for two alleles A, a at one locus, the expected gene frequencies in the nth generation in terms of those in the previous generation are given by

(1) $$p_i^{(n)} = \sum_{j=1}^{k} (1 - \alpha_i) M_{ij} p_j^{(n-1)} + \alpha_i x_i, \quad i = 1 \ldots k$$

Here $p_i^{(n)}$ is the gene frequency in colony i at generation n and α_i is the proportion of individuals coming into colony i from an external population with a constant gene frequency x_i. The parameters α_i and x_i represent all the linear pressures taken into account by the model. To take account of random sampling variations (random genetic drift) we assume that the realized gene frequency in colony i at generation n is the result of a binomial sample of size $2N_i$ with parameter $p_i^{(n)}$ given by equation (1), where the $p_i^{(n-1)}$ are the realized frequencies in the previous generation. N_i is the population size of the ith colony. Using the angular transformation $p = \mathrm{Sin}^2 \theta$, so that $V(\theta)$ is approximately independent of θ, and of p, it is possible to obtain explicit expressions for the variances and covariances in gene frequency in the nth generation for all colonies in terms of the initial frequencies. This amounts to a complete solution of the specified model. Monte Carlo simulations show that the angular transformation is valid so long as all gene frequencies alter at only near 0 or 1. The results are given in terms of the elements of the powers of the migration matrix. This, effectively, limits numerical computations to specimens of about 40 colonies. Larger sets of colonies can of course be reduced, without bias, to smaller sets by simple condensation of the migration matrix. Special features of this approach are:

1) It is not necessary to assume equilibrium. The kinetics of the approach to equilibrium can, in fact, be readily determined.
2) Migration is arbitrarily defined by the migration matrix and need not be isotropic.
3) Linear pressures and population size may vary between classes.

It is clear that as the number of colonies tends to infinity, this discontinuous model must tend to the continuous model. For the specific case of colonies arranged on a circle, and migration occuring only between neighboring colonies, it can be proved that, as the number of colonies tends to infinity, the results tend to those given by Kimura and Weiss (1964) for the simple linear stepping stone model. Bodmer and Cavalli-Sforza (1968) have used the model to investigate the effects of variations in the migration and selection patterns on the observed gene frequency variances and covariances. The limitations of the model are common to those of most previous models. They are mainly the lack of a random element to migration and the ignoring of the population's age structure. This latter can, to some extent, be taken into account by adjusting the effective population size N_i. The migration matrix model does, however, provide a somewhat broader framework than previously proposed models for assessing the effects of a variety of factors on random genetic drift.

Bodmer, W. F. and Cavalli-Sforza, L. L. 1968. A migration matrix model for the study of random genetic drift. Genetics 59:565-592

BARRAI: The possibility of evaluation of an inbreeding coefficient in the presence of a pedigree or of a mating system allows the geneticist to ascertain individual values of inbreeding, and with such an accuracy as is permitted by the number of generations the pedigree fans backward, or by the number of generations the mating system has persisted. No allowance is made, in such an estimate, of the contribution to inbreeding by preferential mating taking place beyond these limits, which could alter the estimate of the probability of homozygosis by descent in an individual, and its estimate in a population.

In the last decade, it has been stressed that the geneticist must consider total inbreeding as made up essentially of two components, the one due to ascertained consanguinity, which can be detected in the present generation or in the past few generations, and the other to undetected, remote consanguinity.

An ingenious method has been devised by Morton and Yasuda to bioassay remote or total consanguinity. In absence of a system of records that reaches far enough in the past, the bioassay method may be used for particular loci, at which there is the possibility to discriminate between the homozygous and the heterozygous genotypes.

Yasuda showed that inbreeding may be estimated more efficiently using mating type frequencies than phenotype frequencies. His method is an extension of Wahlund's principle which applied to marriages taking place in northeastern Brazil, estimated that remote inbreeding was as great as close consanguinity.

It is possible that the recent advances in biochemical genetics, through which the heterozygous state is easily detected for many polymorphs and idiomorphs, will promote a large application of the bioassay method in the study of human population structure and in the estimate of the inbreeding coefficient; however, studies on genetical isolation based on frequencies of consanguineous marriages should not be abandoned on the ground that consanguineous marriage is due to factors partly independent of distance. This objection, advanced by Morton, is based on the finding by Freire-Maia and Freire-Maia that the frequency of consanguineous marriages tends to a rectangular distribution as the mean matrimonial radius becomes greater than one hundred kilometers; such being the case, consanguineous marriages take place preferentially at large distances, so that the concept of Dahlberg of genetic isolates is not valid for recent consanguinity. However, it may well be that the observation by the Freire-Maias depends on the peculiarities of the Brazilian population.

In the past ten years, it has been our effort to list and study the factors that may affect the frequencies of consanguineous marriages, and indirectly, the recent component of inbreeding. We do not pretend that inbreeding as predicted by consanguineous marriage is a substitute for total inbreeding; actually, using existing records, only three or four generations of ancestors may be traced back, and nothing is known from the more remote past.

It is apparent that the remote component may equal the recent one, as Yasuda has found; but it would be most relevant if his findings were confirmed for other populations, and with different methods. We have under study a human population in Val Parma, Italy, whose recent inbreeding has been estimated with some accuracy. For the same population, we have demographical records going back in the past for about ten generations; it is our effort to build the pedigree of this population from the available records, in order to have an estimate of the remote component of inbreeding and to compare it with the value estimated from consanguineous marriages of the last few generations. The project is well under way, and we hope to present some results before long.

YASUDA: I would like to take this opportunity for presenting an isonymy study that I am now carrying on at Mishima, Japan, although the results are still preliminary. This is a simple method to ascertain the effect of random genetic drift in human populations.

Isonymous marriage, that is marriage between persons having the same surname, takes place either due to mere chance or to consanguineous marriage. In the latter case, a quarter will be isonymous (I). Let I be the proportion of observed isonymous marriages, k the expected isonymy frequency due to inbreeding,

q_m the male proportion with a certain surname and q_f the corresponding female proportion. Then we have

$$I = \Sigma q_m q_f + (1-\Sigma q_m q_f)k$$

where $\Sigma q_m q_f$ is the contribution due to random matches, and $(1-\Sigma q_m q_f)k$ to non-random marriages. Since the population is divided into subgroups by surname, $\Sigma q_m q_f$ and k would correspond to the inbreeding coefficients F_{ST} and F_{IS}, respectively. F_{ST} is a coefficient for an individual <u>A</u> whose common ancestor was a person having a different surname from <u>A</u>, while if the surname of the ancestor was the same as that of <u>A</u>, we have F_{IS}.

Therefore, as shown by Crow and Mange (1), the total inbreeding coefficient F_{IT} of the population can be obtained from the relation (2):

$$F_{IT} = F_{ST} + (1-F_{ST})F_{IS}$$

where

$$F_{ST} = \Sigma q_m q_f/4$$

and

$$F_{IS} = (I-\Sigma q_m q_f)/4(1-\Sigma q_m q_f).$$

At Nakazato village of Mishima district, 1,464 couples were sampled from <u>koseki</u>, or household records, in which the parents' surnames as well as the couple's are recorded. Among them, 1,063 males and 614 females were born in the area in question during the last hundred years. Their main occupation is farming. Table 1 summarizes the result, including a study in Tokyo (3). We have found that 95 pairs were isonymous marriages according to koseki, but only nine of them were truly isonymous.

Table 1. Proportion of isonymous marriages and estimated F-statistics at <u>Nakazato</u> and in <u>Tokyo</u>.

Isonymy (I)	F_{IS}	F_{ST}	F_{IT}	α	α/F_{IT}
Nakazato					
9/430	0.00456	0.00276	0.00731	0.0044*	0.60
(0.0209)					
Tokyo					
10,116/363,797	0.00641	0.00051	0.00695	0.0029**	0.45
(0.0278)					

*Shizuoka study (4). **α=0.049/16, first cousin only. Furusho, personal communication.

In this connection, we define the proper surname of an individual as his or her father's surname when he or she was born. The above phenomenon is mainly due to <u>muko-yoshi-engumi</u>, a Japanese custom for maintaining the family name through generations. When a couple does not have a son but has a daughter, a bridegroom-to-be is adopted just before the marriage takes place. Consequently, this results in an increase of isonymous marriages in the koseki.

It is still too early to attribute any significance to the figures reported in Table 1, but they suggest that F_{ST} caused by random genetic drift would be much higher in rural than in urban societies, although it is interesting to note that the total inbreeding coefficients (F_{IT}) are almost the same. The mean inbreeding coefficient (α) from pedigree study would be F_{IT} if the ascertainment of pedigrees is complete, but it is only about one half of F_{IT} in both areas.

Sampling bias on F_{ST} may be checked by a comparison of the quantities Σq_m^2, Σq_f^2 and Σq_{m+f}^2, in correspondence to the proportion of isonymous pairs when two males are drawn at random, to the corresponding proportion in females, and to the frequency of isonymous pairs irrespective of sex. Table 2 shows the results in terms of F_{ST}.

Table 2. Random component (F_{ST}) estimated from various procedures.

	$\Sigma q_m q_f/4$	$\Sigma q_f^2/4$	$\Sigma q_m^2/4$	$\Sigma q_{m+f}^2/4$	$\Sigma q^2/4$
F_{ST}	0.00276	0.00294	0.00444	0.00346	0.00492
N	1677	614	1063	1677	5928

N = sample size

The difference between the second and the third columns may be due to sample size with regard to sex, since koseki is filed by <u>hittosha</u>, or family representative, who is usually a husband; only in sixty of the 1,464 couples was the wife the family representative. The last column is taken from a telephone directory which covers the studied area, and it can be compared with the fourth column obtained from koseki.

An attempt was made to study isolation by distance with a relationship between the frequency (I) of isonymous marriages and the matrimonial distances (r). Adding some other data in Mishima to the above survey, we found that the proportion of isonymous marriages decreases with distance (Table 3).

Table 3. The relationship between the proportion of isonymous marriages and matrimonial distance in Mishima.

Marital distance (r) (km)	0-0.4	0.4-0.8	0.8-1.2	1.2-2.8	2.8-
No. isonymous	5	4	2	3	0
No. couples	70	111	96	212	455
Frequency (I)	0.0714	0.0360	0.0208	0.0142	0.0000

A mathematical form of the relationship is tentatively

$$I = \exp(-kr^{1/4}), \text{ where } k = 2.37.$$

1. Crow, J. F. and Mange, A. P. 1965. <u>Eugen. Quart.</u> 12: 199.
2. Wright, S. 1951. <u>Ann. Eugen.</u> 15: 323.
3. Kamizaki, M. 1954. <u>Seibutu-Tokei</u> 2: 292. (In Japanese.)
4. Tanaka, K. 1963. In <u>The Genetics of Migrant and Isolated populations</u> (Ed. E. Goldschmidt). p. 148.

WRIGHT'S COEFFICIENT OF INBREEDING, F, FOR HUMAN PEDIGREES

Arthur P. Mange
University of Massachusetts
Amherst, Mass.

This program, called "HUMAN INBREEDING, NO DISK STORAGE," does two major jobs for a given person: first, it constructs, stores, and punches out his pedigree for six ancestral generations by repeated references to a stored listing of about 1,000 potential ancestors. Secondly, it computes and punches out his coefficient of inbreeding, F, (Wright, 1922) by systematically searching the stored pedigree for common ancestors. In the F computation the program assumes that certain types or patterns of marriages do not occur. For example, the program will ignore a parent-offspring mating whose issue is the person in question. While it is appropriate for most interrelated human populations with five or six generations of ancestry information, it may not work correctly for species in which very close consanguinity is common. With fewer than five ancestral generations known, machine calculation would usually be unnecessary. The program is written in Fortran II and uses only the 40,000 cores of the IBM 1620 for which it was written. The pedigree construction and F calculation for one individual requires two to three minutes. In a single computer run the program can calculate the inbreeding coefficients of all the stored ancestors and all offspring of the stored ancestors.

Having previously calculated by hand about 800 inbreeding coefficients for complex Hutterite pedigrees (Mange, 1964), the impetus for this program was provided by a confrontation with about the same number of additional pedigrees. The friendly but persistent prodding of Dr. A. G. Steinberg was also needed to overcome my reluctance to cope with computer language and logic. Discussions with him and with Dr. P. M. Conneally were extremely valuable, as was the kind assistance of Dr. A. S. Littell, Director of the Computer Laboratory of Western Reserve University Medical School.

Although I will not describe related programs in detail, at least three modifications of "HUMAN INBREEDING, NO DISK STORAGE" are also available:

1. "HUMAN INBREEDING, DISK STORAGE" can accommodate a much larger ancestral pool from which pedigrees are constructed. This program was the one actually used for the Hutterite population.

2. "ANIMAL INBREEDING, NO DISK STORAGE" allows any mating pattern for a dioecious species however close or complex the relationships. This program has been tested on a few representative cattle pedigrees from the Blaine Experimental Farm, Lake Mills, Wisconsin.

3. "ANCES," designed by Mrs. Alice Martin of Case Western Reserve University, will construct an 8-generation pedigree and calculate certain inbreeding coefficients.

In addition to these programs several other recent computer approaches to the inbreeding tedium are available. One of these has been reported by MacCluer et al. (1967) and is based on the method of Kudo (1962) and Kudo and Sakaguchi (1963). Another has been developed by T. H. Roderick of Bar Harbor and is, to my knowledge, unpublished. The discussants of this paper will present still others.

INPUT

Four-digit code numbers (except 0000 and 9999) are assigned, uniquely but in no particular order, to all contemporary and ancestral individuals. Only one of a group of unmarried sibs, however, need be given a number since they will all have the same F value. Fictitious people can be invented, if desired, to estimate coefficients of relationships between any pair of individuals. These two people would be designated the parents of the fictitious person and their coefficient of relationship would be approximately twice the computed F value.

Having assigned code numbers, it is useful to separate those people who are the ancestors of somebody (i.e., people who are parents) from those who are not. While inbreeding values may be desired for people in either group, only the parents need be included in a list of ancestors from which pedigrees are constructed. This program can accommodate up to 479 males and 479 females who are ancestors and any number of others. For the Hutterites this group of others consisted largely of unmarried children of the most contemporary generation.

One input card is punched for each ancestor and this deck is designated LIST. Repeated searching of LIST by the computer forms the basis of pedigree construction. One input card is also punched for each individual whose inbreeding coefficient is desired and this deck is designated FINDIV. A parent may have duplicate cards, one in each deck. The form of the information on these cards is given in Table 1.

As seen in Table 1 the LIST deck will consist of all cards with a 3 in column 1. This deck is arranged by sex, males first, and, within sex, the cards are sorted by year of birth, the earliest date first. This order saves some search time in pedigree construction since there are progressively more ancestors

Table 1

Punch code for input cards.

Column	Punch
1	3 if this person has progeny; 9 if he has none.
3 - 6	the code number of the individual.
7	1 if the individual is male; 2 if female.
9 - 12	the code number of the individual's father.*
15 - 18	the code number of the individual's mother.*
21 - 23	the last three digits of the individual's year of birth.
24 - 80	any other information.

* The code number 0000 is assigned to any unknown parent.

with earlier birth dates. The subdeck of males is terminated by a card with 9999 in columns 9 - 12, and the subdeck of females is terminated by a card with 9999 in columns 15 - 18. The LIST deck is fed to the computer following the program.

The FINDIV deck will consist of all cards with a 9 in column 1 and a duplicate of any, or all, cards with a 3 in column 1. These cards are arranged by sibships by sequence sorting on the code numbers of the individual's parents. With this card order the program will compute the F value for a group of sibs only for the first one and provide an appropriate reference for the others. The FINDIV deck is fed to the computer following the LIST deck.

GENERAL PROGRAM LOGIC

The machine calculation of F follows essentially the same method of path tracing through common ancestors that one might do by hand. The only important difference relates to situations in which a common ancestor is himself inbred. As we will see shortly the program deals with this correctly as long as a common ancestor of a common ancestor appears within the six generations of the machine-constructed pedigree. For most human populations this limitation is not serious since the program will correctly account for two first-cousin marriages "in tandem" or any sequence of two closer consanguineous marriages. Looser patterns of tandem consanguinity--those which the computer is unable to take into account--would add very little to the computed F value.

The program logic hinges on the variable KIN which designates the members of the pedigree of the individual whose F value is sought, i.e. the FINDIV. It is easiest to conceive of KIN as a triply subscripted variable, KIN(L, M, N), where the subscripts are "coordinates" of the members of the pedigree:

L is the side of the pedigree, 1 indicating the father's side and 2 indicating the mother's side.

M is the number of generations removed from FINDIV, and may take values from 1 to 6.

N is a sequential numbering of ancestors within side and within generation. The maximum value of N depends upon the generation number and equals 2^{M-1}.

The framework of the program consists of systematic changes in these, and other, subscripts. The coordinate system is detailed in Table 2.

Note, in Table 2, that the area of one side of the pedigree is 192 (i. e., 6 x 32) units where a unit is the size of a rectangle occupied by an ancestor in the sixth generation. As will be mentioned later, use is made of these pedigree areas to compute the completeness of the known pedigree. In filling in this pedigree form with ancestors, the computer will always insert a father above his wife so that all males have N numbers that are odd and all females will have N numbers that are even. The one exception to this rule is KIN(2, 1, 1), the mother of FINDIV. And with the exception of KIN(1, 1, 1), the father of FINDIV, the last male in a given generation will have an N number equal to $2^{M-1} - 1$ and the last female an N number equal to 2^{M-1}.

Members of the pedigree are usually referred to in the program by the equivalent, singly subscripted variable, KINS(LT), in order to decrease the running time. This linear subscript, LT, is related to the triple subscripts by the formula

$$LT = L + 2M + 12N - 14.$$

Pedigree construction for the first FINDIV begins after the LIST of potential ancestors, along with each ancestor's parents and sex, is read and stored. LIST is also both a singly and a triply subscripted variable but this need not detain us. It will be sufficient to note that the storage of LIST is such that the memory locations of a potential ancestor's father and mother are determinable from the memory locations of the ancestor himself. The FINDIV's card is then read and his father and mother designated KIN(1, 1, 1) and KIN(2, 1, 1), respectively (see Table 2). The computer then searches LIST, by an appropriate IF statement in a DO loop, in order to match the code number of KIN(1, 1, 1). Some machine time is saved by searching only those members of LIST of the proper sex. When a match is made, the locations of this ancestor's father and mother are determined and their code numbers stored in KIN(1, 2, 1) and KIN(1, 2, 2), respectively, these two people being the FINDIV's paternal grandparents. This search, match, and assign procedure is systematically repeated in order to construct the whole pedigree ending with KIN(2, 6, 32). The variable KIN is wasteful of space since 384 [i. e., (2)(6)(32)] locations are reserved for it by the dimension statement, but only 126 ancestors are sought. KIN(1, 2, 3), for example, is biologically meaningless as are about two-thirds of all locations reserved for KIN.

The general relationship which is utilized in pedigree construction can be deduced from Table 2: the father of KIN(x, y, z) is KIN(x, y+1, 2z-1) and the mother, KIN(x, y+1, 2z). As is clear, the pedigree is completely dichotomous with repetition of an ancestor as many times as he may occur. This follows the path diagram procedures of some animal breeders but is not the usual method of drawing human pedigrees. The computer does not search LIST for an ancestor whose code number is 0000 but immediately assigns 0000 to the presumably unknown parents of this unknown ancestor. If LIST simply does not contain a known

TABLE 2

The coordinate numbering system for KIN(L,M,N) = the ancestors of FINDIV. L gives the side of the pedigree, and M the number of generations behind FINDIV. The figures within the pedigree are the N numbers. It is understood that all odd N numbers are male and all even numbers are female with the exception of KIN(2, 1, 1), the mother of FINDIV.

L = 1						L = 2					
M=1	M=2	M=3	M=4	M=5	M=6	M=1	M=2	M=3	M=4	M=5	M=6
1 = father	1 = father's father	1	1	1	1	1 = mother	1 = mother's father	1	1	1	1
					2						2
				2	3					2	3
					4						4
			2	3	5				2	3	5
					6						6
				4	7					4	7
					8						8
		2	3	5	9			2	3	5	9
					10						10
				6	11					6	11
					12						12
			4	7	13				4	7	13
					14						14
				8	15					8	15
					16						16
	2 = father's mother	3	5	9	17		2 = mother's mother	3	5	9	17
					18						18
				10	19					10	19
					20						20
			6	11	21				6	11	21
					22						22
				12	23					12	23
					24						24
		4	7	13	25			4	7	13	25
					26						26
				14	27					14	27
					28						28
			8	15	29				8	15	29
					30						30
				16	31					16	31
					32						32

ancestor the computer punches out an appropriate message telling the programmer he omitted this person from LIST and where this person is needed in the FINDIV's pedigree.

In addition to compiling the code numbers of the pedigree members with their appropriate coordinates, a count is made within side and within generation of the number of known ancestors using the program variable TALLY(L, M). This is accomplished by asking whether each KIN code number is different from zero and scoring 1 for each non-zero value. Those values of TALLY are eventually combined to give measures of the completeness of the pedigree.

The computation of F is based on another systematic search procedure involving only the code numbers of the members of the constructed pedigree. Each KIN number on the father's side of the pedigree is compared with each KIN number of the same sex on the mother's side of the pedigree. The parents of FINDIV, however, are omitted from this search procedure. In all, 1,922 comparisons are made [(62)(31)] . A match of the same code number on the two sides of the pedigree may indicate a common ancestor who forms a "loop" with the FINDIV and thus contributes to his F value. Such a match is referred to as a valid coincidence. For each such valid coincidence the contribution to F is calculated as $(1/2)^x$ where x is the number of individuals in the loop. The computer finds this value of x by summing the generation numbers (M) on the two sides of the pedigree and subtracting one.

Because of the dichotomous, and therefore repetitious, nature of the pedigree, however, not every match indicates a common ancestor by the usual meaning of the term. All ancestors of a common ancestor will also appear on both sides of the pedigree. The existance of these people will provide additional matches, but most of these will turn out to be invalid coincidences making no contribution to the F value of FINDIV.

FIGURE 1

A comparison of hand versus machine calculation of F_Q.

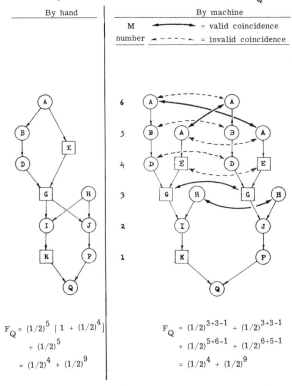

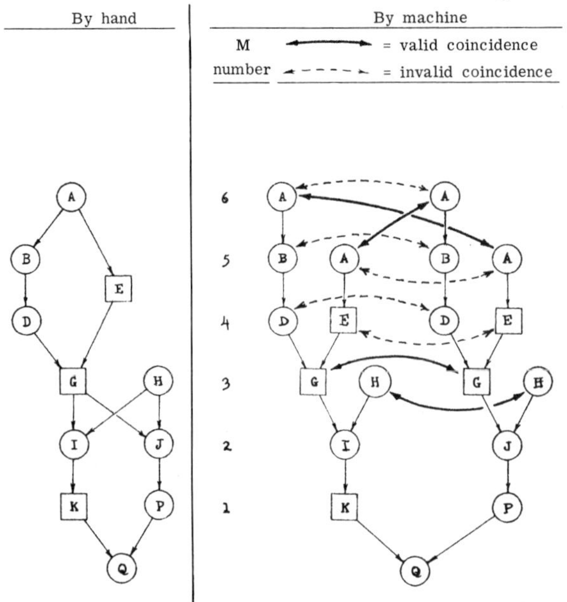

$$F_Q = (1/2)^5 [1 + (1/2)^4] + (1/2)^5 = (1/2)^4 + (1/2)^9$$

$$F_Q = (1/2)^{3+3-1} + (1/2)^{3+3-1} + (1/2)^{5+6-1} + (1/2)^{6+5-1} = (1/2)^4 + (1/2)^9$$

A simple test for the validity of a coincidence was devised which distinguishes "real" common ancestors from spurious ones. Prior to the time when we were considering this test we had decided to ignore the additional contribution to F due to the inbreeding of common ancestors. For the Hutterite population such additional contributions were negligible. But, to our surprise, the test for validity had the side effect of correctly including what we had intended to ignore provided, as mentioned before, the common ancestor of the common ancestor appears within the six generations of the machine-constructed pedigree.

The test for a "real" common ancestor is this: a coincidence is valid when the common ancestor's child on the father's side of the pedigree is different from the common ancestor's child on the mother's side. By way of example, in Figure 1 I have compared the hand calculation with the machine calculation of the inbreeding coefficient of individual Q. This person has a common set of great-grandparents, G and H, one of whom, in turn, has a single common ancestor, A. This is what I would call a first-cousin marriage in tandem with a half uncle-niece marriage. By hand, F_Q is easily calculated as $(1/2)^4 + (1/2)^9$. The machine arrives at the same answer by a somewhat different route: the G-G and H-H coincidences are valid since $I \neq J$ [add $(1/2)^5 + (1/2)^5$] ; the D-D, E-E, and B-B coincidences are all invalid; and two of the A-A coincidences are valid [add $(1/2)^{10} + (1/2)^{10}$] and two are not. In short, the machine, at first, ignores the inbreeding of a common ancestor but picks it up by allowing a more ancestral loop to run through this common ancestor on both sides an appropriate number of times. This method for calculating F has been previously used by Cotterman.

The coordinate numbering system provides the method whereby the machine can pick out the children of the common ancestor needed for this test of validity: if the common ancestor appears on the father's side as KIN(1, y, z) his child is KIN(1, y-1, (z+1)/2) if the common ancestor is male, and KIN(1, y-1, z/2) if the common ancestor is female. Similar expressions exist for the child of a common ancestor in line to FINDIV's mother.

It might be interesting to note that the procedures outlined here are similar to those which were performed by hand prior to the machine program (Mange, 1964). A dichotomous pedigree was drawn on a preprinted form (like Table 2) for the six generations. This pedigree was drawn by couples rather than by individuals. While this cut the pedigree-drawing tedium, it necessitated a separate reference sheet for individuals who had children by more than one spouse. About 1,000 comparisons between paternal and maternal ancestral couples were made by visual scanning, and each valid coincidence was marked by a unique symbol so that it would not be counted twice. Detecting invalid coincidences required normal color vision: the areas on both sides of the pedigree occupied by all couples ancestral to a "real" common ancestor were shaded, a different color being used for each. Then a coincidence formed by a couple who appeared on both sides of the pedigree in an area colored the same was invalid. As long as no ancestor of a common ancestor was inbred, this visual test is equivalent to the machine test outlined above. As explained shortly a measure of the completeness of the pedigree was computed exactly as the machine does it. A number of discrepancies between machine-calculated and hand-calculated F values were all decided in favor of the machine. These hand-calculated errors were always due to overlooking one or more valid coincidences, usually in the sixth generation. This manual procedure is similar to the independently arrived at systematic method of path tracing developed by Kudo (1962).

One general pattern of marriage (at least) will cause the computer to slightly overestimate the F value. This exists when a common ancestor has an ancestor (common or otherwise) with a common ancestor! Two examples of this situation are diagrammed in Figure 2. Pedigrees like these would be vanishingly rare in a six-generation human pedigree, but could easily occur in some animal pedigrees. The animal program, however, employs an extended test for validity which provides a correct computation of F in any situation that I am aware of.

The computer keeps score, by generation, of the increasingly incremented F value. All contributions to F which would be detected if the pedigree were cut off after M generations, but which were not detected on the basis of M - 1 generations, are added to give the inbreeding contribution of the Mth generation. This partitioning of the F value by generation is accomplished by the program variable FPA(M). The total F value for the six-generation pedigree is obtained by summing FPA(M) for M = 2, ..., 6 since FPA(1) must be zero for a dioecious species. For the Hutterites FPA(2) was also zero in all cases (no sib matings), and FPA(3) + FPA(4) due to first- and second-cousin marriages was, on the average, roughly the same as FPA(5) + FPA(6) due to third- and fourth-cousin marriages.

OUTPUT

A sample output, that for Hutterite male 182 and his sibs, is given in Figure 3. The first 11 lines are his pedigree. Mothers are indicated immediately to the right of their husbands in generations 2 - 4, and immediately below their husbands in generations

FIGURE 2

Two path diagrams for which the computer will calculate F_K incorrectly. In each diagram, K has a common ancestor, E, with an ancestor, D, with a common ancestor, A.

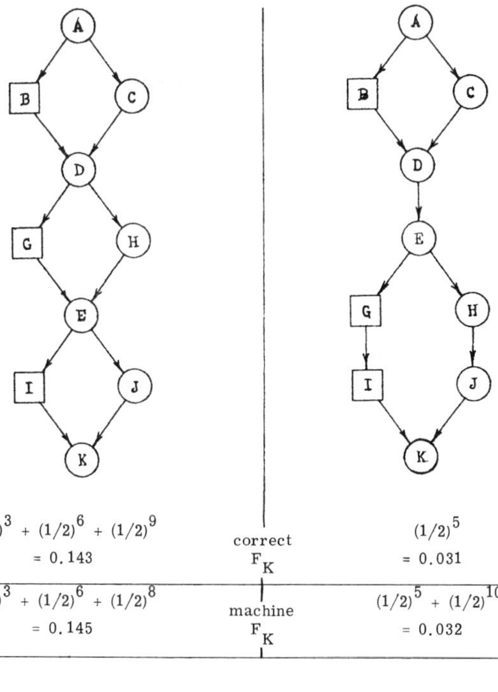

	correct F_K	
$(1/2)^3 + (1/2)^6 + (1/2)^9 = 0.143$	correct F_K	$(1/2)^5 = 0.031$
$(1/2)^3 + (1/2)^6 + (1/2)^8 = 0.145$	machine F_K	$(1/2)^5 + (1/2)^{10} = 0.032$

FIGURE 3

Output for Hutterite male 182 and his sibs

182.1	0	INDIVIDUAL BORN 1923															
182.1	1	501 = FATHER								847 = MOTHER							
182.1	2	1002				951				997				1015			
182.1	3	1080		1019		1081		1021		1085		1003		1061		1022	
182.1	40	1121	1051	1065	1052	1119	1046	1062	1053	1115	1054	1081	1056	1114	1086	1062	1053
182.1	51	1149	1156	1144	1156	1137	1174	1134	1136	1137	1136	1119	1136	1147	1145	1134	1136
182.1	52	1160	1167	1124	1167	1157	1164	1117	1118	1157	1118	1046	1118	1128	1126	1117	1118
182.1	61	1161	1176	1198	1176	1161	0	1157	1159	1161	1159	1137	1159	1159	1186	1157	1159
182.1	62	1180	1168	1197	1168	1180	0	1169	1139	1180	1139	1157	1139	1139	1176	1169	1139
182.1	63	0	0	1165	0	0	0	1184	1178	0	1178	1174	1178	1161	1177	1184	1178
182.1	64	0	0	1190	0	0	0	1138	1136	0	1136	1164	1136	1180	1135	1138	1136

182.1	COINCIDENCE=1081	SEX=1	M=3	N= 3	MM=4	NN= 3	LINKS= 6
182.1	COINCIDENCE=1062	SEX=1	M=4	N= 7	MM=4	NN= 7	LINKS= 7
182.1	COINCIDENCE=1053	SEX=2	M=4	N= 8	MM=4	NN= 8	LINKS= 7
182.1	COINCIDENCE=1137	SEX=1	M=5	N= 9	MM=5	NN= 1	LINKS= 9
182.1	COINCIDENCE=1157	SEX=2	M=5	N=10	MM=5	NN= 2	LINKS= 9
182.1	COINCIDENCE=1136	SEX=1	M=5	N=15	MM=5	NN= 3	LINKS= 9
182.1	COINCIDENCE=1136	SEX=1	M=5	N=15	MM=5	NN= 7	LINKS= 9
182.1	COINCIDENCE=1118	SEX=2	M=5	N=16	MM=5	NN= 4	LINKS= 9
182.1	COINCIDENCE=1118	SEX=2	M=5	N=16	MM=5	NN= 8	LINKS= 9
182.1	COINCIDENCE=1161	SEX=1	M=6	N= 1	MM=6	NN= 1	LINKS=11
182.1	COINCIDENCE=1161	SEX=1	M=6	N= 1	MM=6	NN=19	LINKS=11
182.1	COINCIDENCE=1180	SEX=2	M=6	N= 2	MM=6	NN= 2	LINKS=11
182.1	COINCIDENCE=1180	SEX=2	M=6	N= 2	MM=6	NN=20	LINKS=11
182.1	COINCIDENCE=1161	SEX=1	M=6	N=17	MM=6	NN=19	LINKS=11
182.1	COINCIDENCE=1180	SEX=2	M=6	N=18	MM=6	NN=20	LINKS=11
182.1	COINCIDENCE=1159	SEX=1	M=6	N=29	MM=6	NN=17	LINKS=11
182.1	COINCIDENCE=1139	SEX=2	M=6	N=30	MM=6	NN=18	LINKS=11

FIGURE 3 - continued

182.1	CUTOFF=2	F=0.00000	COMPL= 100.00PCT	PEDFA= 1.0000	PEDMO= 1.0000
182.1	CUTOFF=3	F=0.00000	COMPL= 100.00PCT	PEDFA= 1.0000	PEDMO= 1.0000
182.1	CUTOFF=4	F= .03125	COMPL= 100.00PCT	PEDFA= 1.0000	PEDMO= 1.0000
182.1	CUTOFF=5	F= .04296	COMPL= 100.00PCT	PEDFA= 1.0000	PEDMO= 1.0000
182.1	CUTOFF=6	F= .04687	COMPL= 92.77PCT	PEDFA= .9375	PEDMO= .9895

183.1	0	INDIVIDUAL BORN 1928	
183.1	1	501 = FATHER	847 = MOTHER
183.1	2	SIB OF 182.1 BORN 1923	
182.2	0	INDIVIDUAL BORN 1927	
182.2	1	501 = FATHER	847 = MOTHER
182.2	2	SIB OF 182.1 BORN 1923	
183.2	0	INDIVIDUAL BORN 1931	
183.2	1	501 = FATHER	847 = MOTHER
183.2	2	SIB OF 182.1 BORN 1923	

FIGURE 4

Conventional path diagram of Hutterite male 182.
Bold pedigree symbols indicate common ancestors.

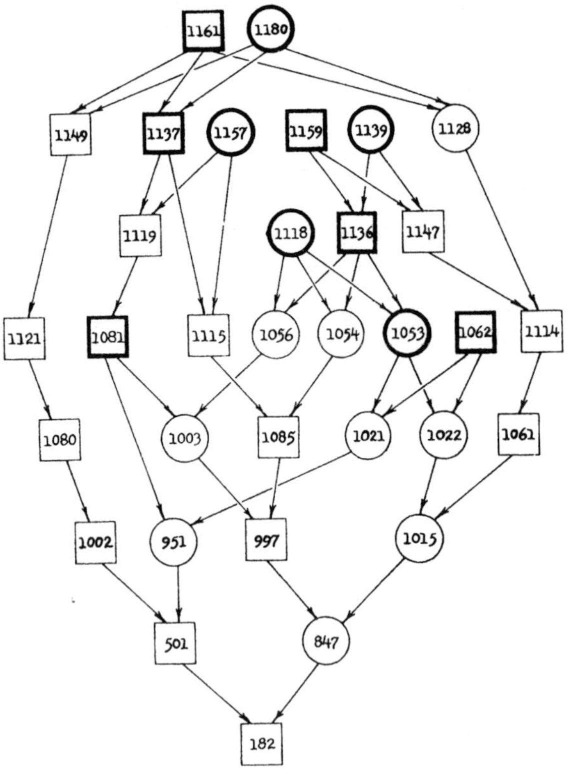

5 and 6. The far left-hand column of ancestors, starting from the fourth generation, should be interpreted as follows: the father and mother of male 1121 are 1149 (line 5-1) and 1160 (line 5-2), respectively; the father and mother of 1149 are 1161 (line 6-1) and 1180 (line 6-2), respectively; the father and mother of 1160 are unknown (lines 6-3 and 6-4).

Next follows a list of valid coincidences (i.e., "real" common ancestors). Using the first coincidence as an example, these items are given: his code number (1081), his sex (1 = male), his coordinates on the father's side (M = 3, N = 3), his coordinates on the mother's side (MM = 4, NN = 3), and the power to which 1/2 should be raised to give his contribution to the inbreeding coefficient of 182 (LINKS = 6). The relevant path diagram, drawn in the conventional way and including only the ancestors needed to calculate 182's F value, has been extracted from the machine pedigree and is presented in Figure 4. The machine output of coincidences will help those who may wish to trace paths by eye and hand.

The five lines beginning CUTOFF in the output give the F value and the completeness of the pedigree if it had been cut off behind the 2nd, ..., 6th generation. The particular measure of pedigree completeness (COMPL) attempts to estimate what percentage of the "true" F value is represented by the F value computed on the basis of the known ancestors. This measure of completeness is calculated by multiplying together the known fractional "areas" on the two sides of the pedigree. "Area" is to be understood literally (as displayed in Table 2) where each side is 192 units of the size of a rectangle occupied by an individual in generation 6. This product of areas should represent the portion of the "true" F which is knowable assuming that the likelihood of a person being a common ancestor does not depend on either his position in the pedigree or on whether he is known or unknown (Mange, 1964). While human populations are not likely to meet these assumptions exactly, for pedigrees that are reasonably complete any other set of assumptions would not give wholly different values for completeness.

In the sample output (Figure 3) the father's side is fully known except for 12 ancestors in generation 6, and the mother's side fully known except for two such ancestors. The percent completeness of the 6-generation pedigree is then:

$$\frac{192-12}{192} \cdot \frac{192-2}{192} \cdot 100 = (.9375)(.9895)(100) = 92.77\% .$$

Had the pedigree been cut off behind the fifth generation, the completeness would be similarly calculated except that there are only 160 units the size of a rectangle occupied by an ancestor in generation 6 (or 80 of the fifth generation units). In the sample output, the pedigree is 100 percent complete back to the fifth generation. This weighting of individuals by area is consistent with their potential contribution to the F value of FINDIV since, for example, a 5-5 coincidence contributes four times as much to F as a 6-6 coincidence. Correspondingly, the product of two fifth generation areas is four times the product of two sixth generation areas since the areas double in size with each generation toward the FINDIV.

Finally, this particular output includes references to three sibs of 182. The FINDIV cards of these sibs immediately followed that of 182. Prior to pedigree construction a match was made by the computer between the parents of 182 and the parents of the sibs. Since they were the same, the computer was programmed to produce the "SIB OF ..." message.

SUMMARY

In summary, the computer accepts a deck of punched cards, one for each ancestral member of an interrelated human population. The essential items of information on these cards are the code numbers of the individual, his father, and his mother. Then the computer accepts similar cards for people whose F value is sought. Out comes, for each such person, his pedigree for six ancestral generations--as far as it is known, a list of all common ancestors with their pedigree coordinates, his inbreeding coefficients--by generation, and measures of the completeness of the pedigrees on which these values are based. This paper attempts to orient a potential user to some of the details and limitations of this program with a brief mention of several modifications of it. An appendix is provided, giving an alphabetical list of the meaning of all program variables.

Acknowledgement

This work was supported in part by Grants HE 03708 (Dr. A. G. Steinberg) and GM 09310 (Dr. A. S. Littell), United States Public Health Service.

REFERENCES

Kudo, A. 1962. A method for calculating the inbreeding coefficient. Am. J. Hum. Genet. 14:426-432.

Kudo, A., and Sakaguchi, K. 1963. A method for calculating the inbreeding coefficient. II. Sex-linked genes. Am. J. Hum. Genet. 15:476-480.

Mange, A. P. 1964. Growth and inbreeding of a human isolate. Hum. Biol. 36:104-133.

MacCluer, J. W., Griffith, R., Sing, C. F., and Schull, W. J. 1967. Some genetic programs to supplement self-instruction in FORTRAN. Am. J. Hum. Genet. 19:189-221.

Wright, S. 1922. Coefficients of inbreeding and relationship. Am. Natur. 56:330-338.

APPENDIX

An alphabetical list of the meaning of all program variables.

Square brackets enclose the statement number where the variable is first used in the program. For example, [nl - 2] means two statements before the statement labelled 1. Also given within some square brackets is an appropriate mnemonic. Some variables defined as subscripts are also used with the same meaning for other purposes.

COMPL(M) [n145] : the percentage completeness of the whole pedigree back to the Mth generation. COMPL(M) = 100· PEDFA(M)· PEDMO(M).

DNOM 1 [n143 + 1; DeNOMinator, 1st factor] : the floating point equivalent of MEXP, the total number of ancestors within side and within generation. Corresponding products of DNOM 1 x DNOM 2 give the total area of the M-generation pedigree.

DNOM 2 [n145 - 3; DeNOMinator, 2nd factor] : the floating point equivalent of M, the generation number.

FINDIV [n12; F of INDIVidual] : the individual whose pedigree and F value is being sought.

F(M) [n80] : the F value of FINDIV based on an M-generation pedigree.

FPA(M) [n48; F, PArtial value] : the contribution to FINDIV's F value detected when including ancestors of the Mth generation that was not detected by all ancestors in generations 1 through M - 1.

I[nl - 2] : the first of three subscripts of LIST indicating the sex of the person: 1 = male, 2 = female.

IEND [n21 + 1] : the non-subscripted equivalent of JMAX(I) needed to end a DO loop.

IFA [n4 - 1] : the code number of the father of the FINDIV who was just previously processed by the computer.

IMO [n4 + 2] : the code number of the mother of the FINDIV who was just previously processed by the computer.

INEE [n15] : the last three digits of the birth year of the FINDIV who was just previously processed by the computer.

IT [n21 + 2] : the linear subscript of LISTS(IT) giving potential ancestors of a FINDIV. IT is related to the triple subscripts of the equivalent variable, LIST(I, J, K), by IT = I + 2J + 960K - 962.

J[nl - 1] : the second of three subscripts of LIST indicating simply a sequential numbering of ancestors within sex.

JMAX(I) [n4] : the actual number of male ancestors (I = 1) or female ancestors (I = 2) read into LIST.

K [nl] : the third of three subscripts of LIST indicating whether the code number in question is the ancestor himself (K = 1), his father (K = 2), or his mother (K = 3).

KEY [n47 + 3] : the line number within a generation in the pedigree output. This is also a dummy variable for the N number of KINS(L, M, N) to provide a neat, systematic pedigree output.

KEY 2 [n44 - 1] : a dummy variable to terminate the N number of KINS(L, M, N) for the pedigree output of generations 5 and 6. See also KEY.

KIN(L, M, N) [n22 + 1] : members of the pedigree of FINDIV; equivalent to KINS(LT). See also L, M, N, and LT.

KINS(LT) [n11] : see KIN(L, M, N).

L [n17 + 2] : the first of three subscripts of KIN(L, M, N) indicating either the father's side (L = 1) or the mother's side (L = 2) of the pedigree.

LBEG [n39 + 1] : a subscript of TALLYS equal to 11 or 12 for scoring known ancestors in the 6th generation within side. LBEG also begins the DO loop for the purpose of this tally.

LD [n17 - 1] : the single linear subscript of TALLYS(LD) indicating generation number: 1, 3, 5, 7, 9, 11 indicating respectively generations 1 through 6 on the father's side, and 2, 4, 6, 8, 10, 12 for the mother's side. LD is related to L and M by: LD = L + 2M - 2.

LEND [n39 + 2] : a dummy variable for ending the DO loop for tallying known ancestors, within side, in the 6th generation.

LINKS [n59] : the number of individuals in a loop through a common ancestor.

LIST(I, J, K) [nl] : potential ancestors of FINDIV; equivalent to LISTS(IT). See also I, J, K, and IT.

LISTS(IT) [n22 - 1] : see LIST(I, J, K).

LT [n10] : the linear subscript of KINS(LT) giving the position of a person in the pedigree of FINDIV. LT is related to the triple subscripts of the equivalent variable, KIN(L, M, N), by LT = L + 2M + 12N - 14.

LT 1 [n48 + 3] : a subscript of KINS referring specifically to a pedigree member on the father's side of the pedigree.

LT 2 [n48 + 4] : as LT 1 but for the mother's side of the pedigree.

LT 1 C [n55] : a subscript of KINS referring specifically to the position of the child of KINS(LT 1).

LT 2 C [n55 + 1] : a subscript of KINS referring specifically to the position of the child of KINS(LT 2).

LTFA [n18 - 1] : a subscript of KINS referring specifically to the position of the father of KINS(LT).

LTMϕ [n18] : a subscript of KINS referring specifically to the position of the mother of KINS(LT).

M [n17 + 3] : the second of three subscripts of KIN(L, M, N) indicating ancestral generations behind FINDIV; M = 1, 2, 3, 4, 5, 6.

MEXP [n18 - 4] : the total number of ancestors within side and within generation; MEXP = 2^{M-1}.

MM [n50] : as M but referring specifically to the mother's side of the pedigree.

MMX [n50 + 2] : as MX but for the mother's side of the pedigree.

MX [n48 + 2] : the total number of ancestors on the father's side of the pedigree in generation M.

N [n17 + 4] : the third of three subscripts of KIN(L, M, N) indicating a sequential numbering within side and within generation. N runs from 1 to 2^{M-1}.

NEE [n12] : the last three digits of the birth year of FINDIV.

NN [n50 - 1] : as N but referring specifically to the mother's side of the pedigree.

NOX [n50 - 3; Number of X's] : sex of a pedigree member; 1 = Male, 2 = Female.

PEDFA(M) [nl45 - 2] : the fraction of the area of the M-generation pedigree which is known on the father's side.

PEDMO(M) [nl45 - 1] : as PEDFA(M) but for the mother's side.

SIBCK [nl5; SIB Check] : the code number of the FINDIV who was just previously processed by the computer.

SIDEFA(M) [n42 + 2] : the area of the M-generation pedigree which is known on the father's side. Area is interpreted literally as displayed in Table 2.

SIDEMO(M) [n42 + 1] : as SIDEFA(M) but for the mother's side.

TALLYS(LD) [nl7] : the number of known pedigree members within side and within generation.

DISCUSSION ON PEDIGREE INBREEDING

T. R. Wolfe, A. W. F. Edwards, and C. C. Cockerham

WOLFE: I would like to compliment Dr. Mange for a well designed and written program. I only wish that more computer programs today were as well documented.

Dr. Mange's program is certainly a practical and useful production program for the calculation of inbreeding coefficients from pedigrees. He has wisely decided that the greatly increased computation required for complete path determination, arbitrarily many generations back, is not worth the small changes in the lower order decimal places. (From my point of view as a programmer, however, the general problem is an enjoyable challenge).

In order that a program of this type be able to analyze an arbitrary pedigree, it is necessary to be able to:

1) specify the relationships among all the individuals in the pedigree
2) specify any prior knowledge we have about inbreeding and consanguinity in the pedigree which cannot be determined from examination of the pedigree alone (for example, the coefficient, F, of an individual who appears only as a parent and common ancestor in the pedigree)

Dr. Mange has dealt with this first problem in perhaps the most convenient fashion: by assigning unique identifiers to each individual in the pedigree and indicating, for each individual, the identifiers of his parent. A unique identifier 0000 has been utilized in this case to allow the program to recognize when a parent is not in the pedigree.

The second requirement, however, is not fulfilled with this program. To take a very simple example, consider the pedigree

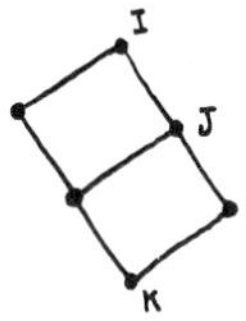

Calculation with this program would indicate that the coefficient of inbreeding of individual K was

$$5/32 = 1/2^5 + 1/2^3$$

whereas, if individuals I and J are also themselves inbred, with approximate F coefficients known but exact pedigree unknown, this answer is not as informative as we would like it to be. Of course, we could attempt to get around this problem by constructing artificial ancestral pedigrees for individuals I and J which would force upon them approximate F coefficients. It would be simpler indeed if we were to build into our program the facility for presetting the F coefficients for certain individuals. Similarly, we might want to indicate that two individuals in the pedigree have a non-zero coefficient of consanguinity despite the fact that we do not know anything about their common ancestry. Here we could either input an approximate coefficient of consanguinity or add a ficticious common parent with the appropriate value of F indicated for him.

This routine should not be used, certainly, for analyzing the effects of systematic breeding, for, as Dr. Mange adequately warns the user, the result will be a higher calculated F than the true value. The two examples given by Dr. Mange where incorrect calculations result do not indicate the severity with which this can occur. The results will certainly be more divergent in highly systematic pedigrees.

I would like to take a moment to point out the simple modifications to this program which would allow the calculation of the F coefficients for sex-linked loci and loci located on the Y chromosome in man. To compute the coefficient for a sex-linked locus we follow the normal procedure, only we do not count the contributions of paths which contain a man and his son. By changing the program so that all males have their fathers indicated with the 0000 code after reading in the data deck, we can use the same deck to produce F coefficients for the sex-linked locus. Similarly, if after reading the data in, we set to 0000 all maternal identifiers, we can process Y chromosome loci.

At this point I would like to discuss a different method of programming this problem of computing inbreeding coefficients given a pedigree. To do this, I must first give a brief description of list processing languages. In normal computer usage, the data are stored in consecutive computer words and access is via single or multidimensional subscripts. For some problems, this has serious drawbacks. Some types of data, for example pedigree structures, are inherently not sequential, but rather are branched. By using lists which are linear sequences of elements, either data values or pointers to other lists, we can represent a branched structure of data values. Algorithms and routines are available for scanning through and analyzing a structure and the data values it contains. Usually storage is dynamically allocated by the list processor, since list structures and data often are constantly changing. The unpublished discussion includes specifications and listings of a program for computing inbreeding coefficients which I wrote using a list processing language, SLIP. I will here summarize its writeup.

Let me give a sharp warning first, however. Because computers are not built to be list processing machines, it must be simulated and hence list processing is often very time consuming. What you gain in simplicity and computing power you pay for in computing dollars.

For some problems in genetics, it may be of value to have a method whereby a high speed digital computer could compute inbreeding coefficients of individuals given their pedigree. It would be quite difficult to write such a program in one of the very common algebraic languages such as FORTRAN and ALGOL because of the complex branching structure of the data which are handled. A list processing language, on the other hand, such as IPL, SLIP or LISP, is ideally suited since the data list structures allowed in these languages may easily be used to represent pedigrees and the arithmetic capabilities are sufficient for the simple computations involved. A listing and thorough description of just such a program written in SLIP (Symmetric List Processor) follows. SLIP is an especially convenient language for this problem since it consists of a set of subroutines which are imbedded in a FORTRAN system. In fact all but a few "primitive" routines are coded in FORTRAN. Thus anyone with a knowledge of FORTRAN may follow the SLIP code easily, and imbedding of SLIP in an existing FORTRAN facility is simply a matter of coding the primitive routines. For a discussion of SLIP and a listing of its FORTRAN routines refer to J. Weizenbaum's article "Symmetric List Processor" in Communications of the ACM for September 1963 (Vol. 6, No. 9, pp. 524+).

The program discussed below is a minimal one but is sufficient for computing the inbreeding coefficients for an autosomal locus with any pedigree. Optional features, such as printing of relationships of pairs of individuals and limiting to sex linkage, have been excluded. They may be added easily but for the purposes of this paper would only serve to obscure the explanation of the program's algorithm.

INPUT The program reads in (in format [I3]) the number N of individuals and (in format [2I3, F10.0]), the three arrays IA(I), IB(I) and BIO(I), I = 1, N. For J fixed between 1 and N, IA(J), IB(J) and BIO(J) completely describe individual J. IA(J) is the integer designating the individual who is one parent (referred to hereafter as the left parent) of individual J and IB(J) designates the other parent (referred to as the right parent). The integer 0 for either one or both indicates that that parent is not in the pedigree. BIO(J) is a floating point value which is the inbreeding coefficient of the value preset by the user from information contained outside of the pedigree.

INITIALIZATION The array KK(I), I = 2, N + 1 is preset to all zeros and KK(1) is set to 1. This array is used in the course of the program to record which individuals have been analyzed and which have not. In particular KK(J+1) is non-zero if and only if the inbreeding coefficient of individual J has been established. By the nature of the method of computing the inbreeding coefficient used by this program, it is necessary that an individual J not be processed until both of his parents have been processed. Thus no attempt is made to analyze the paths to individual J until KK(IA[J] +1) and KK(IB[J] +1) are both non-zero. (Note that KK(1) is preset to 1 so that a parent designated by zero is considered to have been processed.)

METHOD A DO loop scans for and analyzes those individuals which are ready to be processed. The variable ALL is established to stop the scanning procedure when all the individuals of the pedigree have been processed.

The processing of an individual is straightforward. Suppose that the inbreeding coefficient of individual I is being computed. Then three lists are established with names LIST(I), MIST(I) and NIST(I). I is immediately placed on the attribute value list of NIST(I). (In general, K will be on the attribute value list of NIST(I) if and only if individual K is an ancestor of individual I or K=I). These three lists are left empty and the inbreeding coefficient is left at its preset value if K has no parents in the pedigree (if IA(I) = IB(I) = 0). If K has at least one parent in the pedigree, the LIST(I) is the name of a two-celled list containing in its top cell the integer IA(I) designating the first parent and in its second cell a branch pointing to NIST (IA(I)). If there is no second parent in the pedigree, MIST(I) is empty, NIST(I) is identical to LIST(I) and the inbreeding coefficient is left at its preset value. If, though, the other parent is in the pedigree, MIST(I) is the name of a two-celled list, the top cell containing the integer IB(I) and the bottom containing a branch cell pointing to NIST(IB(I)). NIST(I) in this case becomes the name of a 4-celled list, the top two cells identical to MIST(I). Thus for each individual I, LIST(I) names a list structure defining the branching pedigree structure above and including the first parent, MIST(I) that above and including the second parent, and NIST(I) names a list structure which specifies the branching pedigree structure above and including both parents. An example at this point may be helpful. Consider the pedigree

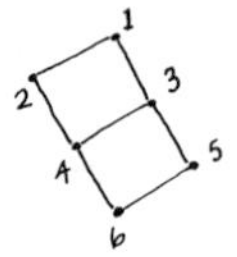

Corresponding to this we make Table 1.

Table 1

I	IA(I)	IB(I)	LIST(I)	MIST(I)	NIST(I)
1	0	0	()	()	()
2	1	0	(1())	()	(1())
3	1	0	(1())	()	(1())
4	2	3	(2(1()))	(3(1()))	(2(1())3(1()))
5	3	0	(3(1()))	()	(3(1()))
6	4	5	(4(2(1())3(1())))	(5(3(1())))	(4(2(1())3(1()))5(3(1())))

PATH DETERMINATION To determine all the paths required to compute the inbreeding coefficient of an individual I, the program uses readers to follow paths into the list structures.

First we note that the attribute value list of the list NIST (IB(I)) contains a tabulation of all the ancestors of I on his right (second parent's) side.

A structural reader KL is sent up through the left side of I, that is through the list structure LIST(I). Whenever it comes upon an individual, say K, that is on the attribute value list of NIST(IB(I)) and hence is an ancestor on the right side, the reader KL pauses and another reader is sent up the right hand side of I to search for paths to K independent of KL as follows:

1) A copy KLL of reader KL is backtracked to determine the path of KL leading from I to K. A tabulation of the individuals in this path excluding I and K is placed in the attribute-value list of a temporary list KEMP. KLL is erased.

2) A reader LL is sent up the right side of I, through MIST(I). If it should reach K, the contribution of the combined paths of the reader KL and LL are added to the inbreeding coefficient (and in this version of the program, subroutine COMPUTE is called to print out this contribution and the path). If, on the other hand, LL reaches an individual other than K, say J, a check is made to see if J is in the left path (made by KL) by checking to see if J is an attribute on the description list of KEMP. If it is not and if K is an ancestor of J (checked by looking in description list of NIST(J)), the reader LL is allowed to continue. Otherwise, however, LL is backtracked and started up a different branch.

Thus in the following situation

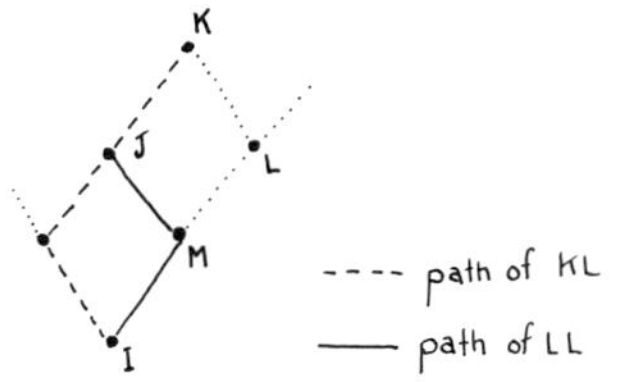

The reader LL is backtracked to M and sent up the branch to L.

3) When the reader LL comes to the end of the list MIST(I), all the inbreeding paths on the right side from I to J not intersecting the path of reader KL will have been found. At this time, control returns to the reader KL which continues its traversal of the list structure LIST(I). When KL comes to the end of LIST(I), all inbreeding paths of individual I will have been found and analyzed. The readers are then erased and the DO loop on I is continued.

When all individuals have been processed, a summary of the results is printed out.

SUBROUTINE COMPUTE This routine determines the contribution and individuals of an inbreeding path and prints them out. It does this by backtracking the two structural list readers (which are its arguments) and tabulating a list of the individuals (in order) in each path. The inbreeding coefficient contribution of the path is computed by referencing the inbreeding coefficient (of the common ancestor) from the array in stock common number 2. This contribution and a list of the path (with a minus preceeding the common ancestor) are then printed out.

A. W. F. EDWARDS: The long, complex, and not invariably general programs for computing the inbreeding coefficient F can be by-passed by appealing to Cotterman's concept of genes identical by descent, and simulating gene-flow through the genealogy.

This simple procedure has the further merit of enabling any pairwise coefficients, such as the coefficients of the additive and dominance components of genetic variance in the covariance of two individuals, to be found simultaneously. The disadvantages of simulation are the time consumed, and the approximate values obtained for the coefficients. Neither of these is critical, for the calculation need only be done once for each population, and there seems little point in having exact values in practice.

I have been developing a series of programs for studying simulated gene flow in genealogies, and one of these, SIMKIN, was readily adapted for finding the required coefficients.

The genealogy, defined by listing the parents of each individual, is first subjected to a preliminary ordering by the program RENUM, which orders the individuals so that each is preceded, though not usually immediately, by his parents in the new list. Those individuals who are founder-members of the genealogy head the list, and SIMKIN allocates two numbered genes to each, so that if there are N founders, genes numbered 1 to 2N are allocated. Straightforward mendelian gene flow is then repeatedly simulated, and gene correspondences accumulated to give the desired coefficients.

Thus, if two individuals possess genes A, B and C, D, the inbreeding coefficient is defined as $P(A \equiv B)$ for the first individual, and $P(C \equiv D)$ for the second. The genetic correlation is $1/2(P\ (A \equiv C) + P\ (A \equiv D) + P\ (B \equiv C) + P\ (B \equiv D))$ and the coefficient of the dominance component is $(P\ (A \equiv C,\ B \equiv D) + P(A \equiv D,\ B \equiv C))$, (see Kempthorne, 1957). Since not all pairwise coefficients are usually required, SIMKIN only prints those requested.

The program has been tested on a genealogy comprising 60 individuals constructed to exhibit various interesting relationships, and all the simpler relationships were elucidated satisfactorily. Thus for sib-sib the two coefficients were 0.5000 and 0.2505, and for grandfather-grandson the correlation was 0.2500. For more complex relationships there was some indication of systematic discrepancies presumably arising through complex serial interactions, possibly of high order, in the generated sequence of random numbers. Thus for sib-sib when the parents are double first cousins, we find coefficients of 0.6286 and 0.2908, whereas the true values, computed by Dr. Cotterman, are 5/8 (= 0.6250) and 37/128 (= 0.2891). That the discrepancies are systematic is shown by the fact that in the former case on each of 24 runs the estimated coefficient exceeded the true value, and in the latter case this occurred on 22 out of 24 runs. The absolute discrepancies are not, however, large, and with more careful random-number generation should not be important.

The simplicity of the program, the generality of the method, and the elucidation of additional coefficients seem to me to outweigh the inherent disadvantages.

Kempthorne, O. 1957. An Introduction to Genetic Statistics, Page 332. Wiley, New York.

COCKERHAM: Dr. Mange's method and program are very straightforward and well documented. He points out the types of errors that are made, which should not be serious in human populations, but may be more serious for other populations. I am not sure what one does with the expansive output of the program. It does list the common ancestors and the number of links for each which is more appropriate for figuring the genetic load for lethal mutations than just the inbreeding coefficient, although in most applications they come out to be very similar. One comparison of interest would be between the coancestries of mates and non-mates. It is not clear how to accomplish this comparison with overlapping generations.

This is a retrospective type program, that is, it starts with individuals of interest and figures back through the ancestors. It is limited in two ways, one being the number of ancestral generations and the other being the total number of ancestors. The limitation in terms of ancestral generations is probably not serious for data available on human populations, but would be for some others. The limitation as to the total number of ancestors would seem to limit the utility of the program to isolate groups.

There is another type of program which is prospective in that one starts with some array of initial individuals and works forward maintaining the coancestries of relevant individuals which leave offspring. This is a very simple process with discrete generations and there should be some way to handle it in terms of overlapping generations. I have no idea about the relative efficiencies of retrospective or prospective procedures or of different methods of programming for the calculation of inbreeding and coancestry, and should like to hear these discussed.

Further, I wonder if pedigree inbreeding is ever sufficiently high in humans to be useful in the analysis of variation.

COMPUTER ANALYSIS OF PEDIGREE DATA

Charles MacLean

National Institute of Dental Research
National Institutes of Health
Bethesda, Maryland 20014

This paper describes a computer program package entitled "Population Searching Program" (P.S.P.), which performs several types of analysis of pedigree data on bisexual organisms. The package includes five basic kinds of calculations:

1. PEDIGREE prints out the entire pedigree for any individual(s) in the population.
2. COEF calculates the coefficient of inbreeding of every member of the population. Either the autosomal or sex linked coefficient can be requested. All common ancestors which contribute to inbreeding may be listed.
3. COEFREL calculates the coefficient of relationship between requested pairs of individuals. This may also be autosomal or sex linked, and the common ancestors can be listed.
4. CLONE traces all the descendants of each or any of a set of individuals requested and lists them together with their parents, sorted into generations.
5. RELATIVES traces all the relatives of any set of individuals and calculates their coefficients of relationship. Common ancestors may be listed.

Any combination of the above five subroutines can be selected by the user through control cards entered with the population data. Complex analyses can be carried out by special use of the subprograms. For example, a population could be sorted into subpopulations by using the coefficient of relationship as a distance function, or the CLONE routine could be used to determine the percentage contribution to the gene pool in any generation for any arbitrary group of "founders."

P.S.P. is written in Fortran and although it can be used by any computer which uses that language, the entire package is best suited to a large, rapid computer. It runs presently on an IBM 360. In addition COEF, the coefficients of inbreeding subprogram, has been designed to run separately on smaller machines such as the IBM 1620.

Part I of this paper is a user's guide which describes input, output and program specifications. Part II is concerned with certain principles of the program design of P.S.P. and is directed at computer programmers.

USER'S GUIDE

Inputs

Data input to P.S.P. consists of one card for every biological sibship in the population. A "biological sibship" consists of all individuals who have the same father and mother (i.e., full sibs) and "population" is used here to denote the entire historical population, past and present. The sibships are numbered from 1 to N (the total number of sibships in the population) and put into the computer in numerical order.

Each card must contain the number of the sibship and the sibship numbers of both parents. Each parent's sibship number must be one of the N sibship numbers. When parents are unknown, their sibship numbers are entered as blanks. All sibships in which both parents are unknown may be combined under a single dummy number if the machine size capacity is a consideration.

When an individual is a parent of more than one sibship, all his subsequent sibships are referenced to his first. This reference is called a link. His first family has no link; it is only his subsequent sibships which are marked.

Thus, each input card must contain the sibship number of both parents, the links (if any) to other sibships having the same father or the same mother, and the number of the sibship itself. The exact format is open to the user and is specified in statement No. 101. The order of the five required items is established by read statement No. 100. The end of data is denoted by a trip card with 9999 in the sibship number.

The data cards are followed by control cards which specify the subroutines to be used and which designate the sibships on which the calculations are to be performed. The first control card must begin in column 1 with any one of the five subprogram titles: 1-PEDIGREE, 2-COEF, 3-COEFREL, 4-CLONE or 5-RELATIVES. If sex-linked genes are to be considered, 1-SEX LINKED is punched in columns 20-31. If common ancestors are to be listed out, 1-COM-ANC is punched in columns 40-48. The PEDIGREE, CLONE, and RELATIVES cards must be followed by cards designating the sibships for which the calculations are to be performed, one sibship per card, in columns 1-5. The COEFREL card must be followed by cards designating the pairs of sibships for which the calculation is to be performed, one pair per card, in columns 1-5 and 6-10.

Except for COEF, which has no following cards, the end of a set of control cards is denoted by a trip card with 9999 punched in columns 2-5. Another control card set may follow. The final control card set is followed by a trip card with 9999 punched in columns 76-79.

Recoding

Population data are almost always prepared according to schemes different from that required by P.S.P. Recoding large amounts of data can be an annoying and time-consuming clerical job, but it is very simple on a computer or electronic accounting machine (EAM). The procedure described below can be applied to any data format by a computer programmer or EAM operator.

- Step 1: Code all individuals in some format related explicitly to their parents.
- Step 2: Reduce to an ordered population of sibships. This can be accomplished on a collator or computer. If the old code has a sibship number imbedded in it, sort by the sibship portion of the code and select the first card after a change in sibship number. If each individual has a code completely independent from that of his siblings, sort by parents' codes, and select the first card after a change in parents.
- Step 3: Remove all sibships with unknown parents, if desired.
- Step 4: Renumber the sibships 1 to N, in this order.
- Step 5: Look up and record the new number of each parent. In a computer this is done by a search through the old codes. On EAM equipment, a new deck is reproduced, sorted by father's old code, merged behind the original deck on a collator, and the new numbers gang-punched into the father's position of the new deck. The new deck is then separated from the original, re-sorted by mother's old code, merged and gang-punched.

Step 6: Mark the links. This is usually a small enough task to be done by hand with the aid of two listings of the data; one sorted by father's old code, one by mother's. In this order multiple families of a parent are contiguous and easy to note.

Outputs

PEDIGREE. The output from PEDIGREE consists of a list of the code numbers of the ancestors of a requested individual in the following order:

1. Propositus
2. Father
3. Mother
4. Paternal grandfather
5. Paternal grandmother
6. Maternal grandfather
7. Maternal grandmother
8. Paternal-paternal great grandfather
9. Paternal-paternal great grandmother
10. Paternal-maternal great grandfather
11. Paternal-maternal great grandmother
12. Maternal-maternal great grandfather

etc.

The organization of this list is discussed in Part II.

COEF. The output from COEF consists of a list of the sibship numbers from 1 to N, each followed by its inbreeding coefficient. The list is headed "SEX-LINKED" if that option is requested. If common ancestors are printed, each line has the format:

PR	=	XXXX	propositus' identification
CA	=	XXXX	common ancestor's identification
FP	=	XXXX	common ancestor's pedigree position through the father's line
MP	=	XXXX	pedigree position through mother's line
COEF	=	O.XXXXXXXX	contribution to the propositus' inbreeding coefficient

COEFREL. Coefficients of relationship are printed one per line, with the identifications of the two sibships followed by their coefficient of relationship.

CLONE. The format of each line of the CLONE output is:

XXXX	sibship
XXXX	parents' sibship
FA or MO	descended through father or mother

Each generation is headed by a line GENERATION XX.

RELATIVES. Output from RELATIVES consists of a list, in the order of input, of all members of the population who have a nonzero coefficient of relationship to the propositus, together with the coefficient. Common ancestors may also be printed according to the format under COEF. If the user selects several propositi they are presented sequentially.

Program Specifications

P.S.P. should be adaptable to any computing system with a Fortran compiler. No tape or disk storage is used. The specifications for P.S.P. on the computers used at NIH should indicate approximate characteristics on similar equipment.

360 P.S.P. The program, written in Fortran II, runs currently on the IBM 360-50 with 50K of memory available for the program. Size restrictions are as follows: (1) 10,000 sibships whose parents are known. This may represent any number of individuals. Any number of sibships with unknown parents in addition to these may be included. (2) The pedigrees may span at most ten generations. If more generations exist for an individual, the earliest are terminated.

PROGRAM DESIGN

The following discussion is directed to computer programmers who are interested in the techniques of pedigree manipulation or in the use of list processing, or who might wish to modify the P.S.P. code. Three subjects are discussed: the use of a binary code for pedigree representation, the location of common ancestors by the technique of list processing, and certain aspects of P.S.P. program code which might be especially confusing.

Pedigree Representation

P.S.P. represents familial relationships by a pedigree "tree" which has two branches at each node and no circuits. A common ancestor is expressed by multiple entry of his identification number, once in each ancestral position which he occupies, rather than by rejoining family lines as in the more familiar pedigree style.

The graphical representation of a pedigree is replaced in a computer by tables and code numbers. There is a natural correspondence between the tree representation and the binary number system, which P.S.P. exploits to construct a Pedigree Table in which each ancestor has a binary code number which exactly specifies the position he occupies in the graphical pedigree. This binary representation greatly expedites programming. The pedigree positions are numbered from left to right across each generation and continued from the most recent to the most ancient generation as shown in Figure 1. Each male ancestral position is assigned an even number, his mate the following odd number.

The Pedigree Table corresponding to Figure 1 is given in Table 1. The coded positions and relationships, e.g., 110 Binary = 6 = maternal grandmother, are the framework of the table itself. It is by assigning particular names from the Population Table to these coded positions that a specific pedigree is constructed. The "name" column of Table 1 constitutes a specific pedigree.

The most important feature of the binary code is the ease with which it is translated into decimal numbers and adapted to the standard indexing techniques of computer programming. In the Pedigree Table, the decimal equivalent of each binary code number specifies the entry number, i.e., the table index, of the corresponding individual. In other words, each ancestor's index in the Pedigree Table defines his binary code uniquely which in turn defines his position in the graphical pedigree.

Simple arithmetic rules are used to calculate the codes for any relationships. For example, the code of an individual's mother in binary notation is the individual's own position number with a 1 appended to the right hand side, i.e., twice his number plus one. The same calculation holds when translated into decimal numbers. The entire pedigree of any arbitrary position of any pedigree is given by such a rule. For example, in Figure 1, the pedigree of position number 2 (10 Binary) consists of 4 (100B), 5 (101B), 8 (1000B), 9 (1001B), 10 (1010B), 11 (1011B), 16 (10000B), etc. By this reasoning the maternal and paternal sides of the pedigree of any subpedigree are distinguished, a requirement in many P.S.P. operations.

Because the number of ancestors in the nth generation is 2^n and the number of ancestors below it is 2^{n-1}, every position in generation n has a binary code n + 1 binits long. This is the characteristic of the base 2 logarithm of their codes. The code for any ancestor can easily be adjusted to yield his code in the pedigree of any other ancestor. In the example above, 10 is the maternal grandfather of 2. In 2's pedigree he would be coded 6. The adjustment is, in general, $B = A - (C-1)\ 2^D$ where C is the code to be adjusted to 1 (2 in the example above), A is the old code of his ancestor (10 above), B is A's new code (6 above), and D is the generation number of A minus the generation number of C (2 = 3 - 1 above).

There is a complication inherent in using a tree pedigree, in which a common ancestor appears in two different places rather than having the lines loop back to the same node. In the tree pedigree the ancestors of a repeated ancestor also appear twice and, in this role, they represent no new information. Considering them multiply is avoided by a simple index operation.

Calculations Based Upon Common Ancestors

One of the functions of P.S.P. is to calculate the coefficient of inbreeding (F) devised by Sewall Wright (Wright, 1922). It is most conveniently considered (Malecot, 1948) as a measure of the probability of genes identical by descent from common ancestors, i.e., ancestors who appear on both the mother's side and father's side of the pedigree.

The calculation performed is

$$F = \sum_{A} 2^{-n_1-n_2+1}$$

where n_1 and n_2 are the number of generations between the common ancestor and the propositus through the father's line and the mother's line, respectively, and the summation is over all common ancestors. One individual may be a common ancestor by several different family lines and is considered as a different common ancestor in each line.

The coefficient of relationship (r_{AB}) between any two individuals (say A and B) is calculated by first establishing a dummy individual whose parents are assumed to be the individuals A and B, and then calculating the inbreeding coefficient (F_D) of this dummy. Then, the coefficient of relationship is given by (Wright, 1923)

$$r_{AB} = \frac{2F_D}{\sqrt{(1 + F_A)(1 + F_B)}}$$

The coefficient for sex-linked genes is complicated by the presence of father-son relationships which exterminate the loop for sex-linked genes, and father-daughter relationships which make transmission of the sex-linked gene certain and thereby have the mathematical effect of reducing the length of the loop by one generation.

Lists and Common Ancestors

Finding common ancestors is the most difficult problem in inbreeding calculations, and it is on this point that P.S.P. differs from programs previously written. Because of the great variety of possible pedigrees, in order to be sure to find all common ancestors, almost everybody on the paternal side of the pedigree must be compared with almost everybody on the maternal side. A single search through a ten generation pedigree (2,048 positions), using no common sense at all, would require 1,024 x 1,024 = about one million comparisons. Sex can be used to cut the number in

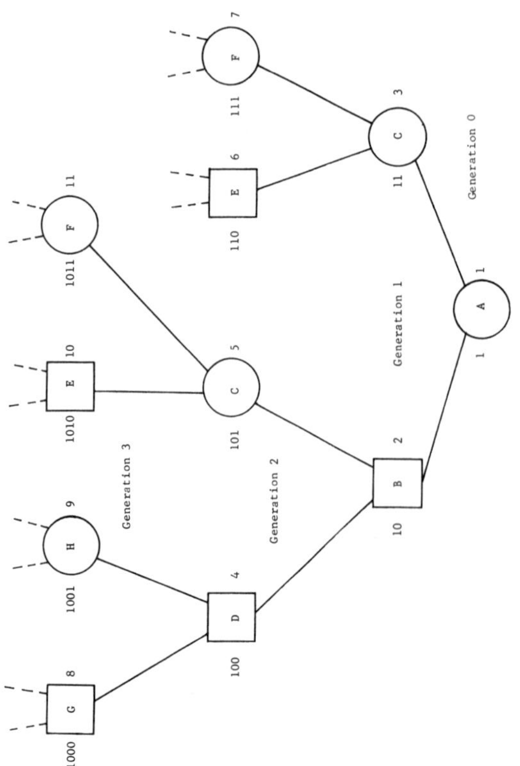

Figure 1. The pedigree is represented in Table 1 by the numbers associated with the positions shown here with binary numbers to the left, their decimal equivalents to the right. The names in the boxes constitute a specific example of a well-known human pedigree. In it A = Antigone, B = Oedipus, C = The Queen, D = The Deposed King, etc.

TABLE 1: Pedigree Table

Binary Code	Decimal Equivalent	Generation Number	Ancestral Position	Ancestor's Name (example)
1	1	0	propositus	A
10	2	1	father	B
11	3	1	mother	C
100	4	2	paternal grandfather	D
101	5	2	paternal grandmother	C
110	6	2	maternal grandfather	E
111	7	2	maternal grandmother	F
1000	8	3	paternal-paternal great grandfather	G
1001	9	3	paternal-paternal great grandmother	H
1010	10	3	paternal-maternal great grandfather	E
1011	11	3	paternal-maternal great grandmother	F
.	.	.	. . . .	.
.	.	.	. . . .	.
.	.	.	. . . .	.

The ancestors' names refer to the pedigree in Figure 1.

TABLE 2: Example of Pedigree List Structure

Population Table				Pedigree Table		
Name	Father	Mother	First Pedigree Link	Position	Name	Link
A	B	C	1	1	A	blank
B	D	C	2	2	B	blank
C	E	F	3	3	C	5
D	G	H	4	4	D	blank
E	..	..	5	5	C	blank
.	.	.	.	6	E	10
.	.	.	.	7	F	11
.	.	.	.	8	G	.
.	.	.	.	9	H	.
.	.	.	.	10	E	.
.	.	.	.	11	F	.
.	.	.	.	.	.	.
.	.	.	.	.	.	.

This is the same example as used in Table 1.

half, and some further improvement can be made by various tricks, but the effort remains enormous in the best of cases. P.S.P. avoids searching through the pedigree by a method called list processing.

List processing is a powerful programming technique which was developed for use in compilers, retrieval systems, etc. (Newell, 1961). In its simplest form, a list is a table each entry of which contains the name of the next entry as part of its information. This device, called linking, makes physical order of the table unnecessary, and thereby allows for different orderings of material simultaneously. In P.S.P. the pedigree is ordered physically by pedigree position, but it is also ordered according to population identity by linking.

The specific pedigrees are created successively from the permanent portion of the Population Table by recording parents and then their parents, etc., in the Pedigree Table. A linking system which includes entries in both the Population and Pedigree Tables is established for each pedigree while it is being constructed. Table 2 shows the list structure of the pedigree in Table 1. All the links in both tables are blank at the beginning of the construction of a pedigree. When an individual is specified for his first appearance as an ancestor in a pedigree, a link to this pedigree position is recorded in his Population Table entry. For example, in Table 2 when B and C are entered into the father's and mother's positions, their first links are marked 2 and 3 respectively. If an individual has been entered previously, a link is recorded in his last pedigree position. For example, when C is entered as B's mother, her first pedigree position, 3, is linked to 5. At the completion of a pedigree, a chain of links extends from each individual's population entry to his first pedigree position (if any) and from his first pedigree position to his next pedigree position and so on to the last pedigree position he occupies in which a blank link indicates the end of his list.

During the search for common ancestors, to find all positions that an individual occupies in the pedigree, P.S.P. simply yanks on the chain of links which starts in his population entry. This requires no searching through the pedigree and is, therefore, literally hundreds of times faster. Whether these pedigree positions are on the father's side or the mother's side of the pedigree is determined by arithmetic on the pedigree indexes as described above.

Special Computer Economies

To facilitate an economic computer program, several important expedients have been used which are of no interest theoretically, but which must be understood by a programmer who wishes to modify P.S.P.

Simply to save storage space, the Population Table does not have an entry for each individual; it has only one entry for each sibship. This table constitutes nearly all the storage requirement of the program, so that economy in it is crucial. To distinguish an individual from his siblings who share his Population Table identification, he is referred to not by his own sibship but by the sibship of which he is a parent. This is convenient in pedigree analysis since his children's sibship number is stored in the pedigree entry which has an index one-half his own.

During pedigree construction, when siblings occur in the same pedigree their links are established from their progeny which, of course, are different. When they themselves are used as the basis for placing their parents into the pedigree, their identical sibship number leads to the same parents twice. No further connection between siblings is of interest. During the search for common ancestors, an individual is recognized only by the links that were established as he was put into the pedigree in more than one place and not on the basis of his code number. It is his children's sibship number under which his first link is stored, and the subsequent links depend only on the links before. This expedient is practicable except when an individual has children by more than one mate. In this case a special input is required to distinguish him from his siblings.

Index calculation is used in place of data movement wherever possible. For example, once a pedigree is constructed, its branches reach out until the knowledge of ancestry fails. Therefore, one pedigree automatically includes the full pedigrees of the father, mother, grandparents, and so on. Data has to be assorted into only this one pedigree in order to calculate the inbreeding coefficients of all the ancestors. Since a pedigree of ten generations has up to two thousand individuals in it, separate pedigrees are constructed for only a small proportion of the population.

NOTE:

3600 COEF. The COEF program written in CDC Fortran IV presently runs on a CDC 3600 with 64K memory. This version packs several items into each word of storage, so that its use on a computer with a different word size would require a modification. The maximal population size allowed is 10,000 sibships spanning at most 12 generations.

Running time depends upon the number of common ancestors in the population. The human populations used by NIH had inbreeding at levels which yielded a calculation rate of about 10-20 sibships per second. One population of 15,000 people with a moderate amount of inbreeding was run in 9 minutes. Another of 50,000 people with a low amount of inbreeding (disclosed) ran in 20 minutes.

1620 COEF. A reduced COEF program, written in Fortran II, without the options is currently used on the IBM 1620 with 40K digits of memory and no auxiliary equipment. Population size is limited to 600 sibships spanning seven generations. The calculation rate is about 10 sibships per minute.

REFERENCES

Hoen, K. and Grandage, A. 1960. "Note: Calculation of inbreeding in family selection studies on the IBM 650 data processing machine." Biometrics 16(2): 292-296.

MacCluer, J. W. 1966. Description of a program to calculate the inbreeding coefficient by Kudo's method. Unpublished. University of Michigan, Department of Human Genetics.

Malécot, G. 1948. Les mathématiques de l'hérédité. Masson et Cie, Paris.

Mange, A. 1964. Fortran programs of computing Wright's coefficient of inbreeding in human and non-human pedigrees. Am. J. Hum. Genet. 16:484-485.

Newell, A. 1961. Information processing language-V. Prentice-Hall.

Rehfeld, C., Bacus, J., Pagels, J., and Dipert, M. 1967. Computer calculation of Wright's inbreeding coefficient. J. of Heredity 58(2): 81-84.

Wright, S. 1922. Coefficients of inbreeding and relationship. Am. Naturalist 56:330-338.

Wright, S. 1923. Mendelian analysis of the pure breeds of livestock I. The measurement of inbreeding and relationship. J. Heredity 14:339-348.

ESTIMATION OF THE INBREEDING COEFFICIENT FROM MATING TYPE FREQUENCY AND GENE FREQUENCY

Norikazu Yasuda

Department of Population Genetics
National Institute of Genetics
Mishima, Shizuoka-ken, Japan

The inbreeding coefficient may be defined as the probability of two genes being identical by descent from a common ancestor if two individuals *I* and *J* are mates, when one of the genes is drawn at random from *I* and the other is drawn from individual *J*. It is a special case of *the kinship coefficient* in which *I* and *J* are not necessarily mates (Malecot, 1967).

This definition of the inbreeding coefficient indicates that couples are useful for extracting information about the inbreeding coefficient. In human genetics it is of special interest to be able to estimate the coefficient from single generation data as has been done in estimating recombination frequency (Penrose, 1935) as well as in determining a mode of inheritance (Cotterman, 1937). A method of *phenotype bioassay*, by which the inbreeding coefficient is estimated as a deviation from Hardy-Weinberg proportion of individuals, however, is unsatisfactory since it gives very unstable estimates with large standard errors (Yasuda, 1968b). Alternatively, *mating bioassay*, based on mating type frequencies which have been described in terms of gene frequency and inbreeding coefficient (Yasuda, 1968a), may give more information (that is, smaller standard errors) about the inbreeding coefficient than phenotype bioassay does when the number of individuals involved is the same.

The purpose of the present technical report is not only to explore a method for estimating the inbreeding coefficient from mating type frequencies, but also to present a use of the electronic computer for estimating gene frequency by phenotype bioassay.

PRINCIPLES AND METHODS

The available data consist of a random sample of couples or pairs whose phenotypes are determined by a mode of monofactorial inheritance. Since the expected proportion of each pair is given in terms of probability, we can apply a method of maximum likelihood for estimating gene frequency (p) and the inbreeding coefficient (α). A summary of the method is given in Appendix 1.

Case 1. Autosomal genes, without dominance. Let *A* and *a* be two alleles at a locus, and p (= 1 - q) and q be the corresponding gene frequencies, respectively. Also let us suppose that mating types, observed numbers, and expected proportions are as follows:

mating type	observed number (n_i)	expected proportion (P_i)
AA x AA	n_1	$(1-q)^4 + 6q(1-q)^3\alpha$
AA x Aa	n_2	$4q(1-q)^3+12q(1-q)^2(2q-1)\alpha$
Aa x Aa	n_3	$4q^2(1-q)^2+4q(1-q)[1-6q(1-q)]\alpha$
AA x aa	n_4	$2q^2(1-q)^2+2q(1-q)[1-6q(1-q)]\alpha$
Aa x aa	n_5	$4q^3(1-q)+12q^2(1-2q)\alpha$
aa x aa	n_6	$q^4+6q^3(1-q)\alpha$

where we assumed that $p > |\alpha|$ and $q > |\alpha|$.

The log likelihood of observation is

$$L = \sum_{i=1}^{6} n_i \ln P_i,$$

and the maximum likelihood scores are

$$U_q = \sum_{i=1}^{6} n_i\left(\frac{\partial P_i}{\partial q}\right)/P_i \quad \text{and} \quad U_\alpha = \sum_{i=1}^{6} n_i\left(\frac{\partial P_i}{\partial \alpha}\right)/P_i.$$

The estimated variances of scores are

$$K_{qq} = \sum_{i=1}^{6} n_i\left(\frac{\partial P_i}{\partial q}\right)^2/P_i^2,$$

$$K_{q\alpha} = \sum_{i=1}^{6} n_i\left(\frac{\partial P_i}{\partial q}\right)\left(\frac{\partial P_i}{\partial \alpha}\right)/P_i^2, \quad \text{and}$$

$$K_{\alpha\alpha} = \sum_{i=1}^{6} n_i\left(\frac{\partial P_i}{\partial \alpha}\right)^2/P_i^2.$$

Under the null hypothesis that $\alpha = 0$, the expected scores and the variances are

$$U_q = [(2c+b)-2(a+b+c)q]/q(1-q)$$

$$U_\alpha = \frac{3(a+b+c)[(4ac-b^2)+\frac{4}{3}(a+b+c)(n_3-2n_4)]}{(2a+b)(b+2c)}$$

and

$$K_{qq} = 2(a+b+c)/q(1-q),$$

$$K_{q\alpha} = 0$$

and $K_{\alpha\alpha} = 3(a+b+c)$,

respectively, where

a $(=2n_1+n_2+n_4)$, b $(=n_2+2n_3+n_5)$ and c $(=n_4+n_5+2n_6)$

correspond to the observed numbers of genotypes *AA*, *Aa* and *aa*, respectively. The maximum likelihood estimate for gene frequency is obtained from $U_q = 0$, or $q = (2c+b)/2(a+b+c)$. The results are remarkable: the amount of information about the inbreeding coefficient is three times as large in mating bioassay as in phenotype bioassay which yields $K_{\alpha\alpha} = (a+b+c)$ (Morton and Yasuda, 1962), indicating that mating types are yielding more information about the inbreeding coefficient. On the other hand, no improvement is gained from the information about gene frequency, denoting that gene frequency estimation is sufficiently made from a sample of individual phenotypes.

The first approximation to α is

$$\alpha_1 = U_\alpha/K_{\alpha\alpha} = [3(4ac-b^2)+4(a+b+c)(n_3-2n_4)]/(2a+b)(b+2c)$$

where the subscript of α indicates the number of iteration. The null hypothesis that $\alpha = 0$ is tested by $\chi^2 = U_\alpha^2/K_{\alpha\alpha}$ with one degree of freedom. If the chi-square becomes significant, then the iterative processes may follow as

$$\alpha_2 = \alpha_1 + (U_\alpha K^{\alpha\alpha} + U_q K^{\alpha q})$$

where *K*'s are elements of estimated covariance matrix when $\alpha = \alpha_1$.

Case 2. Autosomal genes with dominance. Suppose that *A* is dominant to *a*. Mating types, observed numbers and expected frequencies are given as follows:

mating type	observed number(n_i)	expected frequency (P_i)
A- x A-	n_1	$(1-q^2)^2-2q(1-q)(1-3q^2)\alpha$
A- x aa	n_2	$2q^2(1-q^2)+2q(1-q)(1-6q^2)\alpha$
aa x aa	n_3	$q^4+6q^3(1-q)\alpha$

provided that q and $(1-q)$ are greater than $|\alpha|$.

In order either to test the null hypothesis that $\alpha = 0$ or to estimate the inbreeding coefficient, we evaluate the scores and the expected variances at $\alpha = 0$. Namely,

$$U_q = 2[b-(a+b)q^2]/q(1-q^2)$$

and

$$U_\alpha = 2(4n_1n_3-n_2^2)/\ q(1+q)(2n_1+n_2),$$

and the information matrix is given by

$$K = \begin{bmatrix} K_{qq} & K_{q\alpha} \\ K_{q\alpha} & K_{\alpha\alpha} \end{bmatrix}$$

in which

$$K_{qq} = 4(a+b)/(1-q^2),$$

$$K_{q\alpha} = 2(a+b)/(1+q)$$

and

$$K_{\alpha\alpha} = (a+b)(1+7q^2)/(1-q)^2,$$

where $a = 2n_1+n_2$, and $b = n_2+2n_3$ are the observed numbers of A- and aa phenotypes, respectively. The covariance matrix is then

$$K^{-1} = \begin{bmatrix} K^{qq} & K^{q\alpha} \\ K^{q\alpha} & K^{\alpha\alpha} \end{bmatrix}$$

where

$$K^{qq} = (1+7q^2)(1-q^2)/32(a+b)q^2,$$

$$K^{q\alpha} = -(1-q^2)/16(a+b)q^2$$

and

$$K^{\alpha\alpha} = (1+q)^2/8(a+b)q^2.$$

The null hypothesis that $\alpha = 0$ can be tested by

$$\chi^2 = U_\alpha^2 K^{\alpha\alpha} = \frac{(n_1+n_2+n_3)(4n_1n_3-n_2^2)^2}{(2n_1+n_2)^2(n_2+2n_3)^2}$$

with one degree of freedom. The estimated gene frequency from

$$U_q = 0 \text{ is } q = \sqrt{b/(a+b)}$$

so that a first approximation to α is, if desired, $\alpha_1 = U_\alpha K^{\alpha\alpha}$ with the amount of information $I_\alpha = 1/K^{\alpha\alpha} = 8(a+b)[q/(1+q)]^2$.

Table 1. Estimation of the inbreeding coefficient and gene frequency in the MN blood group system from northeastern Brazil (without dominance)

Phenotype Bioassay

phenotype	observed number	expected number ($\alpha = 0$)	expected number ($\alpha = 0.0027$)
M	648	646.58	648.00
MN	1050	1052.84	1050.00
N	430	428.58	430.00
total	2128	2128.00	2128.00
		$\chi^2 = 0.02$	
$U_\alpha = 5.7274$		$\alpha = 0.0027 \pm 0.0217$	
$U_q = 0.$		$q = 0.4488 \pm 0.0076$	
$K_{\alpha\alpha} = 2128$		$K^{\alpha\alpha} = 0.00046992$	
$K_{\alpha q} = 0$		$K^{\alpha q} = 0.$	
$K_{qq} = 17204$		$K^{qq} = 0.00005812$	

Table 1 -- continued

Mating Bioassay

mating type	observed number	expected number ($\alpha = 0$)	expected number ($\alpha = 0.0065$)
M x M	98	98.21	101.33
M x MN	326	319.88	318.72
MN x MN	258	260.45	257.14
M x N	126	130.23	128.57
MN x N	208	212.06	213.01
N x N	48	43.17	45.23
total	1064	1064.00	1064.00
		$\chi^2 = 0.90$ (df= 4)	$\chi^2 = 0.71$ (df= 3)
$U_\alpha = 41.4367$		$\alpha = 0.0065 \pm 0.0125$	
$U_q = 0.0000$		$q = 0.4488 \pm 0.0076$	
$K_{\alpha\alpha} = 6384$		$K^{\alpha\alpha} = 0.00015664$	
$K_{\alpha q} = 0$		$K^{\alpha q} = 0.$	
$K_{qq} = 17204$		$K^{qq} = 0.00005812$	

Although the information about α is small, it is possible to test the null hypothesis that $\alpha = 0$ or to estimate the inbreeding coefficient (or both), while phenotype bioassay cannot do either (Morton and Yasuda, 1962).

Case 3. Sex-linked genes. Since we are little interested in the inbreeding coefficient at a sex-linked locus, no detailed discussion will be given here, but some results are summarized in Appendix 2. One should again notice that in mating bioassay the amount of information about the inbreeding coefficient is larger than that in phenotype bioassay. The gene frequency can be obtained as the solution of $U_q = 0$. However, it should be mentioned that if a population is divided into two groups by sex, phenotype bioassay may be used with a few modifications in which case two situations may occur. If the gene frequency is different in each sex, we apply gene counting for males and the ordinary phenotype bioassay for females. While under the equilibrium where gene frequency is equal in both sexes, a modified gene counting may be useful. Following the notations of Appendix 2, gene frequency is obtained in cases without dominance as

$$q = w\,(2f_3 + f_2)\ /\ 2F + (1-w)m_2\ /\ M$$

or $q = (f_2+2f_3+m_2)/(2F+M)$, and in cases with dominance as

$$q = w[(1-h)f_1+f_2]/2F + (1-w)m_2/M$$

in which $w = 2F/(2F+M)$ is the proportion of genes possessed by females, and $h = (1-q)/(1+q)$ is the probability that a female is an homozygote within the dominant phenotype. A close value of h can be obtained from gene frequency of females estimated by phenotype bioassay so that the population gene frequency is estimated by iterations. In the present case we have the analytic solution of maximum likelihood equation,

$$q = [-m_1 + \sqrt{(m_1^2 + 4(2f_2 + m_2)(2F+M))}\ ]/2(2F+M).$$

Numerical examples. Morton(1964) studied the MN blood groups in a population from Brazil. The data consist of a random sample of couples so that mating bioassay can be applied, whereas phenotype bioassay treats a random sample of individuals that is obtained by ignoring matrimonial relationship. For a case of the system without dominance, the inbreeding coefficients were estimated by both phenotype and mating bioassay (Table 1). For a case of dominance, we assumed that a single serum anti-M was used (Table 2). If the amount of information is compared, there is no doubt that mating bioassay in the system without dominance is the best procedure for estimating the inbreeding coefficient, while phenotype bioassay is sufficient for gene frequency estimation (Table 3). In general, we have a conjecture that the system without dominance (codominant system) affords the maximum information about the inbreeding coefficient by mating bioassay and about the gene frequency for phenotype bioassay (Yasuda, 1968b).

FACTOR-UNION SYSTEM

With increase of the number of alleles, it becomes more complicated to obtain an analytic form of maximum likelihood scores and the variances. A method of gene counting (Ceppellini et al., 1955; Yasuda and Kimura, 1968; Yasuda, 1968c) is useful only for estimating gene frequency. Before we adapt the method of maximum likelihood scoring (Fisher, 1922; Stevens, 1938) by which both gene frequency and the inbreeding coefficient can be estimated, a concept of factor-union algebra should be introduced for grouping genotypes whose phenotype is the same.

Let us consider the A_1A_2BO blood group system for an explanation of the factor-union algebra. There are four main alleles A_1, A_2, B and O at the locus, and A_1 is dominant to A_2 and O; A_2 to O; B to O; but B is codominant with A_1 and A_2. (In addition, genotypes AB and OO are, of course, distinct from each other.) The alleles are detected by reactions with the corresponding sera: A_1 by anti-A_1 and -A; A_2 by anti-A; B by anti-B; and O by no agglutination with any sera. If we assign 1 for positive and 0 for negative reaction (these numbers are called factors) the alleles at the locus are then characterized by an array of factors which has the name of gene vector:

$$A_1 = 110,\ A_2 = 010,\ B = 001 \text{ and } O = 000,$$

where the order of factors is conventional and is anti-A_1, -A and -B in the example.

Since man is diploid, so that two genes are responsible for the phenotype of each individual (except sex-linked characters in the male), the combination of gene vectors should be performed with a logical union of factors, or factor-union algebra (Cotterman, this conference); namely,

$$0 + 0 = 0$$
$$0 + 1 = 1 + 0 = 1$$
and
$$1 + 1 = 1.$$

The additive operator "+" is actually union, one of the binary operators in Boolean algebra (Birkhoff and MacLane, 1965). The genotype vectors are then

$$A_1A_1 = 110 + 110 = 1{+}1\ 1{+}1\ 0{+}0 = 110$$
$$A_1A_2 = 110 + 010 = 110$$
....
$$OO = 000 + 000 = 000.$$

Six of ten genotype vectors are distinct as phenotype vectors;

phenotype vector	genotype
110	A_1A_1, A_1A_2, A_1O
010	A_2A_2, A_2O
001	BB, BO

Table 2. Estimation of the inbreeding coefficient and gene frequency in the MN blood group system from northeastern Brazil (with dominance, assuming that only anti-M was used)

Phenotype Bioassay

phenotype	observed number	expected number (α = 0)	
			$U_q = 0.$
M(+)	1698	1698.00	$K_{qq} = 10668.$
M(-)	430	430.00	$K^{qq} = 0.00009374$
total	2128	2128.00	$q = 0.4495 \pm 0.0096$

Mating Bioassay

mating type	observed number	expected number (α = 0)	expected number (α = 0.0214)
M(+) x M(+)	682	677.34	682.00
M(+) x M(-)	334	343.14	334.00
M(-) x M(-)	48	43.52	48.00
total	1064	1064.00	1064.00

$\chi^2 = 0.75$

$U_\alpha = 35.0472$ $\quad \alpha = 0.0214 \pm 0.0247$

$U_q = 0.$ $\quad q = 0.4495 \pm 0.0118$

$K_{\alpha\alpha} = 2445$ $\quad K^{\alpha\alpha} = 0.00061079$ $\quad I_\alpha = 1/K^{\alpha\alpha} = 1,637.$

$K_{q\alpha} = 2936$ $\quad K^{q\alpha} = -0.00016811$

$K_{qq} = 10668$ $\quad K^{qq} = 0.00014001$ $\quad I_q = 1/K^{qq} = 7142.$

Table 3. Comparison of the estimated inbreeding coefficient and gene frequency in MN blood groups by six different bioassays

Mating Bioassay

	α	I_α	q	I_q
Case 1.	0.0065±0.0125	6,384	0.4488±0.0076	17,204.
Case 2.	0.0214±0.0247	1,637	0.4495±0.0118	7,142.

Phenotype Bioassay

	α	I_α	q	I_q
Case 1.	0.0027±0.0217	2,128	0.4488±0.0076	17,204.
Case 2.	---	0	0.4495±0.0096	10,668.

Case 1. = without dominance. Case 2. = with dominance.
α and q stand for the inbreeding coefficient and frequency of a recessive gene (N), respectively. The amount of information is designated by I.

000	OO
111	A_1B
011	A_2B

It is now clear that the probability and the derivatives of each genotype can be added to those of a phenotype through factor union algebra. Moor-Jankowski et al. (1964) suggested that a binary system for phenotype might prove to be useful for coding. Introduction of factor-union algebra allows not only for the coding of phenotypes, but also for characterizing the mode of inheritance through the logical union of factors. We shall call such a genetic system factor-union system.

The factor union systems are a subset of regular phenotype systems (Cotterman, 1953) which exclude genetic systems with incomplete penetrance; that is, a genotype manifests only a particular phenotype during its life. In addition, none of the homozygotes shows any phenotype of other homozygotes. This is a necessary condition for a factor-union system but it is not sufficient, since phenogram 3-5-4, for instance (Cotterman, 1953), meets the requirements of the case, but the system is not a factor-union system. However, the only example of a regular phenotype, but non-factor-union system in man, as far as I know, is found in the haptoglobin locus, one of the serum protein polymorphisms where alleles Hp^1, Hp^2 and Hp^{2m} generate phenogram 3-4-12 (Cotterman, 1953). The allele Hp^{2m} may be detected only in a heterozygous condition with Hp^1 whose phenotype is Hp2-1 (Mod) (Giblett, 1959).

One more property of binary expression of genes should be mentioned for the use of an electronic computer. It is a fact that the total number of possible phenotypes is computed if alleles at a locus are specified by gene vectors. Further aspects of factor-union systems are reported in this conference by Dr. C. W. Cotterman. In Appendix 3, some examples of factor-union systems are presented. They have been used for estimating gene frequency and the inbreeding coefficient from a population of northeastern Brazil (Yasuda, 1966).

A COMPUTER PROGRAM ALLTYPE

The program ALLTYPE is designed for estimating gene frequencies and the inbreeding coefficient from phenotype or mating type data of factor-union systems by the method of maximum likelihood scoring. It consists of two parts, program G-TYPE and program MATYPE, for phenotypes of individuals and pairs of individuals, respectively.

Program G-TYPE. The program starts from a set of trial values of gene frequencies in a panmictic population (we assume $\alpha = 0$ at the beginning). Let $\underline{n}_\phi$ be the observed number of individuals whose phenotype is $\underline{\phi}$, determined by a locus with alleles $\underline{A}_1, \underline{A}_2, \ldots, \underline{A}_m$ whose frequencies are $p_1, p_2, \ldots, p_m$ $(\sum_{k=1}^{m} p_k = 1)$, respectively. The expected frequency of the phenotype $\underline{\phi}$ is then

$$P(\phi) = \sum_g P_g,$$

where $\underline{P}_g$ is the frequency of a genotype $\underline{g}$. Thus, the likelihood of observation is

$$L = \frac{(\sum_\phi n_\phi)!}{\prod_\phi n_\phi !} \prod_\phi [P(\phi)]^{n_\phi} .$$

The scores designated by U_p^* ($= \frac{\partial \ln L}{\partial p} = \frac{1}{L}\frac{\partial L}{\partial P}$) for instance, are calculated by treating all variables as if they were independent:

$$U_\alpha = \sum_\phi \frac{n_\phi}{P(\phi)} \left[\frac{\partial P(\phi)}{\partial \alpha} \right] = \sum_\phi n_\phi u_{\phi\alpha}$$

and

$$U_p^* = \sum_\phi \frac{n_\phi}{P(\phi)} \left[\frac{\partial P(\phi)}{\partial p} \right] = \sum_\phi n_\phi u_{\phi p} \qquad (p = p_1, \ldots, p_m)$$

where, for instance,

$$u_{\phi p} = \frac{\ln P(\phi)}{\partial p} = \frac{1}{P(\phi)} \frac{\partial P(\phi)}{\partial p} .$$

For a given genotype, we have

$$P_g = p_i^2 + p_i(1-p_i)\alpha \qquad \text{for homozygote}$$
$$= 2p_ip_j(1-\alpha) \qquad (i \neq j) \quad \text{for heterozygote}$$

and

$$\frac{\partial P_g}{\partial \alpha} = p_i(1-p_i) \qquad (i=1,\ldots,m) \quad \text{for homozygote}$$
$$= -2p_ip_j \qquad (i \neq j) \quad \text{for heterozygote}$$
$$\frac{\partial P_g}{\partial p_k} = 2p_i + (1-2p_i)\alpha \qquad (k=i) \quad \text{for homozygote}$$
$$= 0 \qquad (k \neq i) \quad \text{for homozygote}$$
$$= 2p_j(1-\alpha) \qquad (k=i) \quad \text{for heterozygote}$$
$$= 2p_i(1-\alpha) \qquad (k=j) \quad \text{for heterozygote}$$
$$= 0 \qquad \text{otherwise,}$$

so that

$$u_{\phi\alpha} = \frac{1}{P(\phi)} \sum_g \frac{\partial P_g}{\partial \alpha}$$

and

$$u_{\phi p} = \frac{1}{P(\phi)} \sum_g \frac{\partial P_g}{\partial p} .$$

The operation $\sum_g$ is possible through factor-union algebra, while $\sum_\phi$ is an ordinary addition operator. Thus, genotypes whose vectors are the same can be grouped for obtaining both the corresponding probability of phenotype and its first derivatives.

Imposing the restriction that $p_m = 1 - \sum_{i=1}^{m-1} p_i$, the independent scores are now

$$U_p = U_p^* - U_{p_m}^* \qquad (p = p_1, \ldots, p_{m-1}),$$

and the estimated variances of scores are obtained from

$$K_{\alpha\alpha} = \sum n_\phi u_{\phi\alpha}^2 ,$$

$$K_{p\alpha} = \sum n_\phi u_{\phi\alpha} (u_{\phi p} - u_{\phi p_m})$$

and

$$K_{pp} = \sum n_\phi (u_{\phi p} - u_{\phi p_m})(u_{\phi p} - u_{\phi p_m}).$$

The improved estimate is found from

$$\Theta_1 = \Theta_0 + UK^{-1}$$

where the subscript of Θ is the number of iteration, and $\Theta = (\alpha, p_1, \ldots, p_{m-1})$, $U = (U_\alpha, U_{p_1}, \ldots, U_{p_{m-1}})$ and K^{-1} is the covariance matrix.

Further improvement may be obtained by starting from the value Θ_i . The standard error of the estimate is obtained from the square root of the corresponding diagonal element of the covariance matrix. The estimate of the dependent variable is calculated from $p_m = 1 - \sum_{i=1}^{m-1} p_i$, and its variance is the sum of all elements with respect to p in the covariance matrix.

Example. Since the key part of the procedure is concerned with the computation of $P(\phi)$ and u_ϕ's, we illustrate the method for the A_1A_2BO blood group system. For four genes the corresponding gene vectors and trial values of gene frequencies may be given by

gene	vector	frequency
A_1	110	p_1
A_2	010	p_2
B	001	q
O	000	r

where $p_1+p_2+q+r=1$ and $\alpha = 0$. The possible genotypes, the corresponding genotype vectors, the frequency and the derivatives are as follows:

genotype	vector	frequency (P_g)	derivative to: α	p_1	p_2	q	r
A_1A_1	110	p_1^2	$p_1(1-p_1)$	$2p_1$	0	0	0
A_1A_2	110	$2p_1p_2$	$-2p_1p_2$	$2p_2$	$2p_1$	0	0
A_1B	111	$2p_1q$	$-2p_1q$	$2q$	0	$2p_1$	0
A_1O	110	$2p_1r$	$-2p_1r$	$2r$	0	0	$2p_1$
A_2A_2	010	p_2^2	$p_2(1-p_2)$	0	$2p_2$	0	0
A_2B	011	$2p_2q$	$-2p_2q$	0	$2q$	$2p_2$	0
A_2O	010	$2p_2r$	$-2p_2r$	0	$2r$	0	$2p_2$
BB	001	q^2	$q(1-q)$	0	0	$2q$	0
BO	001	$2qr$	$-2qr$	0	0	$2r$	$2q$
OO	000	r^2	$r(1-r)$	0	0	0	$2r$
sum check		1	0	2	2	2	2

By grouping of the same vectors, six phenotypes, the corresponding frequencies and the derivatives are now given by

phenotype	vector	frequency $(P(\phi))$	derivative to: α	p_1	p_2	q	r
A_1	110	$p_1^2+2p_1p_2+2p_1r$	$p_1(1-p_1)$ $-2p_1p_2$ $-2p_1r$	$2p_1$ $+2p_2$ $+2r$	$2p_1$	0	$2p_1$
A_2	010	$p_2^2+2p_2r$	$p_2(1-p_2)$ $-2p_2r$	0	$2p_2$ $+2r$	0	$2p_2$
B	001	q^2+2qr	$q(1-q)$ $-2qr$	0	0	$2q$ $+2r$	$2q$
O	000	r^2	$r(1-r)$	0	0	0	$2r$
A_1B	111	$2p_1q$	$-2p_1q$	$2q$	0	$2p_1$	0
A_2B	011	$2p_2q$	$-2p_2q$	0	$2q$	$2p_2$	0
sum check		1	0	2	2	2	2

It should be $\Sigma P(\phi) = 1$, $\Sigma(\frac{\partial Pi}{\partial\alpha}) = \Sigma P(\phi)u_{\phi\alpha} = 0$, and $\Sigma(\frac{\partial Pi}{\partial\phi}) = \Sigma P(\phi)u_{\phi p} = 2$ (this holds only when $\alpha = 0$), as shown in the sum check. The u-scores are thus calculated by dividing the derivatives by the corresponding phenotype frequency.

Morton(1964) studied the A_1A_2BO blood groups in a population from northeastern Brazil. The data are presented in Table 4. As trial values which were obtained from the formulae of Wellisch and Thomsen (cited in Race and Sanger, 1968), we have p_1=0.1547, p_2=0.0517, q=0.0808, and r=0.7101, and α=0. The genotypes, the corresponding vectors, the frequencies and the derivatives are:

genotype	vector	frequency (P_g)	derivative to: α	p_1	p_2	q	r
A_1A_1	110	0.0248	0.1326	0.3148	0	0	0
A_1A_2	110	0.0163	-0.0163	0.1043	0.3148	0	0
A_1B	111	0.0254	-0.0254	0.1616	0	0.3148	0
A_1O	110	0.2235	-0.2235	1.4202	0	0	0.3148
A_2A_2	010	0.0027	0.0490	0	0.1034	0	0
A_2B	011	0.0084	-0.0084	0	0.1616	0.1034	0
A_2O	010	0.0734	-0.0734	0	1.4202	0	0.1034
BB	001	0.0065	0.0743	0	0	0.1616	0
BO	001	0.1148	-0.1148	0	0	1.4202	0.1616
OO	000	0.5042	0.2059	0	0	0	1.4202
sum check		1.0000	0.0000	2.0000	2.0000	2.0000	2.0000

By grouping genotypes whose phenotype vector is the same we have

phenotype	vector	frequency $P(\phi)$	derivative to: α	p_1	p_2	q	r
A_1	110	0.2646	-0.1072	1.8384	0.3148	0	0.3148
A_2	010	0.0761	-0.0244	0	1.5236	0	0.1034
B	001	0.1213	-0.0405	0	0	1.5818	0.1616
O	000	0.5042	0.2059	0	0	0	1.4202
A_1B	111	0.0254	-0.0254	0.1616	0	0.3148	0
A_2B	011	0.0084	-0.0084	0	0.1616	0.1034	0
sum check		1.0000	0.0000	2.0000	2.0000	2.0000	2.0000

Table 4. The ABO blood groups in northeastern Brazil: observed numbers of mating types and individual phenotypes

O	A_1	A_2	B	A_1B	A_2B		
274	278	84	130	22	11	O	1073
	80	40	68	16	1	A_1	563
		5	22	3	3	A_2	162
			13	7	5	B	258
				1	1	A_1B	51
					0	A_2B	21
				(1064)		total	2128

Note: the observed number of mating type A_1 x B, for example, is 68 in the table, and the observed number of individuals whose phenotype is O is 1073. The total number of couples is 1064.

Therefore, u's are obtained from ratios of the derivatives to the corresponding phenotype frequency:

phenotype	vector	observed number (n_ϕ)	u_α	u_{p_1}	u_{p_2}	u_q	u_r
A_1	110	563	-0.4051	6.9478	1.1897	13.0404	1.1897
A_2	010	162	-0.3206	0	20.0210	0	1.3587
B	001	258	-0.3339	0	0	13.0404	1.3322
O	000	1073	0.4084	0	0	0	2.8167
A_1B	111	51	-1.0000	6.3622	0	12.3937	0
A_2B	011	21	-1.0000	0	19.2381	12.3095	0
total		2128					

Taking r as a dependent variable, we obtain

$$U_\alpha = 563(-0.4051)+ 162(-0.3206)+\ldots+ 21(-1.0000)$$
$$= 0.0583,$$
$$U_{p_1} = 563(6.9478-1.1897)+\ldots+ 21(0.0000-0.0000)$$
$$= -19.8536,$$
$$U_{p_2} = 61.2660$$

and $U_q = -0.9358$.

Apart from α, the information (K) and the covariance (K^{-1}) matrices are

$$K = \begin{bmatrix} 30,000.9507 & 4,863.1128 & 495.2465 \\ 4,863.1128 & 73,164.6471 & 5,354.0472 \\ 495.2465 & 5,354.0472 & 55,991.8296 \end{bmatrix}$$

and

$$K^{-1} = \begin{bmatrix} 0.00003369 & -0.00000223 & -0.00000008 \\ -0.00000223 & 0.00001392 & -0.00000131 \\ -0.00000008 & -0.00000131 & 0.00001799 \end{bmatrix}$$

respectively, where the array of rows and columns are in order of p_1, p_2, and q. Therefore, the improved estimates are

$$(p_1,\ p_2,\ q) = (0.1574,\ 0.0517,\ 0.0808) + UK^{-1}$$
$$= (0.1574,\ 0.0517,\ 0.0808)$$
$$+(-0.0008,\ 0.0009,\ -0.0001)$$

or

$$p_1 = 0.1566 \pm \sqrt{0.00003369} = 0.1566 \pm 0.0058$$
$$p_2 = 0.0526 \pm \sqrt{0.00001392} = 0.0526 \pm 0.0037$$
$$q = 0.0807 \pm \sqrt{0.00001799} = 0.0807 \pm 0.0042$$

and

$$r = 1-p_1-p_2-q \pm \sqrt{\Sigma k} = 0.7101 \pm 0.0076,$$

where

$$\Sigma k = [\ 3369+1392+1799+2(-223)+2(-8)+2(-131)]\ \times 10^{-8} = 0.00005836.$$

Further improvements may not be necessary because of relatively small correction terms, but the next iteration gave

$$p_1 = 0.1566 \pm 0.0058,$$
$$p_2 = 0.0526 \pm 0.0038,$$
$$q = 0.0808 \pm 0.0043$$

and

$$r = 0.7100 \pm 0.0074$$

Although the inbreeding coefficient may be estimated by phenotype bioassay, the ABO system has a singular matrix at $\alpha = 0$ (Yasuda, 1968b). We therefore turn to finding the coefficient from mating bioassay.

Program MATYPE. The principle of mating bioassay is essentially the same as that of phenotype bioassay, but the mating bioassay is rather complicated due to the fact that the data consist of a random sample of pairs, and the probability of mating type is one of seven functionally different types which can be made when one has a set of at least four multiple alleles.

Suppose that n_ϕ is the observed number of pairs whose phenotype is ϕ. Let A, B, C, D,... and p_A, p_B, p_C, p_D,... represent the alleles and the corresponding frequencies, respectively. The frequency of phenotype ϕ and the U-scores are then

$$P(\phi) = \sum_M P_M,$$

$$U_\alpha = \sum_\phi \frac{n_\phi}{P(\phi)} \sum_M \frac{\partial P_M}{\partial \alpha}$$

$$U_p = \sum_\phi \frac{n_\phi}{P(\phi)} \sum_M \frac{\partial P_M}{\partial p} \qquad (p = p_A,\ p_B, \ldots).$$

P_M, $\partial P_M/\partial\alpha$ and $\partial P_M/\partial p$ are given in Table 5 in terms of gene frequencies and the inbreeding coefficient.

Although the genotype vector of an individual is generated from gene vectors through factor-union algebra, the operator $\sum_M$ should be performed by using double comparison of genotype vectors. For instance, both couples $A_1A_1 \times A_1A_1$ and $A_1A_2 \times A_1O$ have the same genotype vector 110, 110 (two vectors are responsible for each pair, of course) so that there is no difficulty in pooling the frequency and its derivatives in a single comparison. However, matings $A_1A_1 \times A_1B$ and $A_1B \times A_1O$, for instance, have the genotype vectors 110, 111 and 111, 110, respectively. The positions of vectors of each individual should be reversed in order that both matings are grouped together.

Example. In the preceding example, gene frequencies are constant (neglecting errors), namely, $p_1 = 0.1566$, $p_2 = 0.0526$, $q = 0.0808$, and $r = 0.7100$. Starting from $\alpha = 0$, we obtain after some iterative processes $\alpha = 0.0081 \pm 0.0099$. This is not significant from zero ($\chi^2 = 0.68$ with df = 1).

DISCUSSION

A computer program ALLTYPE which includes both programs G-TYPE and MATYPE has been written in FORTRAN IV language for the CDC 3100 computer. Details, as well as user's instructions for the program, may be obtained through the Population Genetics Laboratory, University of Hawaii. It is of great advantage that the

Table 5. Frequency of seven functionally different mating types and their derivatives

Mating type		Frequency (P_M)
Incross	AA x AA	$p_A^4 + 6p_A^3(1-p_A)\alpha$
Backcross	AA x AB	$4p_A^3p_B + 12p_A^2p_B(1-2p_A)\alpha$
Outcross	AA x BB	$2p_A^2p_B^2 + 2p_Ap_B(p_A+p_B-6p_Ap_B)\alpha$
Intercross	AB x AB	$4p_A^2p_B^2 + 4p_Ap_B(p_A+p_B-6p_Ap_B)\alpha$
3-way Outcross	AA x BC	$4p_A^2p_Bp_C + 4p_Ap_Bp_C(1-6p_A)\alpha$
3-way Intercross	AB x AC	$8p_A^2p_Bp_C + 8p_Ap_Bp_C(1-6p_A)\alpha$
4-way Intercross	AB x CD	$8p_Ap_Bp_Cp_D(1-6\alpha)$

Table 5 -- continued

$\partial P_M/\partial\alpha$	$\partial P_M/\partial p_A$	$\partial P_M/\partial p_B$
$6p_A^3(1-p_A)$	$4p_A^3 + 6p_A^2(3-4p_A)\alpha$	0
$12p_A^2p_B(1-2p_A)$	$12p_A^2p_B + 24p_Ap_B(1-3p_A)\alpha$	$4p_A^3 + 12p_A^2(1-2p_A)\alpha$
$2p_Ap_B(p_A+p_B-6p_Ap_B)$	$4p_Ap_B^2 + 2p_B(2p_A+p_B-12p_Ap_B)\alpha$	$4p_A^2p_B + 2p_A(p_A+2p_B-12p_Ap_B)\alpha$
$4p_Ap_B(p_A+p_B-6p_Ap_B)$	$8p_Ap_B^2 + 4p_B(2p_A+p_B-12p_Ap_B)\alpha$	$8p_A^2p_B + 4p_A(p_A+2p_B-12p_Ap_B)\alpha$
$4p_Ap_Bp_C(1-6p_A)$	$8p_Ap_Bp_C + 4p_Bp_C(1-12p_A)\alpha$	$4p_A^2p_C + 4p_Ap_C(1-6p_A)\alpha$
$8p_Ap_Bp_C(1-6p_A)$	$16p_Ap_Bp_C + 8p_Bp_C(1-12p_A)\alpha$	$8p_A^2p_C + 8p_Ap_C(1-6p_A)\alpha$
$-48p_Ap_Bp_Cp_D$	$8p_Bp_Cp_D(1-6\alpha)$	$8p_Ap_Cp_D(1-6\alpha)$

Table 5 -- continued

$\partial P_M/\partial P_C$	$\partial P_M/P_D$
0	0
0	0
0	0
0	0
$4p_A^2 p_B + 4p_A p_B (1-6p_A)\alpha$	0
$8p_A^2 p_B + 8p_A p_B (1-6p_A)\alpha$	0
$8p_A p_B p_D (1-6\alpha)$	$8p_A p_B p_C (1-6\alpha)$

Note: A, B,... and p_A, p_B,... are alleles and the corresponding frequencies, respectively ($p_A + p_B + \ldots = 1$). The inbreeding coefficient is denoted by α. It is assumed that $p_A > |\alpha|$, $p_B > |\alpha|$, ...

program will handle any factor-union system. We have used the program for estimating the inbreeding coefficient and the gene frequencies for 16 polymorphic systems in a population from northeastern Brazil (Yasuda, 1966).

As far as gene frequency estimation is concerned, there is little doubt left for further development from phenotype bioassay in combining the method of maximum likelihood with factor-union algebra. However, a selection of trial values is somewhat troublesome, since bad trial values may lead to a divergence of the procedure. To escape the difficulty, either the square root method (e. g. Mourant, 1954) or gene counting (Ceppellini et al., 1955; Yasuda and Kimura, 1968; Yasuda, 1968c) may be used. The latter procedure especially is promising, since it gives maximum likelihood estimate for gene frequency, although the rate of convergence is three or four times slower than Newton-Raphson iteration.

In multi-allelic locus two technical problems have arisen. Firstly, if one of the alleles at a given locus is rare, such as the r^y gene in the Rh system, the convergence by iterative processes then becomes difficult, or it may even fail. In such a case one should build a different genetic system, either excluding the allele in question or pooling it with the other allele making the frequency larger. Secondly, the larger the number of alleles, the more difficult the iterative process. Since analysis is made only for two-allele cases, it is hard to predict the mathematical stability of maximum likelihood equations, namely, whether the solution is unique or even feasible. We have not tried yet to prove rigorously that the likelihood has a unique maximum (except in a few special cases). A practical approach to examining this would be to start with several different trial values of p's to see if the same final estimates are reached. This has been tried, and we are confident that the large sample sizes involved would produce a relatively smooth likelihood function with a single maximum toward which our estimates will converge. Experience also suggests that for any factor union system there is always a unique set of solutions in large samples.

The inbreeding coefficient α obtained from either phenotype or mating bioassay simply indicates a coefficient representing a systematic deviation from a simple product of gene frequencies. The cause of deviation may be sampling bias, misclassification of phenotypes, differential selection, and so on. Differential selection, especially against homozygotes, might tend to give a smaller or even a negative estimate of the coefficient. Sometimes the random sampling does not hold and the effect on the estimate would be considerable. Biased sampling or misclassification has in a statistical sense the same effects on the estimates as selection. Above all, a careful test with large sample may be necessary to estimate a meaningful coefficient of inbreeding by present methods.

Apart from these questions of application, mating bioassay resolves the following disadvantages of phenotype bioassay. Firstly, the inbreeding coefficient cannot be estimated from several genetic systems in which the number of phenotypes is equal to the number of alleles. Secondly, the ABO systems have a statistical singularity at $\alpha = 0$, so that no reliable estimate can be obtained for the inbreeding coefficient. Thirdly, the amount of information about the inbreeding coefficient is in general much larger in mating bioassay than in phenotype bioassay.

Finally, although mating bioassay for a sex-linked locus was not discussed here, the principle presented so far could be applied for estimating the inbreeding coefficient at sex-linked loci.

SUMMARY

An attempt was made for estimating the inbreeding coefficient as a deviation from Hardy-Weinberg proportions based on data obtained from couples. Apart from questions on the biological meaning of the estimate, the present method, which we have named mating bioassay, has two theoretical advantages: namely, it allows an estimate of the inbreeding coefficient and it gives a greater amount of information about the inbreeding coefficient than phenotype bioassay. Furthermore, by applying factor-union algebra to group genotypes whose phenotype is the same, the method of maximum likelihood scoring estimates gene frequencies for any factor-union system, including most of the genetic systems in man. Briefly, phenotype bioassay is sufficient for gene frequency estimation, while mating bioassay is desirable for estimating the inbreeding coefficient. However, a large sample may be needed for obtaining an efficient estimate in mating bioassay.

ACKNOWLEDGEMENTS

The author would like to express his thanks to Dr. Newton E. Morton for his encouragement throughout this work. He also wishes to thank Dr. Ming-Pi Mi for teaching him the computer programming and for discussing the progress during the study.

The work presented here is based on a thesis submitted to the graduate school of the University of Hawaii in partial fulfillment of the requirements for the Ph.D. degree. It is also Contribution No. 672 from the National Institute of Genetics, Mishima, Japan. Support was provided by U.S. Public Health Grants GM 10424, GM 12454 and FR 00247 from the National Institutes of Health.

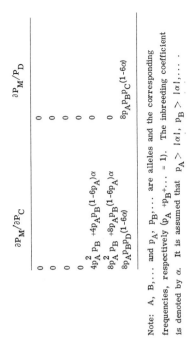

APPENDIX I. A METHOD OF MAXIMUM LIKELIHOOD SCORING

Since the method of maximum likelihood is a well-known technique for estimating the parameters which appear in probability models, it is worthwhile to look at one of the developments, namely, the scoring method that is powerful when likelihood equations are too complicated to obtain the analytic solutions. There are good summaries concerning this method (Rao, 1952; Bailey, 1961). In order to visualize it the description below assumes a single parameter, but the generality is not lost when the number of parameters is arbitrary.

Let us suppose that L denotes a likelihood with a parameter Θ, and we define the score of Θ as the first derivative of ln L with respect to Θ; i.e., $u_\Theta = \partial \ln L/\partial\Theta$. The amount of information on Θ is $k = E(u_\Theta^2) = -E(\partial^2 \ln L/\partial\Theta^2)$, where E is an operator to take expectation. In practice, however, it is convenient to use an estimated variance, $\Sigma n u_\Theta^2$ instead of $E(u_\Theta^2)$, where n is the observed number. Both values should be the same except in the neighborhood of $k = 0$.

For a given set of independent samples, the scores and the amount of information are additive so that the total score $U(\Theta)$ and the information $K(\Theta)$ are obtained as

$$U(\Theta) = \Sigma u_\Theta \text{ and } K(\Theta) = \Sigma k$$

where summation is over all independent samples.

Let Θ_o denote an approximate value of the true one, and e be the correction which must be applied to Θ_o to give the exact value of the solution, so that

$$\Theta = \Theta_o + e.$$

The maximum likelihood equation $U(\Theta) = 0$ then becomes

$$U(\Theta_o + e) = 0.$$

Expanding this in the series form of Taylor's theorem, we have

$$U(\Theta_o+e) = U(\Theta_o) + e\,U'(\Theta_o) + \frac{e^2}{2}U''(\Theta_o + \delta e) \quad (0 \leq \delta \leq 1).$$

Hence

$$U(\Theta_o) - e\,K(\Theta_o) - \frac{e^2}{2}K'(\Theta_o + \delta e) = 0 \qquad (1).$$

Now if e is relatively small, we may neglect the term containing e^2 and get the simple relation

$$U(\Theta_o) - e_1 K(\Theta_o) = 0$$

from which

$$e_1 = U(\Theta_o)/K(\Theta_o) \qquad (2).$$

The improved value of the root is then

$$\Theta_1 = \Theta_o + e_1 = \Theta_o + U(\Theta_o)/K(\Theta_o) \qquad (3)$$

and its variance $\sigma^2 = 1/K(\Theta_1)$. The discrepancy between Θ_o and Θ_1 is tested by $\chi^2 = (\Theta_1 - \Theta_o)^2 K(\Theta_o)$ with one degree of freedom. If the chi-square is significant, then we may repeat the above process until no significance appears. The heterogeneity test between classes is given by $\chi^2 = \Sigma u^2/k - U^2(\Theta)/K(\Theta)$ with degrees of freedom i-p-1, where i is the number of independent classes, and p the number of parameters to be estimated.

It is evident from the formula (3) that the larger the K-value, the smaller is the correction which must be applied to get the true value of the estimate. This means that when the graph of likelihood equation is nearly vertical where it crosses the Θ-axis, the correct value of the root can be found very quickly and with very little labor. If, on the other hand, the numerical value of K-score should be small in the neighborhood of the root, the value of e given by (2) would be large and the computation of the root by this method would be a slow process or might even fail altogether. This method should never be used when the graph of $U(\Theta)$ is nearly horizontal where it crosses the Θ-axis. In such a case regula falsi interpolation might be helpful (Barnett, 1966). The process will evidently fail if $K(\Theta) \sim 0$ is in the vicinity of the root, and such an example in estimating the inbreeding coefficient in the neighborhood of zero has been found in phenotype bioassay of the ABO blood group systems (Yasuda, 1968b).

APPENDIX II. MAXIMUM LIKELIHOOD SCORES AND VARIANCES AT A SEX-LINKED LOCUS WITH TWO ALLELES UNDER THE HYPOTHESIS THAT $\alpha = 0$

Model and Data		Score
Phenotype Bioassay		
No Dominance		
AA f_1	A m_1	$U_q = [f_2+2f_3+m_2-(M+2F)q]/q(1-q)$
Aa f_2	a m_2	$U_\alpha = \dfrac{(M+F)(4f_1f_3-f_2^2) + (f_1m_2^2-m_1m_2f_2+f_3m_1^2)}{(m_1+f_2+2f_1)(m_2+f_2+2f_3)}$
aa f_3		$K_{qq} = (M+2F)/q(1-q)$, $K_{q\alpha} = 0$, $K_{\alpha\alpha} = F$
$(f_1+f_2+f_3=F,\ m_1+m_2=M)$		
Complete Dominance		
A- f_1	A m_1	$U_q = (m_2+2f_2)/q - [(2f_1+m_1)q+m_1]/(1-q^2)$
aa f_2	a m_2	$U_\alpha = f_1-f_2+(f_2-Fq)/q(1-q)$
		$K_{qq} = 4F/(1-q^2)+M/q(1-q)$, $K_{q\alpha} = 2F/(1+q)$, $K_{\alpha\alpha} = F(1-q)/(1+q)$
$(f_1+f_2 = F,\ m_1+m_2 = M)$		
Mating Bioassay		
No Dominance		
AA x A n_1		$U_q = (n_{23}+2n_{45}+3n_6-3Nq)/q(1-q)$
Aa x A n_2		
AA x a n_3		$U_\alpha = \dfrac{3N[(3n_1+n_{23})(n_{45}+3n_6)-(n_{23}+n_{45})^2]}{(3n_1+2n_{23}+n_{45})(n_{23}+2n_{45}+3n_6)}$
Aa x a n_4		
aa x A n_5		$K_{qq} = 3N/q(1-q)$, $K_{q\alpha} = 0$, $K_{\alpha\alpha} = 3N$
aa x a n_6		$(n_{23} = n_2+n_3,\ n_{45} = n_4+n_5,\ N = \sum_{i=1}^{6} n_i)$
Complete Dominance		
A- x A n_1		$U_q = [(2n_2+n_3+3n_4)-(n_1+n_3)q-3Nq^2]/q(1-q^2)$
aa x A n_2		$U_\alpha = [3(n_1+n_2)-(n_1+3n_2)q]/(1-q^2) + (n_3+3n_4)/q-3N$
A- x a n_3		$K_{qq} = N(1+5q)/q(1-q^2)$, $K_{q\alpha} = 2N/(1+q)$
aa x a n_4		$K_{\alpha\alpha} = N(1+3q)/(1+q)$ $(N = n_1+n_2+n_3+n_4)$

Where A and a are alleles with frequencies p and q, respectively, α is the inbreeding coefficient, and f_i and m_i denote the observed numbers of male and females respectively. The observed numbers of mating types are designated by n_i. It is assumed in mating bioassay that p and $q > |\alpha|$.

APPENDIX III. BRAZILIAN SEROTYPES WITH REFERENCE TO THE GENE AND PHENOTYPE VECTORS

Genetic system	gene		phenotype		possible genotype
Secretor	Se	1	Se	1	Se/Se, Se/se
	se	0	se	0	se/se
Lewis	Le	1	Le	1	Le/Le, Le/le
	le	0	le	0	le/le

Lutheran	Lu^a	1	a+	1	Lu^a/Lu^a, Lu^a/Lu
	Lu	0	a-	0	Lu/Lu
Gm	a = 1	10000	a	10000	a/a
	ab = 1, 5	11000	abx	11100	ax/ab
	ax = 1, 2	10100	abcx	11110	ax/abc
	abc = 1, 5, 6	11010	axb^2	11101	ax/b^2
	b^2 = 3, 5	01001	ax	10100	a/ax, ax/ax
			ab	11000	ab/ab, ab/a
			abc	11010	abc/abc, abc/ab, abc/a
			$abcb^2$	11011	abc/b^2
			ab^2	11001	ab/b^2, a/b^2
			b^2	01001	b^2/b^2
Inv	Inv^a	1	a+	1	Inv^a/Inv^a, Inv^a/Inv
	Inv	0	a-	0	Inv/Inv
PTC	T	1	T+	1	T/T, T/t
	t	0	T-	0	t/t
P	P_1	1	P+	1	P_1/P_1, P_1/P
	P	0	P-	0	P/P
ABO	A_1	110	A_1	110	A_1/A_1, A_1/A_2, A_1/O
	A_2	010	A_1B	111	A_1/B
	B	001	A_2	010	A_2/A_2, A_2/O
	O	000	A_2B	011	A_2/B
			B	001	B/B, B/O
			O	000	O/O
Kell	K	100	K	100	K/K
	k	010	Kk	110	K/k
	k^s	011	Kk^s	111	K/k^s
			k	010	k/k
			k^s	011	k/k^s, k^s/k^s
Duffy	Fy^a	1	a+	1	Fy^a/Fy^a, Fy^a/Fy
	Fy	0	a-	0	Fy/Fy
Diego	Di^a	1	a+	1	Di^a/Di^a, Di^a/Di
	Di	0	a-	0	Di/Di
MNSsU	MS	1010	M	1000	M*/M*
	Ms	1001	MS	1010	MS/MS, MS/M*
	M*	1000	MSs	1011	MS/Ms
	NS	0110	Ms	1001	Ms/Ms, Ms/M*
	Ns	0101	MN	1100	M*/N*
	N*	0100	MNS	1110	MS/NS, MS/N*, NS/M*
			MNSs	1111	Ms/Ns, Ms/NS
			MNs	1101	Ms/Ns, Ms/N*, M*/Ns
			N	0100	N*/N*
			NS	0110	NS/NS, NS/N*
			NSs	0111	NS/Ns
			Ns	0101	Ns/Ns, Ns/N*
Rh	cde(r)	00011	r	00011	r/r
	Cde(r')	00101	R_1r	10111	R_1/r, R_1/R_o, R_o/r',
					(R_1/r'^n, R_o/r'^n)
	cdE(r'')	01010	R_1	10101	R_1/R_1, R_1r'
	CdE(r^y)	01100	R_o	10011	R_o/R_o, R_o/r
	cDe(R_o)	10011	R_2r	11011	R_2/r, R_2/R_o, R_o/r''
	CDe(R_1)	10101	R_2	11010	R_2/R_2, R_2/r''
	cDE(R_2)	11010	R_1R_2	11111	R_1/R_2, R_1/r'';
					R_o/R^z, R_o/r^y,
					R_2/r'
	CDE(R^z)	11100			R^z/r, (R^z/r'^n,
					R_2/r'^n)
	Cde^S(r'^n)	00111	r'	00101	r'/r'
			rr'	00111	r/r', (r'^n/r'^n,
					r'/r'^n, r/r'^n)
			R_1R^z	11101	R_1/R^z, R_1/r^y, R^z/r'
			$r'r^y$	01101	r'/r^y
			$r'r''$	01011	r/r''
			R^z	11100	R^z/R^z, R^z/r^y
			R_2R^z	11110	R_2/R^z, R_2/r^y, R^z/r''
			r^y	01100	r^y/r^y
			r^yr''	01110	r^y/r''
			r''	01010	r''/r''
Transferrins					
	B	100	B	100	B/B
	C	010	BC	110	B/C
	D	001	BD	101	B/D
			C	010	C/C
			CD	011	C/D
			D	001	D/D
Hemoglobin					
	A	100	AA	100	A/A
	S	010	AS	110	A/S
	C	001	AC	101	A/C
			SS	010	S/S
			SC	011	S/C
			CC	001	C/C

REFERENCES

Bailey, N. T. J. 1961. Introduction to the mathematical theory of genetic linkage. Clarendon Press, Oxford.

Barnett, V. D. 1966. Evaluation of the maximum-likelihood estimator where the likelihood equation has multiple roots. Biometrika 53:151-165.

Birkhoff, G. and MacLane, S. 1965. A survey of modern algebra (3rd ed.). Macmillan Co., New York.

Ceppellini, R., Siniscalco, M. and Smith, C. A. B. 1955. The estimation of gene frequencies in a random mating population. Ann. Hum. Genet., Lond. 20:97-115.

Cotterman, C. W. 1927. Indication of unit factor inheritance in data comprising but a single generation. Ohio J. Sci. 37:127-140.

Cotterman, C. W. 1953. Regular two allele and three-allele phenotype systems. Part I. Amer. J. Hum. Genet. 5:193-235.

Fisher, R. A. 1922. On the mathematical foundations of theoretical statistics. Philo. Trans. Roy. Soc. Lond. A, 222:309-368.

Giblett, E. R. 1959. Haptoglobin types in American negroes. Nature 183:192-193.

Malecot, G. 1967. Identical loci and relationship. Vth Berkeley Symp. Prob. and Stat. 317-332.

Moor- Jankowski, J., Wiener, A. S. and Rogers, C. M. 1964. Blood groups of chimpanzees: demonstrated with isoimmune serums. Science 145:1441-1443.

Morton, N. E. 1964. Genetic studies of northeastern Brazil. Cold Sp. Harbor Symp. Quant. Biol. 29:69-79.

Morton, N. E. and Yasuda, N. 1962. The genetical structure of human population. Pp. 186-203. In Les Deplacements Humaines, Entretiens de Monaco en Sciences Humaines. (Ed. J. Sutter). Hachette.

Mourant, A. E. 1954. The distribution of the human blood groups. Blackwell, Oxford.

Penrose, L. S. 1935. The detection of autosomal linkage in data which consist of pairs of brothers and sisters of unspecified parentage. Ann. Eugen. 6:133-138.

Race, R. R. and Sanger, R. 1968. Blood groups in man. (5th ed.). Blackwell, Oxford.

Rao, C. R. 1952. Advanced statistical methods in biometric research. John Wiley, New York.

Stevens, W. L. 1938. Estimation of blood-group gene frequencies. Ann. Eugen. 8:362-375.

Yasuda, N. 1966. The genetical structure of northeastern Brazil. Ph.D. thesis. The Univ. of Hawaii, Honolulu.

Yasuda, N. 1968a. An extension of Wahlund's principle to evaluate mating type frequency. Amer. J. Hum. Genet. 20:1-23.

Yasuda, N. 1968b. Estimation of the inbreeding coefficient from phenotype frequencies by a method of maximum likelihood scoring. Biometrics 24:915-935.

Yasuda, N. 1968c. Gene frequency estimation by a counting method. Jap. J. Hum. Genet. 12:226-245. (in Japanese)

Yasuda, N. and Kimura, M. 1968. A gene counting method of maximum likelihood for estimating gene frequencies in ABO and ABO-like systems. Ann. Hum. Genet., Lond. 31:409-420.

DISCUSSION OF ESTIMATION OF GENE FREQUENCIES AND INBREEDING

T. W. Kurczynski, J. H. Edwards, and V. Balakrishnan.

KURCZYNSKI: Computer programs for the maximum likelihood estimation of gene frequencies and other quantities such as inbreeding have become valuable tools for genetic studies, especially in man. Such programs, when written with a general approach, eliminate the necessity for deriving and using the tedious algebraic equations involved in point estimation in complex multiallelic systems. Since maximum likelihood scoring has become a standard procedure for solving likelihood equations, general programs for gene frequency estimation by maximum likelihood are likely to have much in common. Differences, however, are to be expected in the basic logic by which data are coded and such differences in logical design determine the distinctive features and limitations of each program.

To illustrate a different approach I would like to compare some aspects of Dr. Yasuda's program with the program MAXIM reported by Kurczynski and Steinberg (1967). As Dr. Yasuda has described, the basic method of coding data in his program depends on a vector of binary digits or factors representing an allele, a genotype, or a phenotype. With this notation, alleles may be combined into genotypes, or genotypes into phenotypes, by the logical union of factors. This procedure provides a concise representation of secondary units of a genetic system, such as genotypes, in terms of the primary units, the factors, and also simplifies some of the logic of the computations. Unfortunately, as Dr. Yasuda has pointed out, not all genetic systems can be expressed in terms of factor union algebra. Furthermore, and more importantly, the set of all genetic systems in which maximum likelihood estimation is possible includes systems which are not factor union systems. However, before discussing some of these systems, I would like to consider the methods of coding data for the program MAXIM. The logical basis of coding in MAXIM depends on each allele in a k allelic system being assigned an unique integer number from 1 to k. All possible genotypes are then specified by a pair of numbers, one for each allele. These genotypes are arranged in an array, designated the G matrix, such that the position of a genotype and its probability are given by the pair of numbers associated with that genotype, where the first number gives the row and the second the column. For purposes of efficiency, the symmetric elements of the G matrix are pooled into an upper triangular matrix. An example should clarify the preceding points. Consider the ABO system with alleles A_1, A_2, B and O, and suppose the alleles are numbered in the order given. The genotype A_1B for example, is then designated by the number pair (1, 3), and the probability of the genotype A_1B, given by $2p_1q$, is found in position (1, 3) of the G matrix. The entire G matrix would appear as follows:

$$G = \begin{bmatrix} p_1^2 & 2p_1p_2 & 2p_1q & 2p_1r \\ 0 & p_2^2 & 2p_2q & 2p_2r \\ 0 & 0 & q^2 & 2qr \\ 0 & 0 & 0 & r^2 \end{bmatrix}$$

Since each of the alleles is represented by a number, the entire array of alleles is coded as a vector, E, (in the actual program the vector is designated EG) where an element, e (i), is the ith gene frequency estimate. It is simple then to generate the G matrix from the E vector since g (i, i) = e(i) x e(i), g(i, j) = 2e(i) e(j) if $j > i$, and g (i, j) = 0 if $j < i$. The major remaining coding problem is to arrange the genotypes into sets representing each phenotype. This is accomplished by reading the genotypes, i. e., their code numbers as described above, in a particular sequence such that the computer recognizes the sets corresponding to each phenotype. The sequence depends on merely arranging the genotypes so that all genotypes comprising a phenotype set are grouped together followed by the next phenotype set. In order to carry out the gene frequency estimation, it is also necessary to compute the partial derivatives of each genotype. This is accomplished, for all possible genotypes, in the same way as in Dr. Yasuda's program, and the results are stored in the three dimensional array, D, where d (l, i, j) is the partial derivative of the ith and jth genotype with respect to the lth allele frequency. With these fundamental arrays defined, the logical steps involved in the calculation of the gene frequencies and their standard errors are quite similar to those in Dr. Yasuda's program and need not be discussed here. The essential difference in the two programs is the manner in which genes are coded, in one case by a vector of binary factors, in the other case by an integer number. In practice the coding of data in terms of binary factors is likely to become cumbersome with systems having many factors, such as might occur in the Gm or Rh systems. For example, if there were 15 anti-sera, it would require 15 binary digits to represent one genotype, whereas if each allele is given a number, a genotype would be coded by, at most, four digits, two for each allele, provided there were less than 100 alleles. A more important consequence of the difference in coding, at least in theory, is that a binary system of coding precludes the processing of non-factor-union systems which are in principle applicable for solution by maximum likelihood. In terms of Cotterman's phenogram notation (Cotterman, 1953) a number of regular 2- or 3-allele phenograms could be mentioned in this regard. However, here I would like to consider briefly two such phenograms already referred to by Dr. Yasuda.

The simpler of the two phenograms is phenogram 2-2-2, which consists of two alleles and two phenotypes with both homozygous classes phenotypically identical. An example of this and other similar phenograms is found in the wasp Habrobracon in which femaleness is determined by heterozygosity for any of a series of multiple alleles, while maleness follows from homozygosity or hemizygosity for any of the alleles. In the case of two alleles, the maximum likelihood estimate of one of the allele frequencies may be shown to be as follows:

$$p = \frac{1 + r \pm \sqrt{r^2 - 1}}{2(1 + r)}$$, where r is the ratio of the observed numbers of the two phenotypes. To avoid complex numbers, the ratio must be chosen equal to or greater than one, which is always possible due to the symmetry of the system. A number of hypothetical samples for this system were successfully processed by MAXIM. There seems to be only one minor problem, peculiar to this and other similar symmetric systems, and that is, a value of 0.5 always satisfies the derivative of the log likelihood equation. Therefore, trial values of 0.5 will not generate the other possible estimate of the gene frequency.

Perhaps a system of greater practical importance is one like the Hp system mentioned by Dr. Yasuda, in which the alleles Hp^1, Hp^2 and Hp^{2m} generate phenogram 3-4-12 (Cotterman, 1953). In this phenogram the genotypes Hp^2/Hp^{2m}, Hp^{2m}/Hp^{2m}, and Hp^2/Hp^2 are phenotypically the same, and the allele Hp^{2m} is only detected in the genotype Hp^1/Hp^{2m}. To test this system with the program MAXIM, a set of data on Seattle Negroes published by Giblett (1959) was used. The data and results of the computer run are presented in the table below. The 17 individuals with no detectable haptoglobin were omitted from the calculations. Trial values of .5, .4 and .1 were chosen for the frequencies of the alleles Hp^1, Hp^2, and Hp^{2m}, respectively. After two iterations the maximum likelihood estimates of .525707, .377002, and .097271 were obtained. The goodness of fit is very close with χ_1^2 = .010631, and is consistent with the existence of a Hp^{2m} allele suggested by the family studies of Giblett and Steinberg (1960).

TABLE

Haptoglobin types of Seattle Negroes
(Giblett, 1959)

Haptoglobin type	Observed	Expected
Hp 1-1	107	107.507
Hp 2-2	87	87.507
Hp 1-2	155	154.194
Hp 1-2m	40	39.791
Absent	(17)	-------
Total	389	388.999

*χ_1^2 = .010631

*Gene frequencies

Hp^1	.525707±.017902
Hp^2	.377002±.019788
Hp^{2m}	.097291±.014233

*computed with program MAXIM (Kurczynski and Steinberg, 1967)

Since non-factor-union systems seem to be rare, the practical usefulness of a gene frequency program based on the coding of data by a binary notation is not diminished. However, some generality is lost which is not dependent on the actual methodology of maximum likelihood estimation.

Cotterman, C. W. 1953. Regular two-allele and three-allele phenotype systems. Amer. J. Hum. Genet. 3:193-235.

Giblett, E. R. 1959. Haptoglobin types in American Negroes. Nature 183:192-193.

Giblett, E. R. and Steinberg, A. G. 1960. The inheritance of serum haptoglobin types in American Negroes: evidence for a third allele Hp^{2m}, Amer. J. Hum. Genet. 12:160-169.

Kurczynski, T. W. and Steinberg, A. G. 1967. A general program for maximum likelihood estimation of gene frequencies. Amer. J. Hum. Genet. 19:178-179.

EDWARDS: A discussion of Yasuda's paper is difficult, since it appears to be the final evolution of the conventional maximum likelihood estimation with the typographical convenience of matrix notation and the option of estimating inbreeding coefficients; it is particularly valuable in documenting this procedure in such great detail. As the completeness of this leaves little room for criticism, or even comment, I will limit this discussion to aims and alternative methods. As this appears to be but one of three possible procedures, I will attempt to review and compare all three, and also comment on the coding system.

Any estimation procedure must start with a set of data, a coding convention, a model relating the data to some parameters, which I shall call structural parameters, a set of algorithms for the estimation of these parameters on some criterion such as maximum likelihood, and, when necessary, summarizing indices to reduce the number of the results, or to express the closeness of fit of the model. I will consider these in order.

The set of data is defined by numbers of persons, either individuals or pairs of individuals in Yasuda's two problems, responding in various ways to sera, or sharing various of the known pairs of sets of bands in gels, or whatever other criterion of phenotype we may have.

Following Mendel, if we suppose phenotypic expression to be capable of dominance and recessivity, we have the logical "or" relationship. That is, the set of possible phenotypes is logically related to the set of possible genetic factors, and is identical to it when both contain all possible elements. As in

	a	A
a	a	A
A	A	A

or

	0	1
0	0	1
1	1	1

where

a + a = a	or	0 + 0 = 0
a + A = A		0 + 1 = 1
A + a = A		1 + 0 = 1
A + A = A		1 + 1 = 1

This OR set has some interesting properties, in that however large the number of tests used (n), there will be 2^n possible responses or 2^n possible phenotypes. For example with two tests the possible phenotypes are 00, 01, 10, and 11 and the possible genotypes of the same form.

	00	01	10	11
00	00	01	10	11
01	01	01	11	11
10	10	11	10	11
11	11	11	11	11

Where many tests are used octal equivalents are useful. This is equivalent to any set of conventional symbols as in the ABO blood groups. The binary numbers may also be represented by decimal numbers (Figure 1): a great convenience of this notation is that numbers are handled in binary forms in most computers, and the simple "OR" instruction, such as K = I.OR.J in FORTRAN IV, defines the phenotypic responses K related to genetic factors I and J.

Indeed, Mendel's work could be summarized as the discovery that the genetic factors are related to phenotype by "OR" rather than the more obvious "+" relationship, Mendel using lower and upper case symbols as equivalent to 0 and 1 in a binary string.

In decoding the set of phenotypes D, consisting of d_o, d_1 ..... $d_{(2^n-1)}$ to give the frequency of genetic factors inferred,

OR SETS

	Binary					Decimal			
00	00	01	10	11	0	0	1	2	3
01	01	01	11	11	1	1	1	3	3
10	10	11	10	11	2	2	3	2	3
11	11	11	11	11	3	3	3	3	3
	00	01	10	11		0	1	2	3

Conventional	O	A	B	(AB)
O	O			
A	A	A		
B	B	AB	B	
(AB)	AB	AB	AB	AB

Compact integers	1	2	3
1	1		
2	2	2	
3	3	4	3

FIGURE 1

we have to estimate a vector P, consisting of p_o, p_1....... $p_{(2^n-1)}$ for the n phenotypes; $\sum_i p_i$ = UNITY. In order to simplify the argument I shall follow conventions similar to those of the Tensor calculus; that is, summation or equivalence will be implied whenever an index occurs twice on the right hand side. The indices will have values o to (2^n-1), rather than starting at 1. I will variously use w_i and w_{ij} as unit multipliers merely to carry indices, where $w_i = w_{ij} = 1$ for all values of i or j.

For example

$1.0 = p_i w_i$ is equivalent to $1 = \sum_{i=o}^{2^n-1} p_i$

$b_i = a_{ij} w_j$ is equivalent to $b_i = \sum_{j=o}^{2n-1} a_{ij}$ for all values of i.

Maximize $r_i \simeq s_i$ means maximize some criterion of closeness of the vectors R and S; where the indices are present on the left as well, summation will not be implied.

For example

$c_{ij} = a_{ij} b_{ij}$ specifies the products for all pairs of i, j, which is to be distinguished from matrix multiplication.

$c_k = a_{ij} b_{ij}$ where k = i. or .j specifies the phenotypic (k) in terms of the summation being implied for all identical values of k.

$L = a_i \log(e_i)$ is the likelihood of expectations e_i given observation a_i.

Our data consists of the vector integers

$d_o, d^1, \ldots\ldots\ldots\ldots d_{(2^n-1)}$

for the n tests. (This vector is equivalent to Cotterman's G_2.)

The set may be, and usually is, incomplete; i.e., $d_i = o$ for some i, and we have to infer the most likely pair of sets of genetic factors, of known or unknown gametic correlation, sampled from the same universe, of known or infinite size, which would lead to the results observed. The problem is of impractical size for contemporary computers, and it also involves the logical problems of an arbitrary total likelihood function compounded of the likelihood of the dissimilarity of the two sets of gametic contributions, the likelihoods of the inferred phenotypic sets, with a bonus for null genotypes which we may term Occam's bonus.

The conventional procedure of gene frequency estimation (Figure 2) which is an approximation based on large numbers, seems to me to involve undue casualness in the assumption of identical parental contributions. In particular, it would seem to lead to biassed inbreeding coefficients since dissimilarity of the parental sets will be expressed as a heterozygote excess.

Basically, the procedure maximizes the likelihood of the wrong comparison; rather than finding the most similar gametic sets and the most likely compatible genotypic set it invents imaginary parents who shed exactly equal numbers of each type of gamete, and even identical fractions of gametes, which segregate with gene-splitting precision, and compares the ideal children of these ideal couples with the actual phenotypes observed. This may give the right answer, but it is not obvious why it should, and even less obvious why it should give the right variance. However, since the combinatorial solution seems insoluble at present the real-number approximation, with fractional genes and identical gametes, seems the only solution practical.

A full solution should, given the 2^n possible phenotypes by frequency, provide a set of estimates of gametic frequencies over the whole range 0 to 1.0. The solutions so far found fall short of

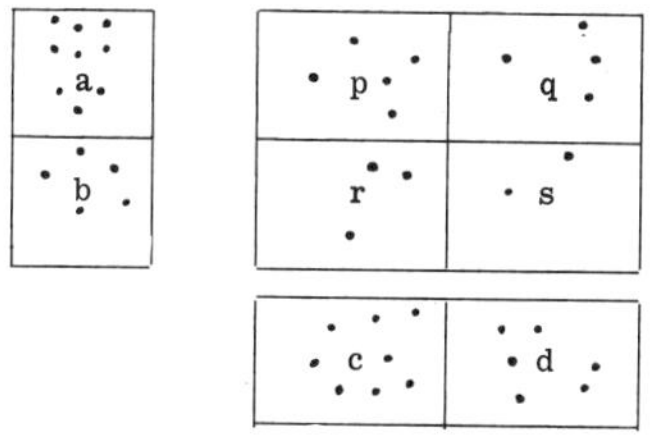

1) Find sets p, q, r, s conforming to restraints $a + b = c + d = p + q + r + s = n$, p as given
2) Find most likely set $a \simeq c$, $ac/n \simeq p$
3) Find $(a + c)/(a + b + c + d)$

FIGURE 2

this; the decision to place some at zero normally has to be made casually, and, where several reasonable decisions are possible, each will involve considerable difference of estimates. No objective criterion for choosing different null sets seems to be available. Occam's razor, conventionally wielded to the limits of statistical credulity, can only pare away at a likelihood. A criterion amounts to defining the null/null contribution $0.\log(o)$ or 0^0 as a determinate positive number. Figures 3 and 4 summarize this.

Yasuda maximizes likelihood by defining null sets, assuming gametic equivalence, and comparing expectation with observation, starting with provisional estimates, and iterating. This may be the fastest method of maximizing likelihood, but it is not the simplest; a computer permits the direct maximation of likelihood by ascending the likelihood hypersurface, and this method largely evades the problems of functions which are difficult or impossible to differentiate by automatic methods. The evasion of differentation seems to me to be one of the major advantages of using computers.

The detection of the maximum likelihood involves a peak climbing algorithm which, ideally, should be fast and simple; consider a man on a foggy mountainside with a compass and an altimeter, and the knowledge that the mountain was parabolic in section and had elliptical contours. If he was intelligent he could walk around any equilateral triangle, find the height at the corners, and predict the bearing and distance of the peak; with less thought and more exercise he could stroll along the vertical planes $x = o, y = o$, and, $x = y$, and solve the conventional variance and covariance matrix. When he reached the predicted point he would be there if his assumptions were correct; if not he could repeat or iterate, introducing the interesting paradox that if iterations are necessary the results are based on erroneous assumptions.

Consider a man who could not invert matrices but could manage quadratic equations; we would advise him to take the height of three equidistant points in a vertical plane, say $x = o$, and predict and go to the highest point, then do the same in the plane at right angles, and carry on doing this, reducing the length of his exploratory walk systematically unless the peak was outside the limits of his walk. An imaginary man of equally limited intellect could change cyclically from dimension to dimension and climb a likelihood hypersurface and once at, or marginally near, the peak, by wandering around an equilateral triangle, or other simplex, he could obtain data to define the information matrix from the constants a_i, b_{ij}, c of the likelihood hyperparabola $L_k = b_{ij}x_i x_j + c$ for each of k values x_i, x_j of sets of parameters. By differentiation we may obtain the information and covariance matrix. On computers it is simpler to measure the second derivatives directly by estimating the curvatures of all orthogonal cuts and one of their bisectors. Preliminary estimation is no problem since, on the assumption of one peak, we will always reach it, and we can put $p_i = 1/m$ where m is the number of non-null parameters. That is, if we imagine a system of homogeneous coordinates in a regular simplex, we always start at the center. The number of iterations is likewise no problem, since the procedure can be programmed to terminate when the increment in likelihood is some small amount, say 0.01 or 0.001 or so. If a log-likelihood hypersurface is convex on any complete set of orthogonal cuts, it seems impossible that it should have any concavities, that is, several peaks.

The output of my program for the Brazilian data quoted by Yasuda is given in the appendix. The results are, of course, identical.

Yasuda suggests terminating the iteration if the discrepancy is not statistically significant. This seems unsatisfactory, since,

ALLELE PROPORTION ESTIMATION

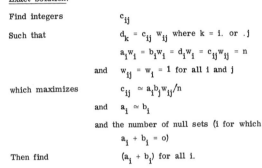

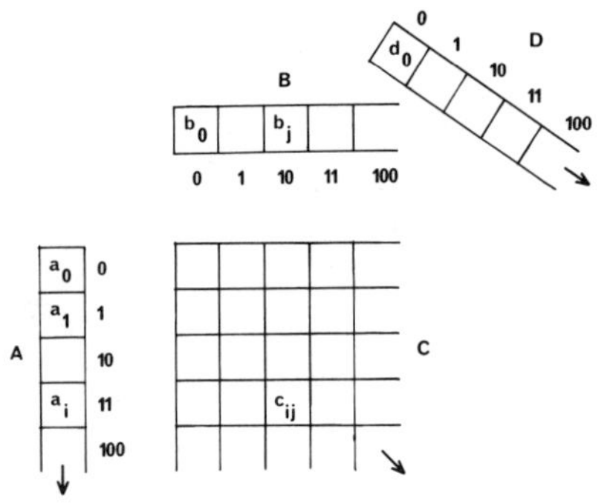

Exact solution.

Find integers c_{ij}

Such that $d_k = c_{ij} w_{ij}$ where k = i. or .j

$a_i w_i = b_i w_i = d_i w_i = c_{ij} w_{ij} = n$

and $w_{ij} = w_i = 1$ for all i and j

which maximizes $c_{ij} \simeq a_i b_j w_{ij}/n$

and $a_i \simeq b_i$

and the number of null sets (i for which $a_i + b_i = o$)

Then find $(a_i + b_i)$ for all i.

FIGURE 3

Figure 3 - continued

Approximate solution.

Designate null sets.

Assume $b_i = a_i$

Find a_i/n for all i

so that if $e_k = na_i a_j w_{ij}$ where k = i. or .j

$L = d_k \log(e_k)$ is maximized.

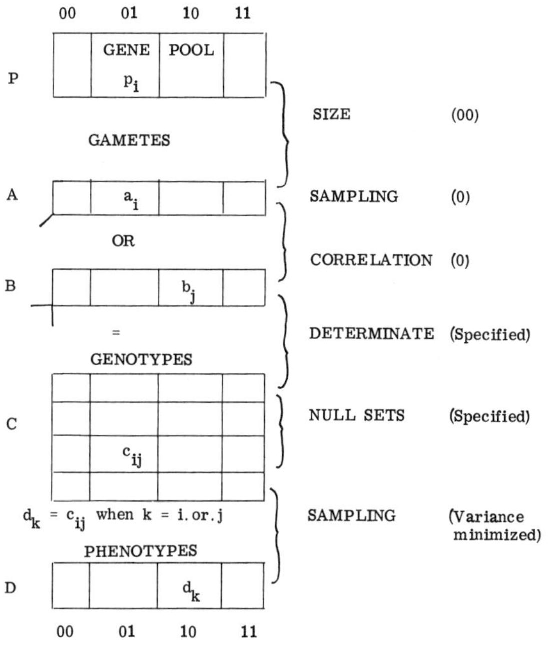

FIGURE 4

while the peak is unique, the routes leading to it are not, but depend on where they come from, that is on the provisional estimators. I see no point in not getting as close as practical to the peak.

Likelihood estimation by iterative orthogonal cuts is equivalent to zeroizing all off-diagonal elements in the information matrix: this simplifies the programming and shortens each cycle, but increases the number of cycles. Since there is only one peak, and computers are fast, the answer will always be the same whether or not covariances are estimated. If they are useless or, as in gene frequency estimation, meaningless, I see little point in estimating them.

Finally, we may consider the problem of specifying the results. Ideally these should be both intelligible and additive, and should allow comparison.

I would like to suggest giving results in terms of integral gene number on the most likely estimate as a simple compromise between practicality and rigor; these are easily conveyed, approximately comparable by n x 2 chi square tests.

The rigorous alternative of finding information or covariance matrices is impractical for large series due to the costs of printing; matrix algebra is unknown to most laboratory workers and anthropologists who handle these data, and it is questionable if acquiring the necessary knowledge would be adequately justified.

In summary, I suggest we need a simple algorithm which analyses

	00	01	10	11	100	etc.
D	d_o	d_i	d_2	. .	d_i	. . . .
	to give					
G	g_o	g_i	g_2	. .	g_i	- - - - - - -

where D is the vector of phenotype number and G (of Cotterman's terminology) the vector of gene numbers. Ideally the null set (g_i = 0) should be defined by the program but, until this is possible, it is necessary to specify the null group intuitively. Where intuition is permitted, I think we should admit to this by expressing results in small numbers of significant figures.

The problem of gene frequencies in randomly defined individuals in a large population is merely one aspect of the problem of total data analysis, where a whole group is analysed.

Ideally a gene frequency program, given any set of phenotypes and genetic relationships, should, as efficiently as possible, infer the gametic correlations, the gametic proportions, and the gametic recombinations. This is important, as random selection, even if it were possible, is no longer the best way of studying populations, and, at a small cost in the precision of gene frequency estimation, anthoropological studies could obtain data on the position of the loci they study and the inbreeding coefficients of the population.

As the value of gene frequency data is proportional to n, and of linkage data to almost n^2, where n is the number of loci studied, a change in collection habits is urgently needed.

BALAKRISHNAN: Dr. Yasuda has presented a comprehensive method for the maximum likelihood estimation of gene frequencies from phenotype and mating bioassays under the asuumption of panmixia and for estimating the inbreeding coefficient α using the gene frequency estimates so obtained.

I am not, however, sure whether the coefficient α estimated in the manner suggested by Dr. Yasuda can really be called the inbreeding coefficient. As Dr. Yasuda himself says, it simply indicates a coefficient representing a systematic deviation from the expectations under the Hardy-Weinberg equilibrium. Such deviations need not necessarily be due to inbreeding only.

In our experience, we have found that even quite large samples (about 3500) have given phenotype frequencies for the ABO system which do not deviate significantly from the Hardy-Weinberg equilibrium even though inbreeding coefficient was estimated to be .008 by survey techniques. In such cases, α estimated by Dr. Yasuda's method is likely to be unrealistically low.

I would think that the efficiency of this method can best be tested on data from some of the Indian communities. We have in India communities differing widely in respect of inbreeding. In some, marriage is not allowed within 4 or 5 degrees of relationship while in others the preferred husband for a girl is her maternal uncle, leading to situations where 5-10 per cent of the marriages are between uncle and niece, and 20-30 per cent of the marriages are between first cousins. It would be interesting to compare estimates of α obtained by this method in these different types of communities.

Another question regards the interpretation of α when different values are obtained from different genetic systems from the same sample. In the example of Brazilian data given by Dr. Yasuda, we have three different values on the basis of mating bioassay -- 0.0065 from the MN system without dominance, 0.0214 from the MN system with dominance and 0.0081 from the ABO system. Here, of course, all three values are insignificant and hence there is no problem. If, however, we get significantly different values from different genetic systems, the question of interpretation comes in.

I would like to mention that in the program for the maximum likelihood estimation of allelic frequencies developed by Kurczynski and Steinberg (1967), the phenotypes are obtained by summing the likelihood scores for the genotypes within each phenotype. This is how the generalization was achieved. This has now been incorporated in our program.

In addition, in our program, no assumption is made regarding the presence or absence of any of the alleles. In the earlier form, this was achieved by using subscripted variables as subscripts which cannot be done with all computers. Now we have modified the program and this difficulty has been removed.

The number of alleles and phenotypes which could be handled by the program can also be increased very much by some adjustment of the way in which the scores for the phenotypes are obtained from those for the genotypes.

LIKELIHOODS AS SUCCESSIVE ITERATIONS

```
ER  F8 -.17564226E+04
    1
    2-108.572
    3   -6.631
    4   -6.294
    5   -6.286
    6   -6.285
    7   -6.285
```

	CLASS	OBS	EXP	OBS. PC	EXP. PC	CHISQ
1	O	1073	1072.36	50.42	50.39	0.00
2	A1	563	560.80	26.46	26.35	.01
3	A2	162	165.01	7.61	7.75	.05
4	B	258	257.87	12.12	12.12	0.00
5	A1B	51	53.86	2.40	2.53	.15
6	A2B	21	18.10	.99	.85	.46
7		2128	2128.00	100.00	100.00	.68

	FACTOR	NUMBER	PERCENTAGE
1	O	3021	70.99
2	A1	667	15.67
3	A2	224	5.27
4	B	344	8.08

$\log_e$ likelihood = -6.28526@ Max Possible = -5.95189

Diff/2 = .16668 D. F. = 3

Inbreeding coefficient (read in; not estimated) = 0.00000

APPENDIX

TITLE	NO	NP	NI	LA	LB	LC	LD	LE	NA	NB	NC	IC
									Control card			
ABO	6	4	9	M					0	0	0	0.000

3

INPUT

	-A1	-A	-B	Computed Decimal Eq.	Computed Serial
O	0	0	0	0	1
A1	1	1	0	6	2
A2	0	1	0	2	3
B	0	0	1	1	4
A1B	1	1	1	7	5
A2B	0	1	1	3	6

3

	-A1	-A	-B		
O	0	0	0	0	1
A1	1	1	0	6	2
A2	0	1	0	2	3
B	0	0	1	1	4

OUTPUT

Map relating genotype to phenotype. Computed.

	O	A1	A2	B
O	O			
A1	A1	A1		
A2	A2	A1	A2	
B	B	A1B	A2B	B

0 DATA FROM BRAZIL (MORTON) QUOTED BY YASUDA

2128.000	1073.000	563.000	162.000
258.000	51.000	21.000	

GENETIC LINKAGE IN MAN

J. R. Renwick
University of Glasgow
Scotland

The past year has brought some exciting developments in human linkage research. The first two autosomal assignments of a locus to a specific autosome have been confirmed. The first three-locus linkage group has been found and the number of autosomal linkages has been increased to a present total of about 13. The number of available polymorphic marker loci is also growing rapidly so the prospects for the future look promising for the mapping of human chromosomes.

THE GENETIC LINKAGE COMPUTER PROGRAM, NU

In order to indicate the role of the computer in some of these and other advances, it is necessary to describe, briefly, the main linkage program written by Jane Schulze that has been running with minor adjustments since 1961 (Renwick and Schulze, 1961). The likelihoods it calculates are interpreted along the lines suggested by Smith (1959) using the theorem of Bayes (1763), but we now use a somewhat different set of prior probabilities, developed from those of Renwick and Schulze (1964).

MAP INTERVALS AND RECOMBINATION FRACTIONS

In mapping chromosomes, we are primarily interested in the map interval between two loci and this is defined as the mean number of crossover events per chromatid strand that occur between the two loci in a single meiotic division. But the data bear only upon the proportion of chromatids that have suffered 1 or 3 or 5 ... crossover events so as to lead to a new combination of parental alleles at the two loci. This proportion is the recombination fraction, θ. The program, in its usual version, computes the likelihood of getting, with respect to the two loci, the phenotypes observed in a series of pedigrees for each of a set of θ values. This θ is related to the map interval, w morgans, between the loci, but which of the proposed θ/w mapping functions is most suitable for man is still uncertain. The function

$$4w = \tanh^{-1} 2\theta + \tan^{-1} 2\theta$$

of Carter and Falconer (1951) reflects a high level of interference and fits the mouse data fairly well. Fig. 1 shows this and other suggested relationships as a plot of w against θ. The property of additivity of true map intervals is not in dispute and we can also probably agree to take the data, bearing on the recombination fraction, at face value. If, as a temporary expedient, we assume that the mapping relationship is constant for all regions of the chromosome complement, then the suitability of the various mapping functions could be assessed by comparing the likelihoods based on them, averaged over all sets of map intervals that obey the additivity rule. The incorporation of prior information about the chromosome length, even a crude estimate from a chiasma count, is important. This is so because this chromosome length imposes restrictions on the order of the mapped loci more stringently for those mapping functions that tend to inflate the map interval, e.g., that of Haldane (1919),

$$\theta = \frac{1}{2} - \frac{1}{2} e^{-2w} .$$

Quite appropriately, then, the likelihood based on such functions is correspondingly reduced. The cancellation of this effect by an assumption of infinite length for the chromosome therefore tends to bias the test in favour of the Haldane type of function. I hope Dr. John Edwards will be able to tell us what light his computer program for optimising a chromosome map has thrown on the problem of the most appropriate mapping function for man.

The question is particularly relevant here since it affects the prior distribution of the recombination fraction. Recent chiasma counts in males (McIlree et al., 1966), confirming previous counts, and evidence to be discussed later that there is more recombination in females than in males make it harder for us to continue accepting the very convenient fiction that for two loci on a chromosome, the recombination fraction is equally likely to be any value from

$$0 \text{ to } \frac{1}{2}.$$

As Morton (1955) pointed out, this simple relationship would hold approximately only for a chromosome of about 1 morgan, but it now seems likely that the average length of a human autosome is at least 1.5 morgans. It is, of course, possible to calculate a suitable prior probability distribution for θ, given the distribution of lengths and given a mapping function, but it is probably easier in practice to take the suggestion of Dr. John Edwards and graph the likelihoods and probabilities against map interval directly.

One of the things we shall need for this is the prior distribution of map intervals of particular length given that the loci are on the same chromosome. For a single chromosome of length, l, the distribution is triangular, the prior probability of a particular length being proportional to l-w. But, normally, when testing for linkage between two loci we do not know which autosome pair they are on, even if they are on the same one. So we need to add up all the 22 triangles of Fig. 2 to get the overall distribution of autosomal map intervals. Part of this summated polygonal distribution is shown in this figure and the whole, on a

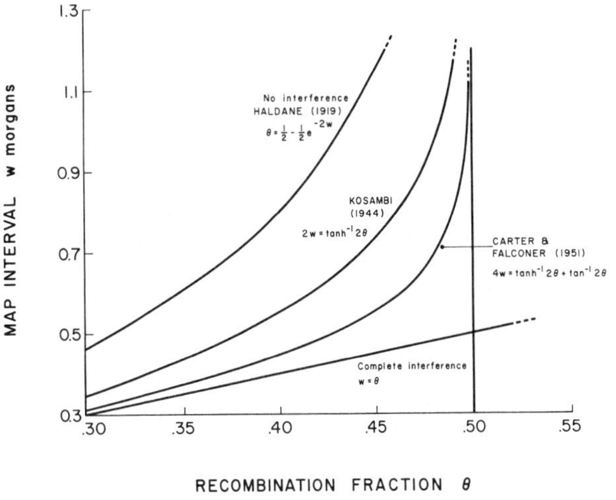

Fig. 1. Relationships between the recombination fraction, Θ, and the map interval w morgans, depending on the degree of interference.

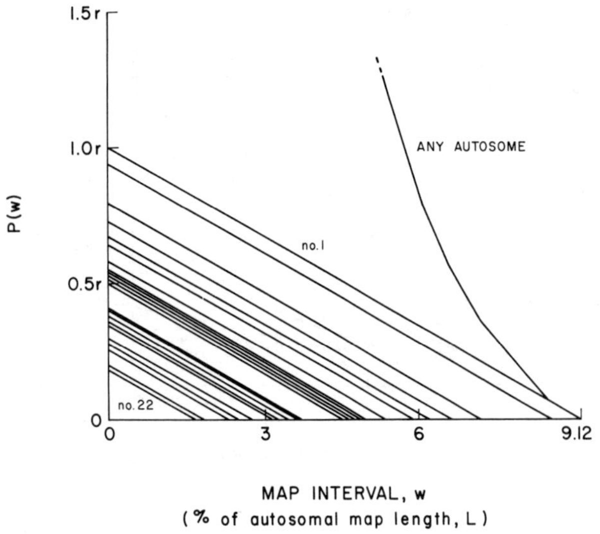

Fig. 2. Distribution or prior probability for particular lengths of map interval on any one of the 22 autosomes. Each triangle has its base (map interval) determined by the length of the autosome and its area by the square of this. Part of the distribution of prior probabilities summated over all autosomes is also shown. Lengths are expressed as percentages of total autosomal map length.

different scale, is shown in Fig. 3. Clearly, very long map intervals could only be on chromosome 1, the largest; and short intervals could be on any chromosome. If m is the maximum number of chromosomes that we therefore need to consider for any particular interval and if l_i and w are both expressed as proportions of total autosomal length, the required distribution is

$$P(w) = \frac{1}{\sum_{i=1}^{22} l_i^2} \sum_{i=1}^{m} \frac{2}{l_i^2} (l_i - w) l_i^2$$

$$= \frac{2}{\sum_{i=1}^{22} l_i^2} \left(\sum_{i=1}^{m} l_i - mw \right) \quad \text{where } m \leq n;\ l_m \geq w$$

This distribution sums to unity. The relative lengths of the individual autosomes, l_i, were taken from Smith, Penrose and Smith (1961). To use the graph, we must choose a particular mapping function and a particular autosomal map length, L, but the graph itself is independent of these.

PROCEDURE FOR ESTIMATING θ

Choosing Carter and Falconer's function, in the absence of a better one, and choosing $33\frac{1}{3}$ morgans as a conservative and arithmetically convenient estimate of total map length averaged over males and females, we translate our set of θ values, for which we have computed the likelihoods, into a set of map intervals. We then multiply the likelihoods (standardised in the usual way to be 1 at $\theta = .5$) by the corresponding prior probabilities of Fig. 3 and graph the resulting final chances of particular lengths of the map interval. For illustration we can take the likelihoods for some data concerning the rare condition of white sponge nevus of the mucosa (Browne, Renwick and Izatt, 1969). The white mouth locus, WM1, was tested for linkage with a series of markers and the standardised likelihoods for the white mouth: MNS pair of loci, for example, are given in Table 1. The graph of Fig. 4 shows the data-modified chances for particular lengths of the interval between WM1 and MNS. The ratio, Λ', of the stippled area under this graph to the area under the prior distribution is the left-hand side of the observational component of the odds and expresses the contribution of the data to the odds that the two loci are on the same chromosome. The initial odds that they are,

$$\Sigma l_i^2 : 1 - \Sigma l_i^2 \qquad \text{or } 1:17.5$$

must also be incorporated by simple bilateral multiplication of the odds as in Table 2.

It will be seen that, had we not standardised the likelihoods to be unity at $\theta = \frac{1}{2}$, the right side of the observational component would have been determined by the likelihood of the pedigrees, given the free assortment ($\theta = \frac{1}{2}$; $w > 3.04M$) which is expected if the two loci are on different chromosome pairs.

If we had used the customary assumptions of equally likely recombination fractions and of equally long autosomes, we should have obtained the smaller value, 0.035, for the probability that the loci are on the same chromosome, instead of the value 0.045. Depending on the data, the difference could be more marked and could be in either direction.

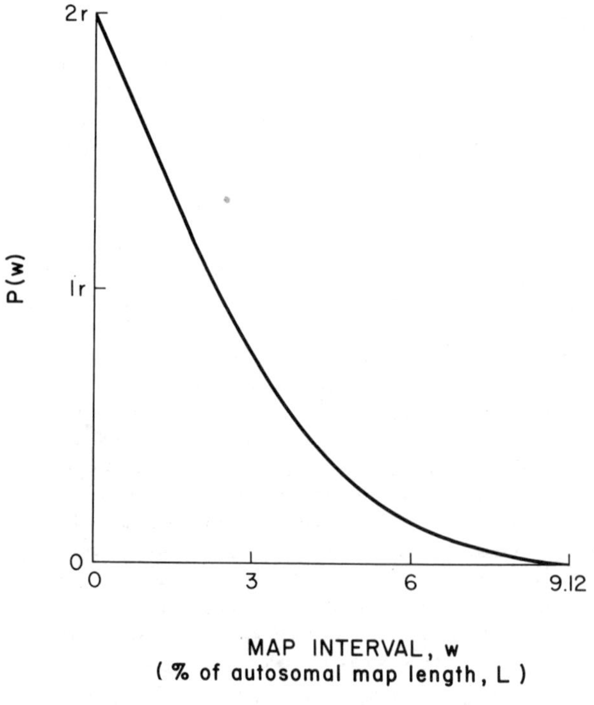

Fig. 3. The complete distribution of prior probability, for particular lengths of map interval, summed over all autosomes. The vertical scale is very different from that used in Fig. 2. Lengths are expressed as percentages of total autosomal map length.

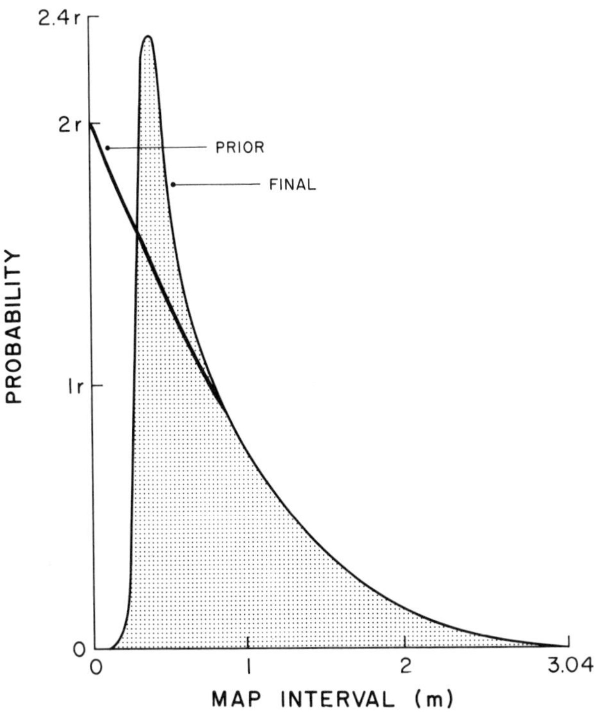

Fig. 4. Graph of the prior distribution of autosomal map intervals as in Fig. 3, together with graph of the distribution modified by data respecting the White Sponge Nevus locus, WM1, and the MNS blood group locus. The ratio of the areas under these graphs, conveniently measurable by planimetry, is the observational component of the odds, Λ':1.

TABLE 2

Combination of the prior and observational components of the odds on two loci being on the same autosome. The findings for the WM1 and MNS pair of loci are given for the purpose of illustration only.

	In general	For WM1:MNS
Prior odds that two autosomal loci are on the same chromosome, $\Sigma\, l^2 : 1 - \Sigma\, l^2$	1:17.5	1:17.5
Observational component	Λ':1	0.82:1
Final odds	Λ':17.5	0.82:17.5

TABLE 1

Standardized likelihoods and prior probabilities for the relationships between the white nevus locus WM1 and the MNS blood group locus. Each likelihood and probability is appropriate to a particular recombination fraction or to the corresponding map interval on the Carter-Falconer (1951) scale. The product of the likelihood and the prior probability is the final data-modified probability distribution for the map interval WM1:MNS and is the basis of the illustrative graph in Figure 4.

Map interval in morgans (w)	0	.10	.20	.31	.44	.55	.69	.95	3.04
Recombination fraction (Θ)	0	.10	.20	.30	.40	.45	.482	.4975	∼.50
Pedigrees WM1ME WM1WE									
a) Standardized likelihood ratio	0	.00004	.072	.914	1.563	1.180	1.050	1.005	1.000
b) Prior probability densities	2.00r	1.87r	1.73r	1.58r	1.41r	1.27r	1.11r	0.81r	0r
c) Final probability densities	0r	.00007r	.125r	1.444r	2.204r	1.499r	1.166r	.814r	0r

FUTURE PLANS

The present computer program for the IBM 7094 gives the likelihoods (which are the real output) and it also gives the area under the final probability distribution of recombination fractions, which were, until recently, taken as equally likely. In the re-writing of this program, which will soon be necessitated by the proposed change of machine in Baltimore, we propose to fit a polynomial to the final distribution of the map interval and have the ratio of the areas, Λ', as part of the printout. We shall also compute the likelihood for zero recombination fraction - a trivial deficiency in the present program - and, possibly, we might try to introduce age as a variable affecting the frequency of crossing over. We can already vary the male and female θ values independently. We shall not attempt to deal with the three-point linkage data other than by estimating the three component intervals independently.

ANCILLARY PROGRAMS

The ancillary programs are written in Manipulator, a high-level, string-handling language.

1) CF - to detect discrepancies between the two hand codings and to detect certain internal inconsistencies in the data.
2) PEDCOD - to prepare these data, which cover several loci, for the pair-wise linkage testing by NU.
3) PRUNE - to eliminate, with variable restrictiveness under user control, the relatively uninformative links between remote branches of a pedigree and to remove branches that are totally uninformative.

4) SELINK - to select suitable batches of predigested data for analysis by NU. A vast data store is being built up in Baltimore and the whole suite of programs is designed (a) to analyse those parts of the data store that can give information on the linkage that looks most promising on the results thus far and (b) to store the results of that analysis and re-cycle with further data if they are encouraging or to turn to the next most promising linkage otherwise. The pairs of loci are ordered according to their current probabilities of being on the same chromosome and data relevant to the leading pair of loci are processed in batches. SELINK also allows the user various other options with respect to the order of processing the data. All programs are now operational but, for various reasons, we have decided to defer the building up of taped results until the central NU program has been rewritten.

This suite of programs has been described elsewhere (Renwick and Bolling, 1967) but the need for analysis on this scale might be re-emphasised here. About 2-3,000 pedigrees are already available, giving information on an average of 10 pairs of loci each and covering, between them, about 1,000 pairs of loci. Only a fraction of these data have been exposed to more than cursory scrutiny and several loose linkages could, presumably, be demonstrated from these data by means of a systematic analysis.

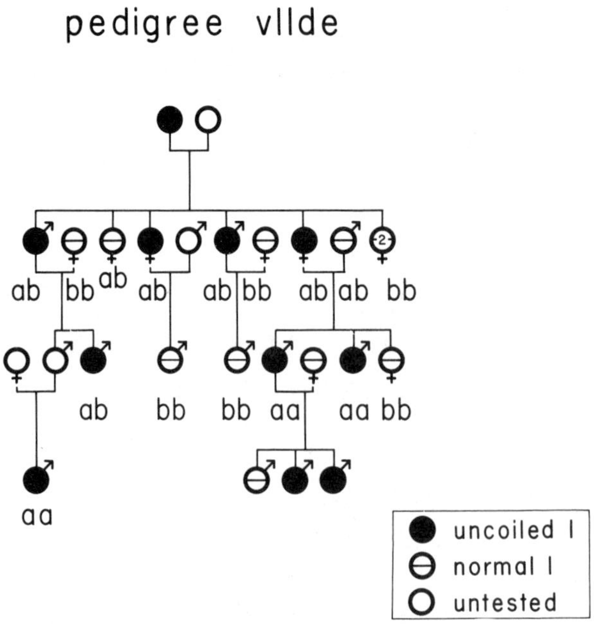

Fig. 5. Pedigree VIIDE showing linkage between the Duffy blood group locus and the Un locus controlling the extension of chromosome 1 (Donahue et al., 1968).

ASSIGNED AND UNASSIGNED AUTOSOMAL LINKAGE GROUPS IN MAN

Table 3 lists the autosomal linkages and chromosome assignments that can be considered to be convincing.

Fy:1 The localisation of the Duffy locus on chromosome 1 was the result of a study of the blood groups in a family in which the harmless extended chromosome 1 variant was segregating as in Fig. 5 (Donahue, Bias, Renwick and McKusick, 1968). The computer analysis included two small pedigrees from the literature (Cooper and Hernits, 1963; Philip, Frydenberg and Lele, 1963) and gave a final probability of 0.90 that the Duffy locus was on the same chromosome as the locus controlling the uncoiling. Since the abnormal allele at this locus is uncoiling only one of the two homologues, it seems reasonable to assume that the uncoiler locus is on chromosome 1 itself, each allele affecting only the homologue on which it sits. The linkage of Duffy to this locus has been shown independently by Ying and Ives (1968) on a similar extensive pedigree. The locus Cae for a zonular, pulverulent type of congenital cataract was shown by Dr. Lawler and myself to be linked to Duffy (Renwick and Lawler, 1963), so this is also on chromosome 1. The data of Crawford et al. (1967) had indicated that Duffy was on chromosome 16, not 1, but an alternative interpretation involving a silent Duffy allele now seems to be the true one.

Thymidine kinase:17 Mouse-man hybrid cells in culture throw out only the human chromosomes. This property, discovered and exploited by Weiss and Green (1967), was used to locate the human thymidine kinase locus. They did this by finding out which human chromosome became obligatory in a hybrid of which the mouse component was genetically deficient for this enzyme. Dr. Barbara Migeon (personal communication) has found that

TABLE 3

Assigned and unassigned autosomal linkage groups in man.

Autosome	Loci	Locus Symbols	Estimated map intervals m=male: f=female	95 percent limits	Source
Chromosome 1	Duffy:Uncoiler	Fy:Un	3	1-21	1
	Duffy:Cae cataract	Fy:Cae	0	0-19	2
Chromosome 17	Thymidine kinase	TK	-	-	3
Other	ABO:Nail-patella syndrome	ABO:Npl	11 (f 14; m 8)	6-19	4
	Nail-patella syndrome:AK	Np:AK	0	0-50	5
	Adenylate kinase:ABO	AK:ABO	16 (f 24; m 8)	5-30	6
	Hemoglobin$_\beta$:Hemoglobin$_\partial$	Hb$_\beta$:Hb$_\partial$	0.1	0- 6	7
	Albumin:Gc protein	Alb:Gc	2	1- 6	8
	Elliptocytosis$_1$:Rhesus	El$_1$:Rh	3	2- 7	9
	Sclerotylosis:MNS	Tys:MNS	4	0.3-19	10
	Lutheran:Secretor	Lu:Se	13 (f 16; m 10)	7-19	11
	Transferrin:Cholinesterase$_1$	Tf:E$_1$	16 (f 19; m 12)	7-28	12

Only the original source is given here:-
1. Donahue et al. (1968)
2. Renwick and Lawler (1963)
3. Weiss and Green (1967)
4. Renwick and Lawler (1955)
5. Renwick (unpublished)
6. Rapley et al. (1968)
7. Ceppellini (1959)
8. Weitkamp et al. (1966)
9. Lawler (1954)
10. Mennecier (1968)
11. Mohr (1954)
12. Robson et al. (1966)

chromosome 17 or 18 is the obligatory one. Such elegant methods do not require a computer.

The suggested assignment of the Hageman locus for one of the clotting factors to the short arm of chromosome 16 (de Grouchy et al, 1968) is not included in Table 3. If it is a true bill, a second child with a deficiency of this arm will be sufficient to establish this but, because of ascertainment bias and a frequency as high as 1 per cent for the Hageman deficiency allele at this locus, and because of imperfection in the laboratory identification of the heterozygotes, the present odds on this assignation are no better than evens, on my assessment. Prior odds, of about 50:1 against, arise from the small size of this arm and have been included in the assessment.

The odds that Hp is on chromosome 13 are somewhat more favorable (perhaps 6:1 on various assumptions about allele frequencies, etc.), but there are alternative explanations of what appeared to be the simultaneous loss of one Hp allele and of part of one chromosome 13 in the data of Gerald et al. (1967). These alternatives, fully recognised by the authors, are an error of paternity and the effect of the rare, silent Hp^0 allele. Other evidence (e.g. that of Bloom et al., 1967) does not conclusively either favor or disprove the assignment of Hp to No. 13.

UNASSIGNED LINKAGE GROUPS

ABO:Nail-patella:AK Rapley et al. (1968) described the linkage of the polymorphic adenylate kinase, AK, locus to the ABO locus which is already known to be 10 units from Np, which is responsible for the nail-patella syndrome in the heterozygote. We have, so far, found six non-recombinants out of six for the Np:AK interval, making the order ABO:Np:AK about four times more likely than all other orders together.

DIGRESSION ON FEMALE AND MALE SUSCEPTIBILITY TO CROSSING-OVER

We have used the ABO:AK data of the Galton laboratory workers and added in our own data in a recalculation to compare the female and male susceptibility to crossing-over. The lods (logarithms of the standardised likelihoods) for pairs of female and male recombination fractions are given in Table 4. These indicate a peak in the neighbourhood of 0.24 for female θ and 0.08 for male θ. The antilods, or likelihood ratios themselves, are points on a probability surface if they are combined with equal prior odds ascribed to each pair of female/male recombination fractions. This reversion to equal odds for all recombination fractions is reasonably safe for this purpose since male and female values will be distorted in a roughly comparable fashion. The relative volumes under the surface on the two sides of the equality diagonal then give the odds on the susceptibility of the female to recombination in this interval exceeding that of the male. These odds are 24:1 and would be higher if it were possible to quantify more realistic prior odds. These would somewhat favour values of the female:male susceptibility ratio greater than 1, by analogy with other species.

Table 5 gives these odds of 24:1 and comparable figures for those other linkages - Lu:Se and $Tf:E_1$ - that are loose enough to give good evidence on sex differences. In sum, an estimated $2\frac{1}{2}$ linkages out of 3 tested show more recombination in females. To establish this tendency as a reality in man will require study of further linkages until the proportion of linkages showing this female excess exceeds $\frac{1}{2}$ to a greater extent than expected under chance deviations alone. The matter is more fully discussed by Renwick (1968).

TABLE 4

Lods for various pairs of female and male recombination fractions, Θ, for the ABO:AK linkage. More than half the data were made available by Prof. H. Harris and his colleagues.

Male Theta Values	Female Theta Values .50	.45	.40	.35	.30	.25	.20	.15	.10	.05	.01
.50	0	-.019	.105	.332	.597	.824	.923	.769	.147	-1.541	-6.278
.45	.290	.279	.414	.647	.914	1.144	1.242	1.088	.461	-1.229	-5.968
.40	.785	.790	.932	1.169	1.437	1.660	1.752	1.589	.955	-.742	-5.490
.35	1.416	1.437	1.588	1.823	2.082	2.294	2.371	2.195	1.548	-.162	-4.926
.30	2.109	2.147	2.303	2.534	2.781	2.975	3.032	2.837	2.171	.440	-4.345
.25	2.810	2.862	3.022	3.246	3.477	3.651	3.686	3.466	2.775	1.021	-3.794
.20	3.466	3.531	3.693	3.910	4.124	4.277	4.286	4.040	3.322	1.536	-3.318
.15	4.022	4.097	4.263	4.471	4.672	4.803	4.786	4.512	3.764	1.938	-2.969
.10	4.391	4.477	4.643	4.846	5.034	5.144	5.105	4.800	4.016	2.140	-2.843
.05	4.360	4.456	4.622	4.820	4.998	5.092	5.028	4.693	3.865	1.916	-3.205
.01	2.600	2.702	2.873	3.068	3.237	3.318	3.233	2.864	1.982	-.081	-5.527

TABLE 5

A count of the linkages showing an excess of recombination in one or other sex in man. Each linkage is partitioned, in the proportion of the odds, between the two classes, using a true susceptibility ratio of 1:1 for discrimination. Np:ABO has been excluded because of the present 4:1 odds (unpublished observations) that it is included in the looser ABO:AK linkage.

	Susceptibility ratio More than 1:1	Less than 1:1	Totals
Lutheran:Secretor Lu:Se	6/7	1/7	1
Transferrin:Cholinestarase$_1$ $Tf:E_1$	3/4	1/4	1
ABO:Adenylate kinase ABO:AK	24/25	1/25	1
Number of linkages	2.57	0.43	3

OTHER UNASSIGNED LINKAGE GROUPS

Of the remaining linkages that have not yet been assigned to an autosome, the one I wish specially to refer to is that between MNS and the locus, Tys, responsible for a rare and newly-described heterozygous condition of the skin with features of both hyperkeratosis and of scleroderma: sclerotylosis seems a suitable name. Dr. Marc Mennecier in Lille described the disease and the linkage this year, using Fisher's u scores (Huriez et al., 1968; Mennecier, 1968). The Bayesian approach outlined above, applied to the computer-determined likelihoods, gives a maximum probability at $4\frac{1}{2}$ units for the map interval and we calculate odds of 186:1 that the loci share the same chromosome - a direct probability of 0.995.

There will be reason later to refer to some of the other known linkages listed in Table 3.

NON-RANDOMNESS OF LOCI ON CHROMOSOMES

Finally, I wish to turn to the interpretation of the expected non-randomness of chromosome maps, a subject which will soon be getting much more attention. Figure 6 shows the X chromosome of the domestic fowl as it was known in 1960 (Hutt, 1960). Dr. Hutt did not comment on this remarkable clustering of color loci but it seems that the few extra loci added to the map since 1960 also fit the same picture. In 1960 there were 8 color loci known in this stretch of 77 units and, in the rest of the chromosome (35 units), there were 6 known loci, not one of them affecting color. Dr. Lewontin and Dr. Bodmer and others here who have worked on the selective pressures on linked and interacting gene loci (Lewontin, 1965; Bodmer and Parsons, 1962) will not be too surprised to see this diffuse clustering over such a long chromosome segment. It is natural to speculate that some of the interaction is that of clashing of certain colors with one another, but there are probably also other types of interaction of a more physiological or biochemical nature. Behavioral geneticists might go even further and might wish to seek an explanation of the sequence of color loci even within the cluster. They might also predict that some of the spaces between the known color loci would contain not only further color loci but also gene loci controlling courtship display. These would be expected to interact strongly with the color loci since there is little point, for instance, in some feather having a bright spot if it is never displayed.

This brings me to a possible genetic technique which might very well become commonplace in the future. Formal genetics used to proceed by, first, a study of the genes, then a mapping of them. Perhaps the reverse order is more appropriate for difficult fields such as behavioral genetics. It might be possible first to predict the location of a behavioral gene locus, say somewhere in the color region of the X, and to detect a polymorphism in it by comparing the two color classes of progeny of a cock heterozygous at a color locus on the X. Of course, this clustering of color loci on the X of the fowl seems to be an extreme case: it will be some time before we can, more generally, predict a gene function from those of its neighbors on the chromosome. But it does direct attention towards a more systematic study of chromosome maps for non-randomness.

COMPUTER AID IN CLUSTER ANALYSIS

Computers can be useful here for the Knox (1959) test for non-randomness or one of the more sophisticated tests discussed by Cox and Lewis (1966). The cluster analysis program discussed by Dr. A. W. F. Edwards at this conference seems particularly promising. Its value will be in drawing attention to possible clusters, perhaps as striking as that on the chicken's X chromosome but at present obscured by the background of other loci. A necessary preliminary will be the scoring of each locus in numerous dimensions of function, as well as in the one dimension of the linear map.

FACTORS INFLUENCING GENE SEQUENCE

Including the interactions discussed above, we can list some of the factors influencing (or possibly influencing) the sequence of loci on chromosomes as follows:

1. Fitness interactions between loci.
2. History: New genetic material is believed often to arise by tandem duplication of the old with subsequent divergence of function (Haldane, 1932; Epstein and Motulsky, 1966). Until the randomising or selection effects of chromosomal rearrangements have had time

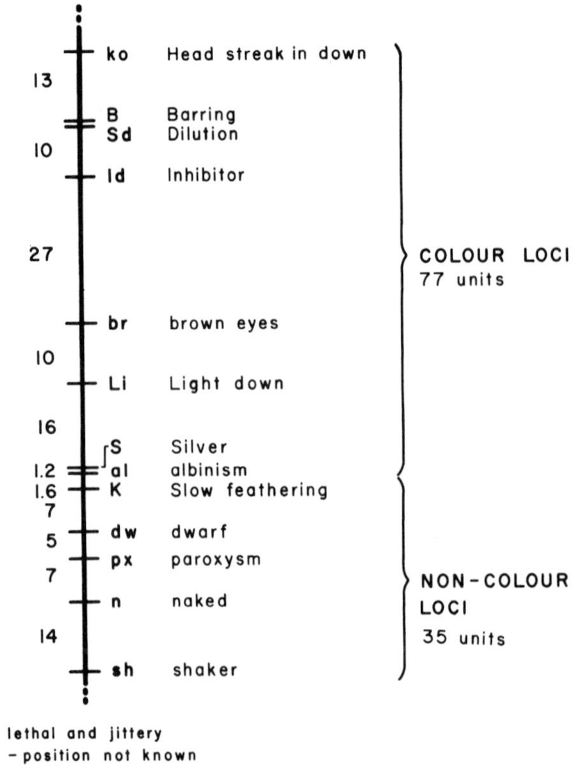

Fig. 6. Striking non-randomness as indicated in the linkage map, published by Hutt (1960), of the X chromosome of the domestic fowl.

to obscure completely this relationship of the old and the new, a degree of 'transient' non-randomness will remain in the gene sequence for a few million years.

3. Economy: If several loci require switching on or off simultaneously or require a particular susceptibility to crossing-over or mutation, there might well be economies if the control operated on these loci as a unit. Arrangement together in a chromosome segment would facilitate such a group control. Operons are examples of the expression of this economy factor in regard to small dense clusters of loci, and X-inactivations are examples in regard to large clusters.

There are probably other examples of the operation of this economy factor on groups of loci between these two extremes. In particular, we could consider a class of loci affecting a particular tissue and thus sharing certain switching requirements. Loci belonging to such a class might be found grouped in a particular chromosome segment - either dense or diffuse clustering.

Similar clustering of loci that are concerned with the same stage of embryogenesis, and that have mutational and switch requirements in common, might also be expected. Since the embryo is partially protected against the fluctuations of the outside environment, and also for reasons of evolutionary antiquity, such loci would probably have less than their share of polymorphisms (in which the commonest allele has a frequency of, say, 0.99 or less). The remaining segments (those concerned with post-natal stages of the life-cycle) would correspondingly show more than their share of polymorphic loci and this can be looked for, insofar as it would lead to an increased expectation of linkage.

The factor of switch-economy in development and in tissue differentiation has previously been given little attention except in T4 bacteriophage in which there is a clear arrangement of loci in large blocks according to function (Edgar and Epstein, 1965). The block of genes for, say, forming the phage head is large in proportion to the total genome and this might correspond in scale to a postulated block of gene loci controlling the differentiation of some tissue in a mammal. Each block in T4 is not an operon as the loci do not share a common messenger RNA, to judge from the paucity of polar mutants in the group. The T4 clustering could be interpreted either on the basis of interacting loci or on the basis of economy of control mechanisms or both. The map of *Streptomyces coelicolor* is also strikingly non-random (Hopwood, 1967).

All the above three mechanisms favouring clusters will tend, for different reasons, to group together the same loci. Historically-related loci are likely to have similar functions for purely chemical reasons and these functions may have switch and control requirements in common. If these loci have similar function, their alleles are likely to interact and be clustered for that reason. Thus, all factors will be operating, more or less, in parallel. This should enhance the total clustering effect and make the clusters more easily detectable. However, the next stage, that of interpretation, is inevitably made more difficult.

DIFFICULTY IN INTERPRETING PARTICULAR GENE CLUSTERS

This difficulty can be illustrated even on the few relevant data available in man. No doubt the following hints of non-randomness from our present meagre knowledge of about a dozen autosomal linkages would be most wisely dismissed as coincidences or due to biased sampling but they will serve for discussion. Against each is listed the possible factors that might favour the linkage; Z denoting some interaction (through intervening loci) too remote to be detectable:

		Possible factors
i.	Transferrin and cholinesterase are both believed to originate in the liver and both are plasma proteins. They are both absent from the majority of other tissues. Their loci, Tf and E_1, are about 16 units apart (Robson et al., 1966).	1, 3, Z
ii.	The same remarks apply to albumin and the glycoprotein known as Gc or group-specific component. The loci for these proteins are 2 units apart (Weitkamp et al., 1966).	1, 2, 3
iii.	The ABO antigens and the enzyme adenylate kinase are both widely distributed but the erythrocyte excels at making both. Their loci are about 16 units apart (Rapley et al., 1968).	3, Z
iv.	The two adult haemoglobin chains β and δ are active only in erythrocytes. Their loci are contiguous (Epstein and Motulsky, 1966). The α chain, whose synthesis spans nearly all stages of development as well as post-natal life has a locus not linked to these two. Its locus would not be expected to share a switching mechanism with the loci of the adult chains β and δ.	1, 2, 3
v.	The Lutheran and Secretor loci are both concerned (in different ways) with blood group antigens and they are about 13 units apart (Mohr, 1851; Renwick, 1968). This functional similarity might possibly imply a general parallelism for the activity of these loci, in terms of timing or of distribution in the tissues, though it can hardly be a close parallelism.	1, 3, Z

None of these linkages can be attributed to a single causative factor unambiguously and, in the case of the haemoglobin β and δ loci, it is very likely that all three factors have been influential. A similar statement could apply to the close linkage of the deuteranopia and protanopia loci on the X chromosome.

All the linkages listed are between loci whose products are present together in one tissue and absent together (or reduced together) in others, as might be expected if factors 1, 3 or Z were operating. It might further be noted that there is, as yet, no example of autosomal linkage between a locus scored on a product in erythrocytes and one scored on a product in plasma in spite of adequate data being available. At least some of the effect might, of course, be due to a specialist interest of an investigator in erythrocyte phenotypes or in plasma protein phenotypes making a linkage more likely to be detected if both loci are within the researcher's specialist interest.

One of the merits of the large-scale computer linkage analysis in Baltimore is that it will minimize, at least at the analysis stage, the factor of unequal thoroughness of the linkage searching. It seems quite possible that interesting clustering of loci in man could be demonstrated in the course of a few years, if the non-randomness is of sufficient degree.

SUMMARY

A variant of Smith's (1959) Bayesian approach to human linkage analysis is outlined which takes into account increasing knowledge relevant to the prior distribution of particular lengths of the map interval between two loci chosen at random on the same (unspecified) autosome. This circumvents the assumption, perhaps inaccurate for man, that recombination fractions are all equally likely, without adding to the difficulty of analysis.

A summary is given of present knowledge of the seven unassigned autosomal linkage groups in man and of the two assigned linkage groups - Fy-Cae, TK on chromosomes 1 and 17, respectively. Where appropriate, the contribution of the 1961 computer linkage program to some of this mapping is indicated. The contribution has been facilitated recently by the employment of ancillary programs constituting a suite for preparing, analysing and storing linkage data and for storing, on a cumulative basis, the corresponding lod scores.

Finally, attention is directed to the need for computer assistance for the study of non-randomness in linkage maps. A striking clustering of color loci on the X chromosome of the domestic fowl is pointed out, which should serve to encourage interest in the factors influencing the sequence of gene loci on chromosomes. The main factors are considered to be (1) interaction between loci with respect to the fitnesses of different combinations of genotypes, and (2) historical factors - a new gene locus will tend to be near the locus from which it arose by duplication. More tentatively, (3) economy of control is believed to play a part - control of mutation rate, control of crossover rate, control of gene activity could be exercised for a large group of linked loci in common, particularly for tissue differentiation or for embryological development. It is predicted that computers will have an increasing role in the making of chromosome maps and in the analysis of non-randomness in them.

REFERENCES

Bayes, T. 1763. An essay towards solving a problem in the doctrine of chances. Phil. Trans. Roy. Soc. 53: 370-418.

Bloom, G. E., Gerald, P. S., and Reisman, L. E. 1967. Ring D Chromosome: A second case associated with anomalous haptoglobin inheritance. Science 156: 1746-1748.

Bodmer, W. F., and Parsons, P. A. 1962. Linkage and recombination in evolution. Adv. Genet. 11: 1-100.

Browne, W. G., Izatt, M. M., and Renwick, J. H. 1969. White sponge naevus of the mucosa: clinical and linkage data. In preparation.

Carter, T. C., and Falconer, D. S. 1951. Stocks for detecting linkage in the mouse and the theory of their design. J. Genet. 50: 307-323.

Ceppellini, R. R. 1959. Discussion. In Biochemistry of Human Genetics, ed. G. E. W. Wolstenholme and C. M. O'Connor. Ciba Foundation Symposium. London: Churchill. p. 135.

Cooper, H. L., and Hernits, R. 1963. A familial chromosome variant in a subject with anomalous sex differentiation. Amer. J. Hum. Genet. 15: 465-475.

Cox, D. R., and Lewis, P. A. W. 1966. Statistical analysis of series of events. London: Methuen.

Crawford, M. N., Punnett, H. H., and Carpenter, G. G. 1967. Deletion of the long arm of chromosome 16 and an unexpected Duffy blood group phenotype reveal a possible autosomal linkage. Nature 215: 1075-1076.

Donahue, R. P., Bias, W. B., Renwick, J. H., and McKusick, V. A. 1968. Probable assignation of the Duffy blood group locus to chromosome 1 in man. Proc. Nat. Acad. Sci. 61: 949-955.

Edgar, R. S., and Epstein, R. H. 1965. Conditional lethal mutations in bacteriophage T4. Proc. XI Internat. Congr. Genet., The Hague 2: 1.

Epstein, C. J., and Motulsky, A. G. 1966. Evolutionary origins of human proteins. Progr. Med. Genet. 5: 85-127.

Gerald, P. S., Warner, S., Singer, J. D., Corcoran, P. A., and Umansky, I. 1967. A ring D chromosome and anomalous inheritance of haptoglobin type. J. Pediat. 70: 172-9.

de Grouchy, J., Veslot, J., Bonnette, J., and Roidot, M. 1968. A case of 16d-chromosomal aberration. Amer. J. Dis. Child. 115: 93-99.

Haldane, J. B. S. 1919. The combination of linkage values, and the calculation of distance between the loci of linked factors. J. Genet. 8: 299-309.

Haldane, J. B. S. 1932. The causes of evolution. New York and London: Harper.

Hopwood, D. A. 1967. Genetic analysis and genome structure in streptomyces coelicolor. Bact. Rev. 31: 373-403.

Huriez, C., Deminatti, M., Agache, P., and Mennecier, M. 1968. Une genodysplasie non encore individualisee: la genodermatose scleroatrophiante et keratodermique des extremites frequemment degenerative. Semaine des Hopitaux de Paris 44: 481-488.

Hutt, F. B. 1960. New loci in the sex chromosome of the fowl. Heredity 15: 97-110.

Knox, E. G. 1959. Secular pattern of congenital oesophageal atresia. Brit. J. Prev. Soc. Med. 13: 222-229.

Kosambi, D. D. 1944. The estimation of map distances from recombination values. Ann. Eugen. (Lond.) 12: 172-175.

Lawler, S. D. 1954. Family studies showing linkage between elliptocytosis and the Rhesus blood group system. Caryologia, Suppl. 6: 26-27.

Lewontin, R. C. 1965. The role of linkage in natural selection. Proc. XI Internat. Congr. Genetics, The Hague 3: 517-525.

Mennecier, M. 1968. Individualisation d'une nouvelle entite: la genodermatose scleroatrophiante et keratodermique des extremites, frequemment degenerative: etude clinique et genetique (possibilite de linkage avec le systeme MNSs). M. D. thesis: Lille.

Mohr, J. 1951. Estimation of linkage between the Lutheran and the Lewis blood groups. Acta path. microbiol. Scand. 29: 339-344.

Morton, N. E. 1955. Sequential tests for the detection of linkage. Amer. J. Hum. Genet. 7: 277-318.

McIlree, M. E., Tulloch, W. S., and Newsam, J. E. 1966. Studies on human meiotic chromosomes from testicular tissue. Lancet 1: 679-682.

Philip, J., Frydenberg, O., and Sele, V. 1965. Enlarged chromosome No. 1 in a patient with primary amenorrhoea. Cytogenetics 4: 329-339.

Rapley, S., Robson, E. B., Harris, H., and Maynard-Smith, S. 1968. Data on the incidence, segregation and linkage relationships of the adenylate kinase (AK) polymorphism. Ann. Hum. Genet. 31: 237.

Renwick, J. H. 1968. Ratios of female to male recombination fractions in man. Bull. Europ. Soc. Hum. Genet. 1:

Renwick, J. H., and Bolling, D. 1967. A program-complex for encoding, analyzing and storing human linkage data. Amer. J. Hum. Genet. 19: 360-7.

Renwick, J. H., and Lawler, S. D. 1955. Genetical linkage between the ABO and nail-patella loci. Ann. Hum. Genet. 19: 312-331.

Renwick, J. H., and Lawler, S. D. 1963. Probable linkage between a congenital cataract locus and the Duffy blood group locus. Ann. Hum. Genet. 27: 67-84.

Renwick, J. H., and Schulze, J. 1964. An analysis of some data on the linkage between Xg and color blindness in man. Amer. J. Hum. Genet. 16: 410-418.

Renwick, J. H., and Schulze, J. 1961. A computer program for the processing of linkage data from large pedigrees. Excerpta Med.: Internat. Congr. Ser. 32: E145 (abstr.).

Robson, E. B., Sutherland, I., and Harris, H. 1966. Evidence for linkage between the transferrin locus (Tf) and the serum cholinesterase (E_1) in man. Ann. Hum. Genet. 29: 325.

Smith, C. A. B. 1959. Some comments on the statistical methods used in linkage investigations. Amer. J. Hum. Genet. 11: 289(2).

Smith, S. M., Penrose, L. S., and Smith, C. A. B. 1961. Mathematical tables for research workers in human genetics. London: Churchill.

Weiss, M. C., and Green, H. 1967. Human-mouse hybrid cell lines containing partial complements of human chromosomes and functioning human genes. Proc. Nat. Acad. Sci. 58: 1104-1111.

Weitkamp, L. R., Rucknagel, D. L., and Gershowitz, H. 1966. Genetic linkage between structural loci for albumin and group specific component (Gc) Amer. J. Hum. Genet. 18: 559-571.

Ying, I. A., and Ives, E. 1968. Personal communication.

Ying, I. A., and Ives, E. 1968. Canad. J. Genet. Cytol. 10: 575.

Present address: Department of Social Medicine
London School of Hygiene and Tropical Medicine
Keppel Street
London, W.C.1., England

Supported by grants from the Hartford Foundation and the National Institutes of Health (GM-10189) to Dr. V. A. McKusick, Baltimore, and from the Medical Research Council (G960/109B). The computing facilities used were those of the Computing Center of the Johns Hopkins Medical Institutions, supported by grant FR-00004 from the National Institutes of Health and by educational contributions from the International Business Machines Corporation.

DISCUSSION ON GENETIC LINKAGE

J. H. Edwards, W. J. Schull, N. E. Morton, and R. C. Elston

J. H. EDWARDS: I will discuss Dr. Renwick's paper covering linkage analysis firstly in general, and secondly in some specific details.

Given a set of phenotypic responses in individuals and the identity of their parents, it should be possible to summarize our observations in terms of genetic factors related by gametic exclusiveness (allelism), their relative abundance (gene frequency) and their location (linkage). The first problem, allelism, appears to be unsolved at the algorithmic level, although up to 5 or so alleles seem capable of resolution intuitively and, even with larger numbers, such as the possible set of 2^{12} of Ceppelini, intuition may make plausible models in which the alleles inferred are far fewer than the number of phenotypes.

The gene frequency problem is solved, given large numbers and intuitive help on designating absent genetic factors. Linkage, although apparently more complicated than either allelic specification or the identity of absent genetic factors, now appears to have been solved.

No doubt there is room for improvement in speed and simplicity, but a programme is now operational which will accept phenotypes and relationships and deliver likelihoods for any pair of male and female recombination fractions, and I will therefore limit the discussion to this programme.

Dr. Renwick adds various **a priori** log-likelihoods to his log-likelihood scores, which lead to different results; these are presumably better results if the **a priori** assumptions are justified. In relation to chromosome number, this knowledge is clearly relevant to linkage detection. In assuming a high correlation between functional and structural chromosomal distances, the assumptions are justified; but the benefits of the correction appear small in relation to the complications, and recent evidence on the length of the human genome only mildly steepens the flat **a priori** likelihood distribution noted by Morton. So far as mapping functions are concerned, I think these should be as simple as possible, or estimated from the linkage data, rather than imposed from the mouse. As there are real objections to **a priori** distortions of the results, I would like to suggest that here, as elsewhere, the corrections are applied at the very end, and both raw and corrected, or otherwise changed, likelihoods or estimates given in the results.

The coding problem is of general importance and in view of the work involved in coding data, I will merely comment briefly on the system I am using of logical operations on two binary strings, which I term markers and masks. The latter, the mask, specifies what has been done: the former, the marker, the result. Both can be abbreviated into octal or decimal numbers, and we may then classify the results into compact integers by considering only the required tests. For example, consider responses a, b, c, d, e

	a	b	c	d	e
MARKER	1	1	0	0	1
MASK	1	0	1	0	1

This codes three tests out of five possible tests. The result of the test not done is never used and is therefore arbitrary.

We now introduce the concept of a necessary marker, NM, which specifies the tests needed for the analysis. A subroutine then ascertains that all the required tests are done, i.e., is NM.AND.MASK = NM

For example

		Decimal	Octal
MARKER	1 1 0 1 1	27	66
MASK	1 0 1 0 1	21	52
NM	1 0 1 0 0	20	50
MASK. AND. NM	1 0 1 0 0	20	50
AND. MARKER	1 . 0 . .	2	2

The advantage of this is that any system of phenotypes can be coded in full, and stored in two integer words, and any subset of tests analysed at high speeds on binary computers. The decimal marker has a 1 to 1 relationship with any other phenotypic symbol. The alleles to be considered can be expressed in similar binary form and their relative frequency estimated from the same data. The procedure has been found quite easy to operate in practice, the data being presented in the form of +, - or blank, blank imposing zero in mask, and + or - being coded as 1 or 0.

The output is a more difficult problem. Ultimately, for every pair of loci we can derive, at a resolving power limited only by cost, a likelihood surface relating likelihood to θ_p and θ_m, the paternal and maternal recombination fractions, and we wish to go from $(n^2-n)/2$ surfaces for n loci to n points both assigned and located on 22 linear structures of arbitrary polarity and approximately known length.

First, we may consider reducing these surfaces, summarized by the computer as matrices, into lines or vectors and then cutting down the number of points or elements.

This can be summarized, in order of meaningfulness, by the mean, the mean and peak-curvature, or by three or more ordinates.

I would like to propose that, initially, we summarize results in the form of "equivalent observations," that is, the number of equivalent total observations such as total number and number of recombinants which would be necessary to give the same mean and variance. These have the simple properties of intelligibility and addivity, and are equivalent to means and weights, although, since in these likelihood surfaces the peak does not specify the foothills, equally approximate.

A pair of numbers, such as these, has the advantage that the total linkage data at any time can be expressed in the form of an n x n matrix, the main diagonal being meaningless and the two other parts containing respectively equivalent numbers and recombinants, or alternately likelihood values for $\theta = 0.1$ and $\theta = 0.3$.

We can represent these results by a figure of nodes and lines, or wires, in which length is proportional to some function of recombination fraction, restricted to $\theta = 0.0 \rightarrow 1.0$, and line thickness represents weight of evidence, say (NR + R) or $(NR + R)^3/(NR \times R)$, between pairs of loci. A wire model of this was once known as "Castle's rat cage."

In view of the complexity of handling the results, I see little advantage in expressing them in unduly complex terms, such as polynomial fits, exact areas, etc. Nor do I see any particular advantage in worrying about safe assignment (certain linkage) rather than most likely assignment. Now that a few secure base-lines are established, this seems less important.

The problem will ultimately reduce to the optimal assignment of the set of equivalent observations on to another set of linear maps. A similar but more complicated model may be envisaged using likelihood vectors for various values of θ.

Initially we will ignore wire thickness and the different functional length of chromosomes. Our problem then is to fit n loci onto m chromosomes. This appears simple intuitively, as we have merely to cut away as much wire as possible to leave m sets of connected points, and then find an orientation in which the order of the points is unambiguous over the widest change of orientation. This is hardly rigorous, but it is capable of more rigorous expression. Clearly both the segmentation and the sequence will be dependent on the mapping function; however, I think this will be a second order effect for many years in relation to human data, and we may as well stay with the simplest function, Haldane's, until we have more evidence of the relative recombination in male and female and of interference. Other species have been well mapped with very little attention to mapping functions, or to the distortions induced by the corrections based on chromosome lengths.

In more rigorous terms, we have one output in the form of a matrix relating to the two basic parameters of recombination fraction and weight, which may be derived directly from the simpler matrix of total numbers (t) and recombinants (r) since $\theta = r/t$ with weight $= t^3/t(t-r)$, and it is necessary to find some sequence of the rows and columns, and some mapping function, which will maximize the likelihood.

This is equivalent, in the one chromosome case, to the travelling salesman problem on a long thin island with a fare-structure in which long journeys cost less per mile, almost reaching a maximum at less than the length of the island. Given the data on approximate fares, we need a minimum-mileage solution. The many chromosome case is equivalent to the travelling salesman problem on an archipelago with long thin islands and very slow ferries going between the ends of any pair in equal time.

Since the data on mapping functions is of the same form as the data on linkage it is necessary to estimate the mapping function from the data. All mapping functions are restrained by the asymptotes $w=\theta$ when θ is small and $w \rightarrow \infty$ as θ approaches 1/2. Haldane's, the simplest, is equivalent to

$$w = (1/2)(2\theta + (2\theta)^2/2 + (2\theta)^3/3 + (2\theta)^4/4 + (2\theta)^5/5 \quad \text{etc.})$$

Kosambi leaves out the even powers, giving

$$w = (1/2)(2\theta + (2\theta)^3/3 + (2\theta)^5/5 \quad \text{etc.})$$

and Carter and Falconer leave out another half giving

$$w = (1/2)(2\theta + (2\theta)^5/5 \quad \text{etc.})$$

Some other function is needed which has a parameter available for estimation, or a pair of parameters for each sex. A simple function is

$$w = (1/2K)((1+2\theta)^K-1),$$

or

$$w = \frac{[(1 + \theta)^k - 1)]}{k}$$

The approach to the asymptote is of little interest in man as bivalents with more than six crossovers are rare in both man and woman.

In the single chromosome case we have the matrix

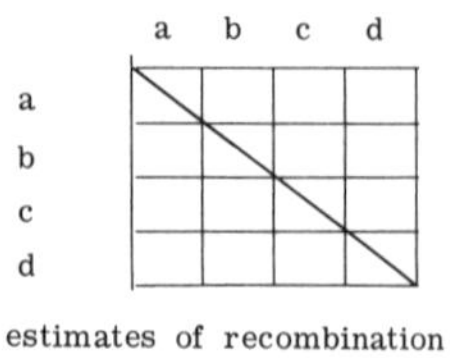

and the various estimates of recombination fractions $\theta_{ij} = \theta_{ji}$ between loci i and j, or, in the diagram

θ_{ab}, θ_{bc} etc.

and we have to find a sequence such that

$$f(\theta_{ac}) \simeq f(\theta_{ac}) + f(\theta_{bc})$$

and, in general,

$$f(\theta_{ik}) \simeq f(\theta_{ij}) + f(\theta_{jk}) \qquad ((\theta_{ik}, \theta_{ij}, \theta_{jk}))$$

where $\simeq$ implies maximization, and the procedure maximizes both the sequence and the parameters of the mapping function f.

A priori evidence of map length is easily incorporated since we have the inequality $(\theta_{1,n})' < e_m$ where e is total genetic length of the chromosome and an **a priori** likelihood, which we may regard as an ascertainment condition, that the distribution of $f(\theta_{1,m})$ is the same as the total length of loose segments in a string of length e_m held at the ends and exposed to m-1 cuts, which, in the absence of interference, are randomly distributed.

In this case the m segments have all an equal expectation of length and a simple distribution. However, it seems that in man the linkage data will soon be better than the cytological data, and e_m should be estimated rather than imposed.

In the multichromosome case, as in the autosomes, it is simplest to regard the n loci as lying on one long chromosome along which are distributed m-1 telomeres, which have the property that the recombination fraction between any loci separated by 1 or more telomeres is 1/2.

We may represent the data in the form of a matrix, when a, b, c, are loci and P, Q, R telomeres and we then have to find the optimal sequence of abcd .... and PQRS so as to maximize the distance functions over all pairs of loci. PQRS have a fixed sequence, but the intervals or the proportional intervals, PQ, QR, RS may be known, or assumed equal.

Finally I would like to refer briefly to the programme I am writing for the analysis of linkage, which appears simpler than Dr. Renwick's, but is disc oriented. I have written all the subroutines, which will just squeeze out an 8k 1130, and would handle four alleles on a 16k machine easily. As repetitive disc seek operations are needed, I do not think a bigger machine would be economical.

Basically we start with a pedigree composed of founders, who have no common ancestors, or ancestors of known phenotype, and others. All data consists of identity, disc address of parents, and phenotype in the MARKER, MASK form, and a working section for the coded phenotype and the gametic vector.

Every founder member receives two gametes and generates a gametic matrix, given the recombination fraction and the phenotype. After assembling these data, the computer then identifies all children and spouses by disc address, the spouses preceding the children, and being identified by a negative disc address, which is made positive before use.

The basic linkage operation is simple, and easily expressed symbolically.

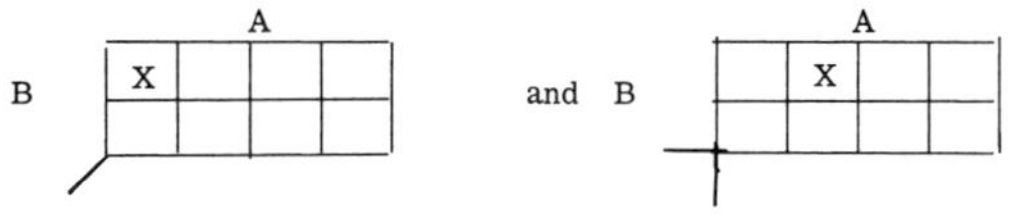

for male and female gametic matrices, where cells refer to the frequencies of all possible allele pairs (in this case 2 and 4) relating to loci A and B.

Founder members receive gametes with allele frequencies defined by the gene frequencies; every individual with children is provided with a gametic matrix by "filtration" and recombination. The gametic matrix formed is the result of all possible pairs, relevant to the parental phenotype, which are then rearranged by phase according to the recombination fraction.

Each sibship, defined to include single sibs when phase is known, or partly known, is considered to be the result of summing the expectations derived from all possible quads of gametes entering the parents, and comparing this with the observed phenotypes of the children. Gametic matrices are restandardized to add to unity throughout.

Filtration is simple, and merely involves defining possibility matrices for A and B, which only "transmit" possible gametes. For example, in the ABO system, where a parent is A the possibility matrix for two gametic vectors is

O	A	B

	O	A	B
O	O	1	O
A	1	1	O
B	O	O	O

Filtration may be regarded as a particular form of ascertainment, which is likewise conveniently specified as a zero-integer vector or matrix, or, in more complicated cases, by a real-number matrix.

Recombination is also simple. The problems of ascertainment are ignored, and largely annulled, through the inclusion of all sibships without reference to certainty of heterozygosity or other criteria of informativeness. There is no limit on family size, excepting the size of the disc, nor on the number of alleles, excepting the size of the core, analysis proceeding in the same sequence as the events being analysed. With this logic, however, it may be difficult to allow overlapping generations.

W. J. SCHULL: Most of the remarks to follow are directed primarily toward the detection of autosomal linkages and the estimation of the recombination fractions associated therewith, and only secondarily toward sex linkage. The latter, since it presumably involves only estimation and not the detection of linkage, presents a somewhat simpler problem at least in theory. As Dr. Renwick has pointed out, past studies of linkage have generally fallen into one or the other of two different but not necessarily mutually exclusive categories. One type of study tended to focus on a small number of families, possibly but a single one, each generally fairly large and involving the reproductive performances of a number of related couples. Often a specific analysis was based upon pedigrees mostly culled from the literature whose manner of ascertainment was apt to be unrecorded. In most instances, a study was a one-time affair with the investigator rarely motivated to continue collection of data for further periodic analyses. Calculation of linkage probabilities under these circumstances can be a formidable undertaking as Dr. Renwick's presentation attests. The theory to guide such calculations seems, however, fairly well advanced (see, e.g., Norton, 1949; Smith, 1959; Renwick and Bolling, 1967).

The second type of study involves a larger number of families each usually represented by a single couple whose reproductive performances constitute the basis for both detection and estimation. All of the information was as a rule collected by a single person or laboratory, and collection of the data was motivated, at least in part, by the prospects of linkage studies. Again, however, analysis was a one-time affair with no effort to augment that which had once been collected and analyzed. A variety of estimation procedures of varying efficiencies have been used to analyze such data. Most assume fixed sample sizes but at least one presumes sampling to be sequential. Among the former are such early suggestions as those of Bernstein (1931), Wiener (1932), Haldane (1934), and the u-statistics advocated by Fisher (1935a, b), Finney (1940; 1941a, b; 1942a, b, c; 1943) and Bailey (1951a, b). Morton's (1955, 1957) method remains the only commonly used procedure which utilizes as a matter of course a sequential stop rule to sampling. A more recent proposal, which neither imposes a sequential or a fixed stop rule, is the Bayesian method suggested by Cedric A. B. Smith (1959); it neatly combines detection and estimation of linkage in probability form. The results of this approach can, therefore, be used for further calculations where and when indicated.

Dr. Renwick now holds out to us the prospects of what might be viewed as a third approach, namely, the creation of one or more centers which would serve as repositories for linkage data and would undertake periodically to collate, colligate and analyze the data in their charge. Hopefully as each set of data arrived it would be possible to reassess those potential linkages to which these new data might contribute. Each new analysis would serve to sharpen previous estimates of recombination probabilities. The prospects of this degree of centralization and automation of a process which has previously been both disjointed and tedious are exciting, but patently problems must exist in implementation. What are they? And are they remediable?

A moment's reflection suffices to establish that implementation of these objectives of greater centralization and automation poses a wide variety of problems too numerous to treat exhaustively here. We propose, therefore, to ignore such matters as authorship, the funding of such an effort, etc. and to restrict our remarks to a few general observations on the control of data quality, the incorporation of ancillary information into the estimation process, and several strategic considerations. Chief among the problems associated with quality control of the data are errors in the specification first, of the biological antecedents of a given individual and second, in the phenotype or genotype of a particular person. Illegitimacy per se is of no moment, of course, if the father and mother of the child in question are properly identified. If, however, the parents are incorrectly specified, the prior probability that a particular child will be of a given phenotype may be markedly different from that presumed. Some notion of the possible magnitude of this problem in the United States is afforded by the data of Schacht and Gershowitz (1961) who found 1.48 per cent of the children of 367 Caucasian families of Detroit to be extra-marital on the basis of the secretor system and/or some seven blood group systems. Among 96 negro families drawn from the same city, 8.89 per cent of the children were extra-marital on the basis of the same tests. The estimated overall frequency of extra-marital issue, as judged by the MN system, was 1.49 and 10.10 per cent in the two groups. It might thus seem that with seven or eight "marker" systems one would detect the bulk of cases where the father or mother have been erroneously identified in white families but not negro; unfortunately those cases not detected would not be randomly distributed over mating types. They would be concentrated in matings involving heterozygous individuals -- the very ones which contribute linkage information. In the present instance, the conclusion seems inescapable that the frequency of extra-marital issue is so high among negro families of Detroit as to impair seriously their value in linkage studies; it seems probable that this also holds true elsewhere in the United States.

No less important than the errors just cited are errors of phenotypic or genotypic classification. Increasingly it is becoming evident that reputable laboratories can differ substantially in the serotype they assign to the same individual. Discrepancies may be as infrequent as one per cent in systems such as the ABO and Rh, and as common as 41 per cent in the P system (Osborne, 1958; but see also Ellis, Cawley, and Lasker, 1963; Neel, Salzano, Junqueira, Keiter, and Maybury-Lewis, 1964). While numerous explanations can be adduced to account for such disagreements, the fact remains that serotyping may be so variable as to render valueless certain blood group systems for linkage studies. Even with the ABO system, where errors are relatively infrequent, they are apt to be systematic within a particular laboratory. It is quite possible that misclassification can both lead to serious errors in the estimation of the recombination fraction when a linkage is detected, or to a spurious linkage heterogeneity. The approach implicit in Dr. Renwick's remarks places a heavy burden upon the constancy of laboratory standards which most certainly does not now exist and may not be readily obtainable.

A second area of difficulty involves those parameters which may be viewed as nuisances in the detection and estimation of linkage. This includes reduced or partially manifesting characteristics, selective differentials, and so on. Possibly the most widespread and potentially troublesome of these are selective disadvantages which may be mediated through the clinical or pathological characteristic in question, the "markers," or both. It seems reasonable to assume, however, that the selection associated with the "markers" is apt to be small, possibly only one to two per cent, and hence not necessarily actively misleading. "Reproductive compensation" is another eventuality which may be more widespread than heretofore believed. The effects of these parameters on the spatial relationships which are estimated have been inadequately studied thus far.

In this same general vein are problems of ascertainment. While considerable attention has been directed towards the kinds of biases which can be introduced by the manner in which one's data come to attention, all of these efforts assume an orderliness or directedness, a plan if you will, to the sampling which in my experience rarely if ever exists. Opportunism more than any other element has characterized data acquisition. Erroneous assumptions about ascertainment can significantly alter the estimates of the spatial relationships which one presumes to exist between two linked loci. This was first pointed out by R. A. Fisher (1936) in connection with Haldane's data on partial sex linkage. Both assumptions of complete and single ascertainment could and did lead to exaggerated notions of the closeness of linkage in these data. While partial sex linkage as a phenomenon is now in disrepute, the effect of assumptions respecting ascertainment on the estimation of recombination fractions remains an important issue.

Finally, there exists the problem of incorporating into the prior probabilities of linkage information which may be derived from other, possibly less classical methods for the detection and estimation of linkage. Specifically, I have in mind the kind of evidence which can be deduced from a study of certain chromosomal aberrations. Thus, complete monosomies such as the monosomy G recently reported (Al-Aish et al., 1967) can, if they arise in the issue of an appropriate mating, serve to localize particular "marker" loci. While instances of complete monosomy may be rare, partial detections of a chromosome are more common and can also provide evidence on spatial relationships of genes. The same can be said of the trisomies, and in fact the latter have been used in precisely this connection on several occasions already. The results are, as yet, inconclusive, but it has been suggested, for example, that the ABO locus may be on chromosome 21 (Shaw, 1962). There are pitfalls to this approach, of course. Witness the elevated levels of glucose-6-phosphate dehydrogenase, an enzyme controlled by X-linked genes, encountered in Down's syndrome (Mellman et al., 1964). While in this speculative vein, it is not beyond the realm of reason to believe that cell culture, granted further developments, may contribute importantly to the mapping of human chromosomes.

Thus far we have concerned ourselves with issues which are only indirectly related to computing. Let us now examine some strategic problems. It seems to me that one of the most important of these involves the choice of programming level since this is ultimately related to machine efficiency and hence to machine time and costs. I take it as an article of faith that we all agree that the implementation of linkage studies on the scope discussed by Dr. Renwick will be both time-consuming and costly, and that economies are in order. Clearly, the most efficient programming would be in machine language, but since this may vary greatly from computer to computer, it obliges one to specify a particular machine on which the analyses are to be done. This would seem unwise in view of the past pace of machine development; a less efficient but certainly less hazardous decision would be to use assembly languages. An equally great economy might be achieved through the use of more efficient list processing procedures in collating the data. I presume that at the present time a very substantial investment in machine time occurs in this area, and that economies here would be particularly desirable. Clearly, an ever present danger in programs of the scope and complexity of the one described to us by Dr. Renwick, is that one becomes wedded to a technology which may be grossly out-of-date. This is not a particularly happy prospect for the one who has labored long, and presumably diligently in the development of the program, but then neither is it a pleasing prospect to those who must ultimately pay for our endeavors.

---

1. Al-Aish, M. S., de la Cruz, F., Goldsmith, L. A., Volpe, J., Mella, G., and Robinson, J. C. 1967. Autosomal monosomy in man. New Eng. J. Med. 277:777-784.
2. Bailey, N. T. J. 1951. On simplifying the use of Fisher's u-statistics in the detection of linkage in man. Ann. Eugen. 16:26-32.
3. Bailey, N. T. J. 1951. The detection of linkage for partially manifesting rare "dominant" and recessive abnormalities in man. Ann. Eugen. 16:33-44.
4. Bernstein, F. 1931. Zur Grundlegung der Chromosomentheorie der Vererbung beim Menschen, mit besonderer Berücksichtigung der Blutgruppen. Ztschr. f. induct. Abstamm. u. Vererbungslehre 57:113-38.
5. Ellis, F. R., Cawley, L. D., and Lasker, G. W. 1963. Blood groups, hemoglobin types and secretion of group-specific substance at Hacienda Cayalti, North Peru. Hum. Biol. 35:26-52.
6. Finney, D. J. 1940. The detection of linkage. I. Ann. Eugen. 10:171-214.
7. Finney, D. J. 1941. The detection of linkage. II. Ann. Eugen. 11:10-30.
8. Finney, D. J. 1941. The detection of linkage. III. Ann. Eugen. 11:115-35.
9. Finney, D. J. 1942. The detection of linkage. IV. J. Hered. 33:157-60.
10. Finney, D. J. 1942. The detection of linkage. V. Ann. Eugen. 11:224-32.
11. Finney, D. J. 1942. The detection of linkage. VI. Ann. Eugen. 11:233-45.
12. Finney, D. J. 1943. The detection of linkage. VII. Ann. Eugen. 12:31-43.
13. Fisher, R. A. 1935. The detection of linkage with dominant abnormalities. Ann. Eugen. 6:187-201.
14. Fisher, R. A. 1935. The detection of linkage with recessive abnormalities. Ann. Eugen. 6:339-51.
15. Fisher, R. A. 1936. Tests of significance applied to Haldane's data on partial sex linkage. Ann. Eugen. 7:87-104.
16. Haldane, J. B. S. 1934. Methods for detecting autosomal linkage in man. Ann. Eugen. 6:26-65.
17. Mellman, W. J., Oski, F. A., Tedesco, T. A., Maciero-Coelho, A., and Harris, H. 1964. Leucocyte enzymes in Down's syndrome. Lancet 2:674.
18. Morton, N. E. 1955. Sequential tests for the detection of linkage. Amer. J. Hum. Genet. 7:277-318.
19. Morton, N. E. 1957. Further scoring types in sequential linkage tests with a critical review of autosomal and partial sex-linkage. Amer. J. Hum. Genet. 9:55-75.
20. Neel, J. V., Salzano, F. M., Junqueira, P. C., Keiter, F., and Maybury-Lewis, D. 1964. Studies on the Xavante Indians of the Brazilian Mato Grosso. Amer. J. Hum. Genet. 16:52-140.
21. Norton, H. W. 1949. Estimation of linkage in Rucker's pedigree of nystagmus and color blindness. Amer. J. Hum. Genet. 1:55-66.
22. Osborne, R. H. 1958. Serology in physical anthropology. Amer. J. Phys. Anthrop. n.s. 16:187-195.
23. Schacht, L. E. and Gershowitz, H. 1961. Frequency of extra-marital children as determined by blood groups. Abstract 89. II Int. Conf. Hum. Genet., Rome.
24. Smith, C. A. B. 1959. Some comments on the statistical methods used in linkage investigations. Amer. J. Hum. Genet. 11:289-304.
25. Wiener, A. S. 1932. Method of measuring linkage in human genetics with special reference to the blood groups. Genetics 17:335-50.

MORTON: Now that there is reason to be dissatisfied with a uniform distribution of prior probabilities, there seems to be even less incentive to emphasize Bayesian methods which do not lead to significant increase in power over a sequential test against a reasonable fixed alternative θ.

ELSTON: I do not think it is right to compare a Bayesian procedure with another procedure on the basis of power or precision of the estimate (as measured by variance). Whether we wish to make a significance test or estimate a parameter (by a point or an interval), the problem can be considered as one of making a "best" decision when we have defined a loss function that measures the loss incurred for each wrong decision. A Bayes' procedure is "best" in that it minimizes the expected value of this loss function, so that in the long run a minimum loss is incurred. The expectation of the loss function is taken not only over all possible decisions for a given true state of nature (i. e., for the case of point estimation, over all possible estimates for a given value of the parameter), but also over all possible true states of nature (all possible values of the parameter). Of course, in order to do this, the prior distribution of the true states of nature (parameter) must be known, and a Bayesian procedure will be "best" only when this is so--I am leaving out of consideration empirical Bayes' procedures. To the extent that the prior distribution is approximated, because it is not really known, the resulting procedure will be an approximation to this "best" procedure. When there is some, but not complete, knowledge of the prior distribution, then one fair comparison of two procedures would be to compare the expected losses they lead to over the whole range of what might be reasonably possible prior distributions--though this has, of course, a subjective element in it.

GENETIC AND SEROLOGICAL ASSOCIATION ANALYSIS OF THE HL-A LEUKOCYTE SYSTEM

Walter Bodmer, Julia Bodmer, Dan Ihde and Stephen Adler

Department of Genetics
School of Medicine
Stanford University
Stanford, California 94305

There are two main reasons for an association between the reactions of a pair of sera with a random population sample. Either (1) the sera contain antibodies directed against antigens which are associated in the population, or (2) one or both of the sera contain more than one antibody and at least one of the antibodies is common to both sera.

A population association between antigens, or, more generally, between two genetically determined traits, may be caused by one or more of the following factors:

(a) The traits may be multiple effects of the same gene. An interesting example of this is the association between red cell ABO compatibility and skin graft survival (Ceppellini et al., 1966; Dausset and Rapaport, 1966).

(b) The traits may be the result of epistatic interaction between two or more genes. A striking example is the interaction between the Lewis and secretor loci, such that Le^b is only found on the red cells of individuals with at least one of the dominant alleles at each locus (see Race and Sanger, 1962).

(c) There may be selective interactions between the loci. Probably the best example of this in man is the association between G6PD deficiency, thalassemia and resistance to malaria in certain areas of the Mediterranean (see e.g. Motulsky, 1964).

(d) Departures from random mating due to inbreeding, assortative mating or population stratification can lead to non-random associations between genes. The effects of inbreeding in human populations are likely to be quite small, and assortative mating with respect to cryptic genetic characters, such as blood group antigens, is unlikely. Population stratification on the other hand, particularly as a result of recent racial admixture, may be a very significant general source of non-random association between unlinked polymorphic loci. For example, in a random sample of the American population one should expect a negative association between the allele R_o (cDe) of the Rhesus system and the Duffy allele, Fy^a, and a positive association between R_o and the Kidd allele, Jk^a. This is because R_o and Jk^a occur with a relatively higher frequency in African, than in Caucasian populations, while Fy^a has a much higher frequency in Caucasians than Africans. The magnitude of the associations found in the American sample will depend on the proportion of individuals with African ancestry in the sample. The associations will, presumably, eventually disappear if random mating with respect to racial ancestry is established on a more or less permanent basis.

(e) The last, and from our point of view most important, cause of non-random association between genes is allelism or very close linkage. The negative association between the antigens A and B of the ABO system was, of course, the basis for Bernstein's interpretation of the system in terms of three alleles and the association between the antigens S, s and M, N was the basis for assigning all these antigens to one system, consisting of the four 'alleles,' MS, Ms, NS, Ns (see Race and Sanger, 1962).

It is this last cause of genetic association which has been emphasized especially by red cell and white cell groupers. This is because the other causes can, mostly, be readily recognized or controlled and are in any case likely to have small effects relative to that resulting from the very close linkage between genes controlling antigens which are part of the same system. Mainly because of technical problems as will be discussed below, the initial genetic analysis of the antigens of the human HL-A leukocyte system was based on an analysis of their population association. (Payne, Tripp, Weigle, Bodmer and Bodmer (1964); Dausset, Ivanyi and Ivanyi (1965); Bodmer and Payne (1965); Bodmer, Bodmer, Adler, Payne and Bialek (1966).

The second main reason for association between serum reactions, namely due to shared antibodies, has played a key role in the initial identification of many of the antigens of the HL-A system. Most available sera are multispecific and, dependon their source, often have low titres. This fact, together with the originally relatively poor reproducibility of the assays used to detect the antigens, the limited availability of tissue for absorption and the complexity of the genetic system, hindered the application of conventional cross absorption techniques to purify the sera. The analysis of 2 x 2 associations between serum reactions for antigen definition in the HL-A system was pioneered by Van Rood (1962), and led to the initial description of the antigens 4a and 4b. The method was further developed by Payne et al. (1964) in their definition of the antigens LA1 and LA2. The main principle involved is the recognition of a group of associated sera which share an antibody, and hence can be used to define the corresponding antigen. One of the main difficulties in the application of this methodology to the HL-A system is the fact that it may often be difficult to disentangle associations between sera due to shared antibodies from those due to antibodies directed against closely associated antigens.

In this paper we first review the theory underlying genetic and serological associations between serum reactions. We next describe our computer programs for the analysis of such associations and lastly, discuss their application to the HL-A system.

THEORETICAL ANALYSIS OF GENETIC ASSOCIATIONS DUE TO ALLELISM OR CLOSE LINKAGE.

We shall limit our discussion to the case of two linked loci each with two alleles (A, a, B, b) in a random mating population. This provides an adequate basis for interpreting 2 x 2 interactions due to allelism or close linkage. We assume that alleles A and B determine the hypothetical antigens A and B and that the four gametes AB, Ab, aB, ab occur with frequencies x_1, x_2, x_3 and x_4 respectively. The relation between the gametic association $\Delta = x_1x_4 - x_2x_3$ and the 2 x 2 association between antigens A and B in the population has been discussed previously by Bodmer and Payne (1965) and Mi and Morton (1966). It can readily be shown that the frequencies of the four phenotypes AB, A, B and - can be expressed in the form

$$AB : \theta_1 = (1 - q_1^2)(1 - q_2^2) + \bar{\Delta} \qquad \text{------ 1a}$$

$$A : \theta_2 = (1 - q_1^2)q_2^2 - \bar{\Delta} \qquad \text{------ 1b}$$

$$B : \theta_3 = q_1^2(1 - q_2^2) - \bar{\Delta} \qquad \text{------ 1c}$$

$$- : \theta_4 = q_1^2 q_2^2 + \bar{\Delta} \qquad \text{------ 1d}$$

where $q_1 = x_3 + x_4$ and $q_2 = x_2 + x_4$ are the population frequencies of alleles a and b respectively and $\bar{\Delta}$ is defined by

$$\bar{\Delta} = \Delta(\Delta + 2q_1q_2) \quad \text{------------------} \quad 2$$

From equations 1a - d it can easily be shown that

$$\Delta = \sqrt{\theta_4} - \sqrt{(\theta_2 + \theta_4)(\theta_3 + \theta_4)} \quad \text{-----------} \quad 3$$

since

$$\theta_4 = q_1^2 q_2^2 + \Delta^2 + 2\Delta q_1q_2 = (q_1q_2 + \Delta)^2$$

which corresponds to the expression given by Ceppellini (1967) for estimating the gametic association Δ from observed phenotypic proportion θ_1, etc. The statistical significance of the estimate of Δ is the same as that given by the appropriate 2 x 2 contingency chi-squared.

When $\Delta = 0$, $\bar{\Delta} = 0$ and the phenotype frequencies are those expected in the absence of any association between the loci. The quantity $\bar{\Delta}$, which always has the same sign as Δ, is a measure of phenotypic association which is analogous to Δ, since as is well known (see e.g., Bodmer and Parsons, 1962), the gametic frequencies can be expressed in the form

$$x_1 = (1 - q_1)(1 - q_2) + \Delta,$$

$$x_2 = (1 - q_1)q_2 - \Delta,$$

$$x_3 = q_1(1 - q_2) - \Delta \quad \text{----} \quad 4$$

and

$$x_4 = q_1q_2 + \Delta.$$

A commonly used convenient measure of the association between the phenotypes is the 2 x 2 'binary' correlation coefficient, defined by

5a $$r = \frac{\theta_1\theta_4 - \theta_2\theta_3}{\sqrt{(\theta_1 + \theta_2)(\theta_3 + \theta_4)(\theta_1 + \theta_3)(\theta_2 + \theta_4)}}$$

$$= \frac{\bar{\Delta}}{\sqrt{q_1^2 q_2^2 (1-q_1^2)(1-q_2^2)}}$$

or

5b $$r = \frac{\Delta(2 + \Delta/q_1q_2)}{\sqrt{(1-q_1^2)(1-q_2^2)}}$$

as given by Bodmer and Payne (1965). A significant phenotypic association is, therefore, on the basis of our simplifying assumption an indication of a value of Δ which is different from zero.

There are two basic reasons why Δ may not be zero. Either (1) the population is not in equilibrium with respect to the gametic frequencies, or (2) selective interactions exist which are enough to maintain Δ different from zero even at equilibrium. In the first case it is a well-known result of population genetics that Δ tends to zero at a rate $(1 - y)^t$, where y is the recombination fraction between the two loci and t the time measured in generations. Thus, the smaller the value of y, the greater the probability that Δ is not yet near zero. The conditions under which selective interaction can maintain Δ different from zero have been amply discussed (for review see Bodmer and Parsons (1962) and Bodmer and Felsenstein (1967)). In general, the magnitude of the selective interaction must be greater than the recombination fraction, y. The smaller y, therefore, the more likely it is that selective interactions exist which are large enough to prevent Δ from approaching zero. Thus in any case a value of Δ different from zero is generally an indication of relatively close linkage between the relevant loci. As emphasized by Bodmer et al. (1966), a genetic association analysis can therefore to some extent be used to detect close linkage.

A series of antigens belong to a 'system' if y, for the genetic determinants of all pairs of antigens, is small (e.g., Ceppellini et al., 1967). The effective limit on y is set by the resolution of human pedigree studies for the detection of low recombination frequencies, and so, in most studies, is unlikely to be much less than 1 per cent. There may of course be many, perhaps even hundreds of cistrons between two loci separated by a one per cent recombination fraction. It seems likely, however, that antigens belonging to a system will be controlled by a block of contiguous cistrons (see e.g., Shreffler, 1967). Thus, unless such a block is exceptionally large, the appropriate values of y might be much less than one per cent, perhaps even as low as 10^{-4} or 10^{-5}. The final answer to these questions will always depend on a detailed chemical understanding of the gene-antigen relationships, as in the case of the ABO-Lewis-Secreter-Bombay system (see Race and Sanger, 1962). Based on these considerations, however, Δ, and so r, for most pairs of antigens belonging to a system should be different from zero. The signs of Δ should help in predicting the most prevalent allelic combinations (phenogroups, or haplotypes, following Ceppellini et al., 1967). It is of course clear that a value of Δ not significantly different from zero by no means proves two determinants are not part of the same system.

Mi and Morton (1966) defined the six basic types of logical relationship that can exist between a pair of loci according to the values of the gametic frequencies x_1 etc., and so Δ, as follows:

1. Permuted $\Delta \neq 0$ and none of x_1, x_2, x_3, or x_4 zero. This is the general relationship expected for a pair of 'associated' loci.

2. Segregant $x_1 = 0$, corresponding to a three allele system with one silent allele, such as the ABO system. In this case

$$\Delta = -p_1p_2, \quad \bar{\Delta} = -p_1p_2(p_1p_2 + 2p_3)$$

and --------- 6

$$r = \frac{-p_1p_2(p_1p_2 + 2p_3)}{(p_1+p_3)(p_2+p_3)\sqrt{(1-(p_1+p_3)^2)(1-(p_2+p_3)^2)}}$$

where p_1, p_2 and p_3 are the frequencies of the three alleles, A, B, and O for example. The association between A and B is always negative (see Payne et al., 1964).

3. Complementary $x_1 = x_4 = 0$, corresponding to a two allele system, such as the MN blood groups, ignoring S, s and other related antigens. Now

$$\Delta = -pq \qquad \bar{\Delta} = -p^2q^2$$

and

$$r = \frac{-pq}{\sqrt{(1 - p^2)(1 - q^2)}} \quad \text{--------------} \quad 7$$

where p and q are the allele frequencies. Once again, of course, r is always negative.

4. Codominant $x_4 = 0$, corresponding to the absence of a 'null' allele. In this case

$$\Delta = -q_1 q_2 \qquad \bar{\Delta} = -q_1^2 q_2^2$$

and

$$r = \frac{-q_1 q_2}{\sqrt{(1-q_1^2)(1-q_2^2)}} \qquad \text{------------- } 8$$

giving, again, a negative association.

5. Subtypic Either $x_2 = 0$ or $x_3 = 0$ corresponding to an operational 'inclusion' of one antigen within the other. If, for example, $x_3 = 0$ we have

$$\Delta = p_2 q_1 \qquad \bar{\Delta} = p_2 q_1^2 (1 + q_2)$$

$$\text{------------- } 9$$

$$r = \frac{p_2 q_1 (1 + q_2)}{q_2 \sqrt{(1-q_1^2)(1-q_2^2)}}$$

where $p_2 = 1 - q_2 = x_1$ is the frequency of the genetic determinant for the 'included' antigen. The association in this case is always positive.

6. Identical $x_2 = x_3 = 0$ and $r = 1$.

Subtypic and identical pairs are clearly identified by the corresponding phenotypic patterns, namely 'inclusion' and identity. Codominant and complementary pairs both lack the 'null' phenotype and are distinguished only by the existence of the AB gamete. Segregant pairs are recognized by the absence of the AB gamete. The phenotypic patterns and the sign and magnitude of r do, therefore,

Table 1

Minimum absolute correlation which, on average, will be significant at the 5 per cent and 1 per cent levels for various values of n, the total number of observations.

	n	100	225	400	625	900	1225	1600
$\lvert r \rvert$ for	5 per cent level	.196	.131	.098	.078	.066	.056	.049
	1 per cent level	.258	.172	.129	.103	.086	.074	.065

The values of $\lvert r \rvert$ are calculated from equation (10) using $\chi^2 = 3.84$ and 6.63 for the 5 per cent and 1 per cent levels respectively.

give an indication of the prevailing pattern of gamete frequencies, though the final elucidation can, of course, only come from family studies.

Given observed phenotype proportions θ_1 for AB, etc., based on a total of n observations it is well known that

$$\chi^2 = n r^2 \qquad \text{------------- } 10$$

where χ^2 is the 2 x 2 contingency χ^2 testing the association between antigens A and B. As discussed by Bodmer and Payne (1965) and Mi and Morton (1966), equation (10) enables one to predict the minimum correlation which, on the average, can be detected at a given significance level for a given number of observations. Some representative values of r for n ranging from 100 to 1,600 and for 1 per cent and 5 per cent significance levels are given in Table 1. Clearly, large numbers of observations are needed to detect small correlations. This is, of course, a major limitation to association analysis, since even quite small correlations may be very significant indicators that two antigens belong to the same system. Mi and Morton (1966) have analyzed the pairwise associations between a large number of blood factors (blood groups, Gm groups, and secretor status) in a population sample based on families drawn from northeastern Brazil. In their data, the r value for the A and B antigens of the ABO system was - 0.14, while those between D and E, and D and e, of the Rhesus system were 0.18 and - 0.05, respectively. The latter value based on 1,991 individuals was not significantly different from zero at the 1 per cent level. However, only two out of the 10 associations between the five antigens D, E, e, C and c of the Rhesus system were not significant. The third lowest r value was 0.12 for the association between D and C, which from Table 1 would have been detected in a sample of about 400 individuals. Thus, in this case, all the antigens would clearly have been assigned to the same system, based only on phenotypic association analysis. However, as Mi and Morton (1966) point out, the estimates of the association between some antigens have a high standard error, making them more difficult to detect without family studies. The Brazilian sample has a mixed Caucasian, African and American Indian origin. Mi and Morton (1966) emphasize that the associations between systems which must have existed in the ancestral population because of racial stratification are no longer significant. This suggests that random mating with respect to racial origin has been occurring for a sufficient number of generations to remove essentially all racial stratification. The application of association analysis to the HL-A system will be discussed later.

THE ANALYSIS OF ASSOCIATIONS BETWEEN THE REACTION OF MULTISPECIFIC SERA DUE TO SHARED ANTIBODIES

Following Payne et al. (1964) consider first the case of two dispecific sera containing antibodies anti-X, anti-Y_1 and anti-X, anti-Y_2, respectively. The expected frequencies of the four corresponding phenotypes, on the assumption that the antigens X, Y_1 and Y_2 are not associated, are shown in Table 2, where x, y_1 and y_2 are the population frequencies of the antigens X, Y_1 and Y_2, respectively. The correlation between the reactions of the two sera is given by

$$r = \frac{x(1-x)(1-y_1)(1-y_2)}{\sqrt{[x+y_1(1-x)](1-x)(1-y_1)[x+y_2(1-x)](1-x)(1-y_2)}}$$

which can be rewritten in the form

$$r = \frac{x}{\sqrt{(x+\frac{y_1}{1-y_1})\,(x+\frac{y_2}{1-y_2})}} \qquad \text{-----------} \; 11$$

The association due to a shared antibody is always positive. The value of r increases as x, the frequency of the shared antibody, increases but decreases as y_1 and y_2, the reaction frequencies of the unshared antibodies, increase. For given x, r tends to 1 as y_1 and y_2 tend to zero. Thus, as expected intuitively, the association between two sera due to a shared antibody decreases as the frequencies of the unshared antibodies increase. Given a group of sera all sharing one antibody, anti-X, any pair of them should be associated, though to varying extents depending on the frequencies of the unshared antibodies. The probability that an individual reacts with all the sera, but does not have the antigen X is $(1 - x)\Pi y$, where the product is over the frequencies of reaction to the unshared antibodies in each of the sera. This quantity will tend to zero as the number of sera in the group increases. The presence of antigen X is then reliably determined by a reaction with all the sera in a group. In principle, therefore, an antigen can be identified by a group of multispecific sera, all pairs of which have a significant r value. This was the basis on which Van Rood first identified the antigens 4a and 4b and on which many of the other antigens of the HL-A system were first identified.

One of the main difficulties with this approach has been that many of the antibodies present in the white cell typing sera are directed against associated antigens of the HL-A system. This leads to associations between sera due to the genetic association between the antigens, as well as to shared antibodies and so invalidates the simple formula for association due only to a shared antibody. However, in practice, if a group of associated multispecific sera divides people into two categories according to whether or not they react to most of the sera in the group, it can be used to define an antigen. An example of such a "reaction distribution" is shown in Figure 1. The histogram gives the number of people who reacted with a given number of the 6 sera originally used to define the HL-A antigen LA1. The distribution is clearly bimodal with a minimum at 3 to 4 sera out of 6. Individuals reacting to 4, 5 or 6 of the sera are LA1+, the remainder LA1-. Technical errors account for the fact that some LA1+ individuals may not react with all the sera of the group. People reacting to 1, 2 or 3 sera are, presumably, reacting to antibodies in these sera other than anti-LA1. The upper limit to the level of misclassification is set approximately by the 3 out of 86 individuals who reacted to 3 or 4 sera. In this example, the definition of the antigen LA1 is unambiguous and is not obscured by associations between antibodies. It was confirmed by absorption studies and the identification of monospecific sera (Payne et al., 1964).

The method used to classify for LA1 in the above example is equivalent to assigning each serum reaction a score of 1 for a positive, and 0 for a negative, and then counting as LA1+ all individuals whose total score exceeds 3. This approach effectively ascribes equal 'weight' to each serum in the group. It is, however, clear that the weight given to a serum should vary both according to its reproducibility and the reaction frequencies of antibodies other than that directed against the antigen defined by the group of sera. A serum with poor reproducibility should clearly receive a relatively low weighting, while a good monospecific serum should be assigned a very high weighting. The weight, also, may vary according to whether the reaction is positive or negative. Thus, a positive reaction to a reproducible multispecific serum may be of little value, while a negative reaction may be a reliable indication of the absence of the antigen. Ideally, positive and negative weights for each serum should be chosen which, on the average, maximize the differences between the scores observed for people proven to have and not to have the relevant antigen. This can be done using the techniques of discriminant analysis, given a sample of individuals whose antigen types have been reliably ascertained. In practice the typing has been done using intuitive empirical criteria to evaluate the various serum reactions. Use is often made of knowledge of other antigen types together with the specificities of secondary antibodies; the results generally seem to agree with those obtained from a simple 'reaction distribution' (Figure 1.)

Knowing the specificity of one of the antibodies in a multispecific serum helps in the identification of remaining specificities. The best approach is to try and absorb out the known antibody and then study the residual serum. A simple preliminary substitute is to analyze the reactions of the serum only with those people who lack the identified specificity. The frequency of reaction to the residual antibodies can also sometimes be a useful guide. Assume, for simplicity, the serum contains two associated antibodies, anti-A and anti-B, determined according to the genetic scheme giving rise to equations 1a-d. The frequency of reaction to the serum is

$$F = \theta_1 + \theta_2 + \theta_3 = 1 - \theta_4 = 1 - q_1^2 q_2^2 - \bar{\Delta} \qquad \text{-------}12$$

from equation 1d. If A is the known specificity, then the frequency of reaction to anti-A, namely $t = 1 - q_1^2$, is known. The frequency of reaction, $1 - q_2^2$, to the remaining antibody is therefore given by

Table 2

Expected phenotype frequencies for the reactions of two dispecific sera with a shared antibody.

		anti-1 +	anti-1 -	
anti-2	+	$x + y_1y_2(1-x)$	$y_2(1-y_1)(1-x)$	$x + y_2(1-x)$
	-	$y_1(1-y_2)(1-x)$	$(1-y_1)(1-y_2)(1-x)$	$(1-x)(1-y_2)$
		$x + y_1(1-x)$	$(1-x)(1-y_1)$	1

Serum 1 contains anti-X and anti-Y_1, serum 2 contains anti-X and anti-Y_2. The antigens X, Y_1 and Y_2 have population frequencies x, y_1 and y_2 respectively and are not associated.

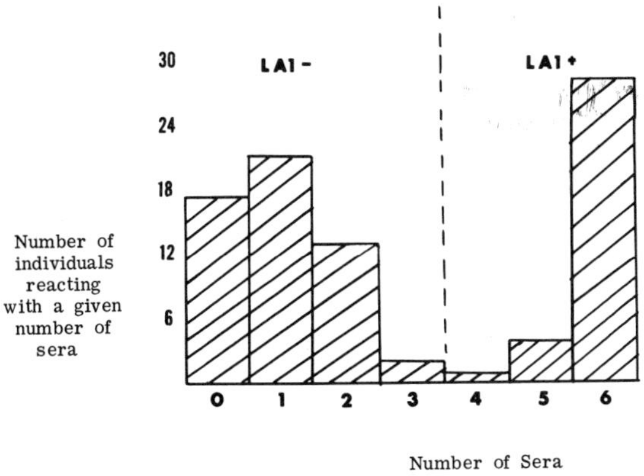

Figure 1. Definition of an antigen using a group of multispecific sera.

$$1 - q_2^2 = 1 - \frac{1 - \bar{\Delta} - F}{q_1^2} = 1 - \frac{1 - F}{1 - t} + \frac{\bar{\Delta}}{1 - t} \quad \text{-------} \quad 13$$

When there is no association between the antibodies, and so $\Delta = \bar{\Delta} = 0$, equation (13) reduces to the formula given by Bodmer and Payne (1965). Since only F and t and not Δ are known, this is the only basis on which an estimate of $1 - q_2^2$ can be obtained. A positive association increases this estimate of the frequency of the remaining antibody while a negative association decreases it.

The known antibody, anti - A, can only be absorbed out by people who are A but not B. The proportion of A individuals who are not also B is given by

$$\frac{\theta_2}{\theta_1 + \theta_2} = q_2^2 - \frac{\bar{\Delta}}{1 - q_1^2} \quad \text{--------------} \quad 14$$

from equations 1(a) and 1(b). This proportion decreases as $\bar{\Delta}$ increases, namely as the positive association between the sera increases. It, of course, approaches 0 as $\bar{\Delta}$ approaches q_2^2 $(1-q_1^2)$ and θ_2 approaches zero, corresponding to a subtypic pair with A completely 'included' in B. Thus, as pointed out by Bodmer and Payne (1965) and Walford and Troup (1967), many cell donors may be needed for the resolution by absorption of a serum containing antibodies directed against closely associated antigens.

COMPUTER ANALYSIS OF ASSOCIATIONS

Seranal is our basic program for the analysis of 2 x 2 associations. The construction of 2 x 2 tables and calculation of χ^2 and r values is a trivial problem from an analytical point of view. As is so often the case, the bulk of the program is concerned with easing the problem of data manipulation. Each serum is given an alphanumeric identifier and assigned a definite position in a two-dimensional array, called the "master list." The rows of the master list are generally used to correspond to some meaningful sub-classification of sera according, for example, to their sources or the assay with which they were used. Different dilutions of a serum, different bleedings from the same donor and reactions using the same serum with different assays are all given separate, but systematically related identifiers. Serum identifiers are added to the master list without changing its previous composition so as to maintain compatibility between old and new data. An individual's reactions are recorded in an arbitrary sequence that corresponds to a given sequence of serum identifiers. This sequence is, of course, generally held constant for a reasonable number of individuals tested. The data input can also be arranged in a two dimensional array, analogous to, though generally different from, that of the master list. This greatly facilitates the analysis of data from a variety of sources, each of which can be assigned to a different row of the data input array. Numerical scores are converted to positive and negative reactions by a series of defined rules which may vary from one serum to another, mainly according to the assay. Zero is always reserved for 'not done' so that comparisons involving two or more sera can readily be omitted if any sera were not tested for, according to whether the product of the relevant reaction scores is zero. The pairwise reactions to be analyzed are defined by two arbitrary sequences of identifiers. The 2 x 2 tables, χ^2 and r values are then computed and listed for all possible pairs including one identifier from each of the two sequences. All significant associations are, for convenience, also listed together at the end of the output. Since antigen types are included in the analysis, this provides a rapid identification of the specificities of new sera. The distributions of scores, and negative and positive reactions for each serum can also be calculated. Data are accumulated from one individual at a time, so that there is no limit on the number of individuals which can be analyzed. The number of pairwise associations which can be analyzed in one run is, however, limited by the core size of the computer. Repeated use of the program in principle allows all possible associations among an arbitrarily large set of sera to be computed. Whenever possible, all data are stored in some form of rapid access memory, such as a magnetic disk. If required the pairwise analysis can be done only with people who reacted in a defined way with a given set of sera. This, as indicated above, can be very helpful for the identification of further specificities in multispecific sera containing one or more known antibodies.

The main auxiliary programs which have been written are for

(a) Selecting people who react in a defined way with an arbitrary set of given sera.

(b) Printing a matrix of χ^2 or r values for arbitrary defined sequences of sera in the rows and columns of the matrix.

(c) Computing the 'reaction distribution' (see Figure 1) for an arbitrarily defined set of sera.

The first of these programs is used to select suitable individuals for absorption studies while the other two are used as aids in identifying meaningful groupings of multispecific sera.

The basic principle of grouping as discussed above is the identification of groups of sera all, or most, pairs of which have significant 2 x 2 associations.

Our program Grouping carries out this search in a systematic way for a given set of sera, using as input data the matrix of all pairwise associations between the sera. The measure of association used is either the 2 x 2 contingency χ^2 or the correlation coefficient r = $\sqrt{\chi^2/n}$,

computed using Seranal. A systematic search is made for groups of given size by scanning all possible combinations of sera in a systematic increasing order and rejecting any combination which does not meet the grouping requirement. This requirement is that, for each group size, the number of pairwise combinations that fall below the critical association level, or "misses," must be less than some given value. These values, as well as the critical association level, are input program parameters. Allowing for a given level of misses gives the program much greater flexibility in searching for groups. The allowed number of misses, of course, decreases with the group size. Two simple devices are used to minimize the output of redundant groups. In the first place, the program always starts searching for groups of the maximum desired size first and then checks each newly found group to see whether it is contained in any previous group, in which case the new group is rejected. Secondly, if any newly found group 'overlaps' with any previous group by more than a given number of sera, the new group is not listed separately. Instead the sera by which it differs from the pre-existing group are listed as 'additions.' Sera are, of course, only listed once as additions. Thus, only those sera not already present as additions are in fact added. The level of overlap allowed is set as a program input parameter for each group size. Each new group, or addition to a group is listed as it is found. In the case of an addition, the actual group found is listed together with the identifier of the group with which it overlaps and the added sera. The final listing is a summary of all groups in order of discovery, with the additions. Because of the large amount of searching involved there are two major computer time limiting parameters to this procedure, the total number of sera and the maximum group size sought for. In practice we have limited the number of sera in any one run to about 80 and the maximum group size to 10. Larger numbers of sera can be dealt with by running the program with overlapping sets of sera. New sera can simply be tested, one at a time, to see whether they belong to any existing group. The 'overlap' device helps in identifying groups which are larger than the set maximum and preventing redundant output of closely related, but not identical groups. Suppose, for example, there existed a group of size 15 while the maximum sought for was 10. Without the overlap check, all the 560 possible selections of 10 out of the 15 sera would be listed. With the overlap set at 6, so that only groups of size 10 which differed by 5 or more sera were counted as new, only two groups would be listed, the remaining sera turning up as additions to these. If the number of antisera is large and group sizes are large, there will, however, always be some duplication in the final output, which is a list according to the size of all the groups, distinguishing the original members from the additions. The tightness of the grouping depends on the critical association level and the maximum numbers of misses and minimum overlaps set for each group size. Once groups have been determined, the reaction distribution program can be used to check the validity of any given group. It is also often helpful to list the matrix of χ^2 or r values with sera belonging to a group next to each other. This facilitates the identification of associations that may exist between groups, which should, of course, reflect the associations between the antigens defined by the groups.

Recently, all our programs and data input procedures have been adapted for use on the Stanford Medical School's IBM 360/50 based time sharing system, ACME. Serum reactions are input directly to a magnetic disk using a typewriter console. The data are edited and checked during input and stored in the standard order corresponding to the serum identifier master list. The use of a time sharing system greatly facilitates data input and editing, as well as providing an almost immediate response for small scale checks on serum associations and the selection of individuals to be used for absorption studies. The system is generally too slow to be used routinely for larger computations. The programs for ACME are written in PL-1 and incorporate prompts for input in such a way that they can readily be used by persons with essentially no previous experience with the use of computers.

ANALYSIS OF THE HL-A SYSTEM*

The initial identification of many of the antigens of the HL-A systems by Van Rood, Bodmer, Payne and colleagues, and Dausset and colleagues using association analysis was done 'by hand,' before the development of grouping programs. The application of our program to Payne and Bodmer's data fully confirmed the original groupings and showed that they were, in fact, exhaustive, as no other groups were identified by the program. The addition of a few further sera lead to the provisional identification of a new specificity, called LA4 because of its 'allelic' or segregant (see above) relationship with each of the three allelic antigens LA1, LA2 and LA3 (see Histocompatibility Testing 1967, pp. 185-6 and the appendix). Similar, if not quite identical, specificities have also been described by Ceppellini et al. (1967), Walford et al. (1967) and Dausset et al. (1968). The existence of the new group was readily confirmed by the grouping program. An example of the output for an analysis of the associations between 80 sera is shown in Figure 2. The correlation, r, was used as the measure of 2 x 2 association. The number of individuals on the basis of which it was calculated ranged from about 50 to 150. The data used for this analysis were obtained in collaboration with Dr. Rose Payne and will be described in more detail elsewhere. The known antigen classifications (F4d, F4d2, FLA2, FLA3, F4A, F4B, F4C, FLA1) were included in the analysis. The prefix F indicates the test used for defining the antigens. The numerical parts of the other identifiers refer to sera, while the prefix letters A and F define the test with which the sera were used. 3001 F and 3001 I are different bleedings from the same immunized donor, who had received a further stimulus between the bleedings. A critical association level of r = 0.4 was used and the maximum size group sought was set at 10. No groups of size 10 or 9 were, in fact, found. The maximum number of allowed misses was 3 for sizes 10 and 9; 2 for sizes 8, 7 and 6; 1 for 5 and 4; and none for 3. Eighteen groups of size 3 were found, which are not shown in Figure 2. All of these were included in the other groups, if no distinction was made between the primary group and the additions. The antigens which the various groups identify are shown in the left margin of Figure 2. All the previously described antigens are clearly separated. The antigen, $4d_2$, defined effectively by the single serum 1052, was known to be closely associated with 4d. The new group for LA4 shows up very clearly. It includes two sera (3001F, 3001I) which are also in the LA1 group. This accounts for the 7th listed group, which is a combination of LA1 and LA4 sera. The 8th and 9th groups are, apart from A1060, included in the 3rd, which defines LA3. The 11th group is included in the 10th

*By international agreement (Bull. Wld. Hlth. Org. 1968, 39, 483-486), some of the antigens discussed in this paper are now defined as follows: LA1=HL-A1, LA2=HL-A2, LA3=HL-A3, 4d=HL-A7.

which defines 4c. Thus, all the significant groupings are accounted for in terms of the previously known antigens together with LA4. Lowering the critical association level to 0.35 or 0.3 did not affect the pattern of groups obtained.

Reaction distributions for four of the LA4 sera are shown in Table 3. The second column shows the distribution without any restrictions while the third shows it among people who are LA1-. This restriction removes the 6 people in the middle of the distribution, as might be expected, since two of the sera are known to contain anti-LA1 activity. The six LA1- individuals who react to only three of the sera are explained by the fact that F4005, which is Walford's serum Hunt (Walford et al. 1967) is known not to be exactly the same as LA4, but to be included in it (subtypic association).

A matrix of the 2 x 2 contingency χ^2s for representative LA1, LA2, LA3 and LA4 sera, together with the antigens themselves is shown in Table 4. The majority of the χ^2s for associations between groups are negative (49/63), as expected from their pairwise segregant associations (see equation 6). The positive association of the LA4 serum 3003J with LA3 is of interest, even though it is not significant, since it is known that this serum contains a second specificity which is probably closely associated with LA3. The negative χ^2s between groups are much lower than the positive values within groups, because the positive association due to shared antibodies is generally much stronger than the negative associations caused by a segregant relationship.

An interesting example of the interaction between the two types of serum associations, namely between antigens or due to shared antibodies, is illustrated by the data shown in Table 5. Given in the table are the 2 x 2 contingency χ^2s for sera containing the specificities 4b and 4d. The antigen 4d is known to be almost included within 4b (Bodmer and Payne, 1965), which accounts for the relatively high positive χ^2s for associations between the two groups. If the critical association level is set too low, the two groups are not separated.

Table 3

Reaction distributions for the LA4 sera F3001F, F3001I, F3003I and F4005

Number of positive reactions	Number of people: Types Unrestricted	Number of people: Only LA1-
0	10	10
1	2	1
2	4	0
3	7	6
4	17	12

Figure 2

Sample output of grouping program illustrating the definition of the new antigen LA4

(See text for detailed explanation)

FINAL LIST OF ALL GROUPS

Antigen	
4d and	I IS 1 GROUP F4D A1013 A1041 F1047 A1052 F4D2 F1034 F1052 F4007
4d2	ADDED
LA2	I IS 2 GROUP FLA2 F1028 A1028 F1043 F4025 F30021 F4052
	ADDED A1002 F4006
LA3	I IS 3 GROUP FLA3 F1045 F4022 F1047 F4019 F4048
	ADDED F4051 F4053 F4054 A1065
4a	I IS 4 GROUP F4A A1042 F1050 A1050 F4050
	ADDED A1060
4b	I IS 5 GROUP F4B A1026 A1046 A1061 F1064
	ADDED F4013
LA1	I IS 6 GROUP FLA1 A1002 F1014 F3001F
	ADDED F4026 F3001I
	I IS 7 GROUP FLA1 F3001F F300I F4005
	ADDED F3003I

Figure 2 -- continued

Antigen	
	I IS 8 GROUP FLA3 F1045 F4051 F4053
	ADDED F4054 F1047
	I IS 9 GROUP F1045 A1060 F1047 F4054
	ADDED F4048 F4053
4c	I IS 10 GROUP F4C F1027 A1033 A1062
	ADDED A1079 F4047
	I IS 11 GROUP F4C F1027 A1079 F4047
	ADDED
LA4	I IS 12 GROUP F1089 F300I F3003I F4005
	ADDED F3001F

Table 4

LA Chi-Square Table

	LA1	1014	4036	LA2	1043	3002H	4025	LA3	1045	4022	LA4	1089	3003J
LA1													
1014	65.												
4036	37.	37.											
LA2	-4.4	-4.4	-7.8										
1043	-4.4	-4.4	-7.4	65.									
3002H	-5.4	-5.4	-4.7	31.	31.								
4025	-5.8	-5.8	-6.2	56.	56.	35.							
LA3	-2.2	-2.2	-0.6	-4.3	-4.3	-4.3	-2.2						
1045	-3.3	-3.3	-1.0	-6.0	-6.0	-6.7	-3.5	57.					
4022	-0.8	-0.8	-1.1	-4.4	-4.4	-8.0	-4.3	53.	53.				
LA4	-0.8	-0.8	0.1	0.1	0.1	-1.1	0.8	-2.0	-3.0	-3.8			
1089	-0.3	-0.3	0.1	0.4	0.4	0.0	0.4	-1.5	-2.6	-1.7	48.		
3003J	-1.4	-1.4	0.0	-0.1	-0.1	-5.8	-0.1	1.4	1.2	1.7	28.	23.	

Negative signs indicate a negative pairwise association.

Total proportion of negative associations between groups = 49/63

(χ_1^2 for 1:1 ratio is 19.3)

Table 4 -- continued

Among LA1 vs. LA1, 3/3 $\chi^2 > +35$
LA1 vs. LA2, 12/12 $\chi^2 < -4$
LA1 vs. LA3, 9/9 $\chi^2 < 0$
LA1 vs. LA4, 6/9 $\chi^2 < 0$
LA2 vs. LA2, 6/6 $\chi^2 > +30$
LA2 vs. LA3, 12/12 $\chi^2 < -2$
LA2 vs. LA4, 4/12 $\chi^2 < 0$
LA3 vs. LA3, 3/3 $\chi^2 > +50$
LA3 vs. LA4, 6/9 $\chi^2 < 0$
LA4 vs. LA4, 3/3 $\chi^2 > +20$

Table 5

Chisquare table for 4b and 4d sera

Group		4B	A1003	A1010	A1026	A1039	A1046	A1061	A1064	A1018	4D	4D2	A1013	A1041	A1047	A1048	A1052
4b	A1003	35.															
4b	A1010	92.	41.														
4b	A1026	140	16.	50.													
4b	A1039	60.	70.	33.	29.												
4b	A1046	110.	42.	74.	72.	54.											
4b	A1061	78.	15.	45.	56.	27.	57.										
4b	A1064	54.	14.	31.	51.	14.	30.	46.									
4b and 4d	A1018	66.	8.3	52.	23.	10.	22.	10.	30.								
4d	4D	20.	4.3	16.	14.	1.6	11.	5.5	14.	48.							
4d	4D2	.04	1.8	5.2	.02	.08	.00	.38	4.4	10.	38.						
4d	A1013	19.	4.4	15.	19.	1.3	13.	4.1	11.	47.	230.	30.					
4d	A1041	10.	.47	14.	2.6	5.7	10.	5.3	6.6	42.	86.	19.	73.				
4d	A1047	3.8	.32	1.1	.44	1.4	2.9	4.4	2.1	16.	53.	12.	39.	42.			
4d	A1048	9.8	2.5	6.8	2.0	2.0	7.5	5.9	8.9		51.	18.	43.	27.	34.		
4d	A1052	25.	7.4	27.	12.	10.	18.	14.	17.	39.	52.	32.	50.	44.	17.	30.	
4d	A1056	16.	6.9	25.	8.1	8.0	20.	10.	10.	32.	50.	16.	45.	80.	21.	17.	45.

Column groups: 4b (4B – A1064); 4b and 4d (A1018); 4D; 4d (4D2 – A1052)

b vs. b 4/28 $\chi^2 < 28$
d vs. d 7/28 $\chi^2 < 28$
b vs. d 64/64 $\chi^2 < 28$

DISCUSSION

It seems clear that, in spite of its ultimate limitations, association analysis can be a very useful guide to the analysis of multispecific sera and to the genetic interpretation of associations between the antigens they define. Largely for technical reasons, this approach has played a major role in the development of the HL-A leukocyte antigen system. Other methods of association analysis based on using the correlation coefficients between quantitative serum reactions for factor analysis, have been discussed by Terasaki et al. (1965) and Elston (1967). Using this approach, Terasaki and colleagues were able to define clearly five groups of sera which corresponded basically to the HL-A antigens LA1, LA2, 4a and 4c. Application of our grouping program to Terasaki's data, which he kindly made available to us on punch cards, gave essentially the same grouping, indicating a good correspondence of the results obtained by the two approaches. The direct association analysis, as we have described it, seems to provide, however, more logical insight into the basic nature of the grouping problem.

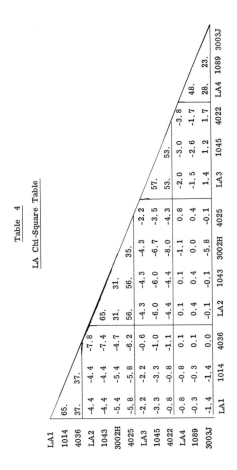

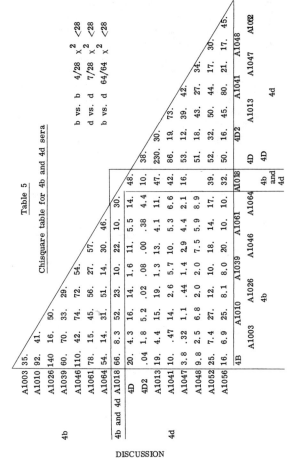

The first estimates of allele frequencies in the HL-A system were based on association analysis (Bodmer et al. 1966). In spite of the subsequent greatly increased complexity of the system (see Histocompatibility Testing 1967), these estimates have turned out to be quite reliable, as far as they went. The major gap in the analysis was the lack of any association, in our data, between the antigens LA1 and LA2, and 4a and 4b. Family data have since then shown clearly that all these antigens belong to the same system (see Histocompatibility Testing 1967). This lack of association between certain antigen sets within the HL-A system suggests that there may be sub-regions, or groups of cistrons, within the major region which have different properties. The characteristics of the respective sera lend some support to this suggestion, though the final interpretation can only really come from a chemical analysis of the relation between gene and antigen. The lack of an association between two antigens within a system means either that there has been enough time for the respective genetic determinants to equilibrate (Δ approaching zero) or that there are no selective interactions between the antigens. Both of these factors may, of course, be correlated with the genetic distance between the determinants.

Higher order interactions between antigens will not, in general, be detected by a 2 x 2 analysis. The relatively large number of antigens of the HL-A system and the very large corresponding number of allelic combinations (certainly greather than 30 or 40) allows plenty of scope for the existence of complex patterns, though these may not be easy to detect. An example of such a complex pattern involving the antigen LA4 is shown in Table 6. The data were obtained from the workshop held in Turin in 1967 and are based on an analysis of 11 families (see Histocompatibility Testing 1967). The antigens TO11 and TO12 are described by Ceppellini et al. (1967). TO11 is known to be "included" in 4a, while TO12 is allelic to LA1, LA2 and LA3 and is "included" in LA4. TO12 is essentially the same as Walford's serum Hunt, our F4005 (see Figure 2, Histocompatibility Testing 1967). From the data given in Table 6, which refer only to the antigens LA4, 4d$_2$, TO11 and TO12, it appears that there are, in Caucasian populations, two main types of LA4 carrying gametic combinations, one combining LA4 with TO11 (type 2) and the other combining LA4 with 4d$_2$ and TO12 (type 1). One combination (type 3) was found which was the logical sum of these two types. All the TO11 combinations are associated with 4a, while 4a occurs more or less at random among the other combinations. These patterns may, of course, vary from one population to another, as is the case to some extent for the Rhesus system. Such a change in pattern could sometimes change the apparent specificity of a serum, if, for example, it contains two antibodies which are closely associated in one population but not in another (Ivanyi and Dausset, 1966; Rubinstein et al., 1967). Preliminary results we have obtained from a study of African pygmy and Bantu populations suggest that the frequencies and associations of the HL-A antigens may be quite different in African from Caucasian populations. Some elements of a complex pattern can, of course, be discerned from the 2 x 2 associations. A complete analysis must, however, depend on the direct identification of the gametic combinations from family studies.

Association analysis is a valuable guide to the initial identification of specificities in multispecific sera. As has been emphasized before, however, the final confirmation of the groupings must be based on absorption of the sera. In the absence of cross reactions and other serological complexities, absorption of a serum should allow a clear-cut definition of the specificities of the antibodies it contains. Three major serological anomalies may, however, complicate the interpretation of absorption data.

1) A serum may absorb for a specificity with which it does not react. Such reactions have been called ANAP by Van Rood et al. (1965) for agglutination assays, and CYNAP by Ferrone et al. (1967) for cytotoxicity assays. We have found them to be quite common and believe they reflect a situation where, for example, an antibody can bind with an antigen, but for some reason the resulting complex cannot bind complement, which is needed to mediate the cytotoxic effect. This may reflect a property of the antibody, or perhaps some form of steric hindrance to complement binding, caused by an adjacent antigenic specificity.

Table 6

Gametic combinations of the HL-A antigens LA4, 4d$_2$, TO11 and TO12

Type	Antigen combination					
	LA4	4d$_2$	TO11	TO12	Number	Proportion which are 4a
1	+	+	-	+	4	2/4
2	+	-	+	-	7	7/7
3	+	+	+	+	1	1/1
4	-	+	-	-	3	1/3
5	-	-	+	-	1	1/1
6	-	-	-	-	28	12/28
				Total	44	24/44

Data are from the Turin Workshop 1967 (see Histocompatibility Testing 1967). TO11 and TO12 are antigens defined by Ceppellini et al. (1967). The data refer only to the specified antigens. Subsequent work has shown that our original LA4 classification confounded, in a number of cases, the differences between TO 11 and TO 12. Thus, many individuals classified as LA4+ could have been either TO 11 <u>or</u> TO 12. This is because some of the LA4 sera included antibodies to both these specificities. More careful analysis of the new specificity LA 4 showed that it could be equated with TO 12, and also Lc 11 (see Walford et al. 1968. <u>Vox Sang.</u> 15:338) which is closely related to Walford's serum Hunt. The specificity 4D2 has, in addition, been shown to be the simple sum of 4d (HL-A7) and LA4 (=TO 12, Lc11). All the reaction patterns shown in Table 6 can be explained on this basis. The results show how association analysis can sometimes, in the absence of further serological manipulation, fail to identify clearly the simple relationships between a set of "primary" specificities. Our data are now consistent with LA1, LA2, LA3 and LA4 forming a strictly mutually exclusive (segregant) set of specificities. The antigens 4d (HL-A7), 7d (HL-A8), 4c and TO 11 form a second mutually exclusive set in line with the suggested existence of sub-regions. (See also Kissmeyer-Nielssen et al. 1968. <u>Nature</u> 29:1116-1119)

2) An antibody may cross-react with two specificities in the sense that it will react with either and be absorbed by either. This phenomenon, also, has been observed by us with HL-A sera. It poses the question whether, in fact, such an antibody should be assigned a new specificity, since by the criteria of absorption it may be a 'pure' serum. Given identification of the two constituent antigens, the cross-reacting serum provides no new information with respect to the definition of allelic combinations, or haplotypes.

3) A serum may interact with two specificities in the sense that it reacts only when both the primary specificities are present. Such a serum, also, gives no new information which can be used to distinguish haplotypes.

It is clear that the definition of a 'pure' specificity by absorption, may be difficult in the face of these complications. The primary aim of the serological analysis is, generally, the definition of the **haplotypes** or, in other words, the recognition of different gene products. Specificities should presumably be defined in such a way that the number needed for the definition of all haplotypes is a minimum. This is a logical problem which can, formally, be expressed in terms of Boolean algebra and the factorization of Boolean matrices (Bodmer, unpublished notes). Ironically enough, this brings us back to an essentially **statistical** definition of an antigenic specificity. The final answers to the interpretation of such complex serological patterns will only come from a chemical understanding of the gene-antigen relationships.

SUMMARY

There are two reasons for an association between serum reactions. Either sera contain antibodies directed against associated antigens, or one or both of the sera contain more than antibody and at least one of the antibodies is shared by the sera. Population associations between genetic traits may be due to multiple effects of the same gene, epistatic interactions, selective interactions, departures from random mating and, finally, allelism or very close linkage. This last is the most important for the analysis of antigenic systems on red and white cells. Assuming a model of two antigens determined by two co-dominant alleles and their silent alternatives, leading to four gametic combinations, and assuming random mating, it is possible to express the association between antigen phenotypes in terms of the association between genes in gametes. This association may be different from zero either because a population is not in equilibrium with respect to gamete frequencies, or because selective interactions exist which are enough to maintain an association. Both of these causes of association are likely to be an indicator of close linkage. Thus, in general, many antigens belonging to a "system," for which the genetic determinants are closely linked, will be associated in the population. Six basic types of pairwise logical relationships are reviewed and measures of association for each of them given. The magnitude of an association which can be detected in a sample of a given size is restricted. Very large numbers of observations may be needed to detect small associations. The association between the reactions of multispecific sera with shared antibodies depends on the frequencies of the shared and unshared antibodies. It increases as the frequency of the shared antibody increases, but decreases as the frequency of reaction of the unshared antibodies increase. An antigen may be identified by a group of multispecific sera, all pairs of which are significantly associated. If such a group of sera divides people into two categories, then these can objectively be used to identify the antigen type. More complicated statistical procedures are needed to take into account differences in reliability between sera. Knowing the specificity of one of the antibodies in a serum, formulae which give the frequency of reaction to the remaining antibodies can be derived.

Seranal is a basic program for the analysis of 2 x 2 associations. The main features of the program involve flexible handling of data input and output. Auxiliary programs select people of given type, print matrices of χ^2 values for arbitrary selections of sera and compute "reaction distributions." The identification of groups of sera, all or most pairs of which have significant 2 x 2 associations, is carried out by means of the program grouping. This searches systematically for groups of given size by scanning all possible combinations of sera in a systematic increasing order and rejecting any which do not meet the grouping requirements. Special devices reduce the number of redundant groups which are put out.

Initial application of the program to data on HL-A leukocyte types confirmed that the original groupings, done "by hand" were exhaustive. Addition of further sera led to the identification of a new specificity LA4, which is allelic to LA1, LA2 and LA3. The overall pattern of associations between LA1, LA2, LA3 and LA4 sera follows the pattern expected from the theory, assuming each pair have a "segregant" relationship. It may sometimes be difficult to distinguish associations between sera due to shared antibodies and those due to presence of antibodies reacting to associated antigens. Higher order interactions may not, of course, be readily discernible using a 2 x 2 analysis and these may be quite common in the complex HL-A system. ANAP and CYNAP reactions with sera (absorption for a specificity for which there is no reaction), cross-reactions and interactions may complicate the confirmation of serum groupings by absorption. Association analysis can certainly be a useful guide to the analysis of multispecific sera and the genetic interpretation of associations between the antigens they define.

ACKNOWLEDGEMENTS

We are much indebted to Dr. Rose Payne who first introduced us to white cell typing and whose data have provided the raw material for many of our analyses. We are also indebted to many colleagues and to the NIH Serum Bank for making sera available to us for our studies. Stimulating discussions with Professor Ceppelini have also helped shape a number of the ideas presented in this paper. The work reported on was supported in part by a Public Health Service Research Career Program Award (to Walter Bodmer - GM 35002) and by USPH Research Grants GM 14650 and GM 19452.

REFERENCES

Bodmer, W. F. and Parsons, P. A. 1962. Linkage and recombination in evolution. Adv. in Genetics 11:1-100.

Bodmer, W. F. and Payne, R. 1965. Theoretical consideration of leukocyte grouping using multispecific sera. "Histocompatibility Testing, 1965." Munksgaard, Copenhagen, Denmark. p. 141-149.

Bodmer, W., Bodmer, J., Adler, S., Payne, R. and Bialek, J. 1966. Genetics of 4 and LA human leukocyte groups. Ann. N.Y. Acad. Sci., Seventh International Transplantation Conference. (F. T. Rapaport, ed.) 129:472-489.

Bodmer, W. and Felsenstein, J. 1967. Linkage and selection: Theoretical analysis of the deterministic two locus random mating model. Genetics 57:237-265.

Ceppelini, R. 1967. Genetica delle immunoglobuline. Atti X Riunione Associazione Genetica Italiana, Parma, 28-30 Ottobre 1966. A.G.I. 12:3.

Ceppellini, R., Curtoni, E. S., Mattiuz, P. L., Leigheb, G., Visetti, M. and Colombi, A. 1966. Survival of test skin grafts in man: Effect of genetic relationship and of blood groups incompatibility. Ann. N.Y. Acad. Sci., Seventh International Transplantation Conference. (F. T. Rapaport, ed.) 129:421-445.

Ceppellini, R., Curtoni, E. S., Mattiuz, P. L., Miggiano, V., Scudeller, G. and Serra, A. 1967. Genetics of leukocyte antigens: A family study of segregation and linkage. "Histocompatibility Testing, 1967." Munksgaard, Copenhagen, Denmark. p. 149-183.

Dausset, J., Ivanyi, P. and Ivanyi, D. 1965. Tissue alloantigens in humans: Identification of a complex system (Hu-1). "Histocompatibility Testing 1965." Munksgaard, Copenhagen, Denmark. p. 51-62.

Dausset, J. and Rapaport, F. T. 1966. The role of blood group antigens in human histocompatibility. Ann. N.Y. Acad. Sci., Seventh International Transplantation Conference. (F. T. Rapaport, ed.) 129:408-420.

Dausset, J., Colombani, J., Colombani, M., Legrand, L. and Feingold, N. 1968. Un nouvel antigene du systeme HL-A (Hu-1): l'antigene 15, allele possible des antigenes 1, 11, 12. Nouv. Rev. Fr. D'Hemat. 8:398-406.

Elston, R. C., 1967. Genetic analysis of white cell blood groups. Amer. J. Human Genetics 19:258-269.

Ferrone, S., Tosi, R. M. and Centis, D. 1967. Anticomplementary factors affecting the lymphocytotoxicity test. "Histocompatibility Testing, 1967." Munksgaard, Copenhagen, Denmark. (Curtoni, E. S., Mattiuz, P. L. and Tosi, R. M., eds.) p. 357-364.

Histocompatibility Testing, 1967. Munksgaard, Copenhagen, Denmark. (Curtoni, E. S., Mattiuz, P. L. and Tosi, R. M.)

Ivanyi, P. and Dausset, J. 1966. Allo-antigens and antigenic factors of human leukocytes. A hypothesis. Vox Sang. 11:326-331.

Mi, M. P. and Morton, N. E. 1966. Blood factor association. Vox Sang. 11:434-449.

Motulsky, Arno G. 1964. Current concepts of the genetics of the Thalassemias. Cold Spring Harbor Symp. Quant. Biology Vol. XXIX:399-413.

Payne, R., Tripp, M., Weigle, J., Bodmer, W. and Bodmer, J. 1964. A new leukocyte isoantigen system in man. Cold Spring Harbor Symp. Quant. Biol. 29:285-295.

Race, R. R., and Sanger, R. 1962. "Blood Groups in Man". Blackwell Scientific Publications Led.

Rood, J. J. van 1962. Leukocyte grouping. A method and its application. Thesis, Leiden.

Rood, J. J. van, Leeuwen, A. van, Schippers, A. M. J., Vooys, W. H., Frederiks, E., Balner, H. and Eernisse, J. G. 1965. Leukocyte groups, the normal lymphocyte transfer test and homograft sensitivity. "Histocompatibility Testing, 1965." Munksgaard, Copenhagen, Denmark. p. 37-50.

Rubinstein, P., Costa, R., Leeuwen, A. van, and Rood, J. J. van 1967. The leukocyte antigens of Mapuche Indians. "Histocompatibility Testing, 1967." Munksgaard, Copenhagen, Denmark. (Curtoni, E. S., Mattiuz, P. L., Tosi, R. M., eds.) p. 251-255.

Shreffler, D. C. 1967. Genetic control of cellular antigens. Proc. Third Int. Congr. Human Genetics. (Crow, J. F. and Neel, J. V., eds.) The Johns Hopkins Press, Baltimore, Maryland. p. 217-231.

Terasaki, P. I., Mickey, M. R., Vredevoe, D. L. and Boyette, D. R. 1965. Serotyping for homotransplantation IV. Grouping and evaluation of lymphotoxic sera. Vox Sang. 11:350-376.

Walford, R. L., Shanbrom, E., Troup, G. M., Zeller, E. and Ackerman, B. 1967. Lymphocyte grouping with defined antisera. "Histocompatibility Testing, 1967." Munksgaard, Copenhagen, Denmark. (Curtoni, E. S., Mattiuz, P. L., and Tosi, R. M., eds.) p. 221-230.

Walford, R. L. and Troup, G. M. 1967. Monospecific lymphocytotoxic antisera. An absorption study. Vox. Sang. 12:173-185.

DISCUSSION ON GENETIC ASSOCIATION

ELSTON: We should all be very thankful for this clear account of association analysis and how it has been applied to the HL-A leukocyte system. As I have not had long to study the paper I shall not attempt to discuss it in detail here, but merely make a couple of comments about association analysis in general.

My first point is that association can be due to many factors, and it is good to have listed for us all these factors. In particular I want to emphasize the possibility of population stratification, or nonhomogeneity, as it seems to be so often overlooked. It is common to find investigators who observe the numbers, 0, of individuals falling into several classes, calculate (on the basis of some null hypothesis) the numbers, E, expected to fall in these classes, and then compare the statistic $\Sigma(0-E)^2/E$ with a χ^2-distribution to test their null hypothesis. But this statistic follows a χ^2- distribution, asymptotically on the null hypothesis, only if the population sampled is homogeneous.

I myself have been considering the analysis of leukocyte systems when the measure of reaction between serum and cell type, e.g., agglutination, is quantitative rather than qualitative; i.e., I have been considering whether one can make use of this further information in the analysis. This brings me to my second point, which concerns the detection of association between two quantitative variables in general. The correlation coefficient, which is often used, has the disadvantage that it detects only linear association. The other method commonly used is to divide the scales on which the traits are measured into class intervals, and form a two-way table containing the number of observations falling into each subclass; a χ^2 test for association is then performed on this table. This has the disadvantage that a large sample size is required for the resulting statistic to be approximately distributed, on the null hypothesis, as χ^2. I should merely like to report here that, in collaboration with Dr. John Stewart of Cambridge University, I have this summer been examining a test for association that overcomes both of these disadvantages. So far it appears that the test we have devised is very powerful for detecting many types of association, and our findings will be published in Biometrics.

SEGREGATION ANALYSIS

N. E. Morton
Population Genetics Laboratory
University of Hawaii

The mechanisms of inheritance have as their end result the generation of phenotype frequencies characteristic of a given mating type in a specified environment. We shall refer to these as *segregation frequencies,* including not only classical mendelian frequencies like 1/2, 1/4, and 9/16, but also modifications by differential mortality and partial manifestation, and even empirical frequencies, the genetic basis of which is unclear. For example, if 1/4 of the children from a particular mating type are expected to be of a certain genotype, but only 80 per cent of them develop a characteristic abnormal phenotype, we will say that the segregation frequency is $(1/4)(.80) = .20$. Similarly, if 20 per cent of the children from a particular mating type have a certain phenotype, we shall say the segregation frequency is .20, even though the mechanism of inheritance may be unknown.

Although segregation frequencies are specified either by genetic hypothesis or empirically from analysis of a series of matings, the actual proportions observed in any particular mating are dependent on gene frequencies, chance, the way the data are collected, and other factors. *Complete selection* is defined as *random sampling of families through the parents, without consideration of the phenotypes of the children* (Morton, 1958). This method of sampling is usually employed with polymorphs and common traits of uncertain etiology, although sometimes additional criteria of selection are imposed based upon the pattern of segregation observed in the children. Such *incomplete selection* is discussed later. Idiomorphs, not recognized in heterozygotes, are rarely studied by complete selection because of the large proportion of families in which the idiomorphic allele is absent or undetected because of recessivity.

For simplicity we shall consider only two segregant phenotypes, which may be called *normal* and *affected.* This loses no generality, since n phenotypes may be examined pairwise in $n - 1$ independent ways. Let there be r affected children in a sibship of size s (the terms *sibship, family,* and *mating* will be used interchangeably to denote the set of children classified as normal or affected). A mating will be called *nonsegregating* if there are no affected children ($r = 0$); *segregating* if there is at least one affected child ($r>0$); and *doubly segregating* if there are both normal and affected children ($0<r<s$). The *segregation frequency* p is the expected proportion of affected children in a given mating type under complete selection, if this proportion is considered to be uniform or continuously variable. However, if the expected proportion of affected children is considered to be sharply discontinuous, with a low value in some matings and a higher value in others, p will be used to denote the segregation frequency in the latter *high-risk* group of families. We shall naturally be concerned to determine which model is appropriate in any particular body of data.

UNIFORM SEGREGATION FREQUENCY (COMPLETE SELECTION)

Suppose that the segregation frequency p is uniform, with no families that cannot segregate, then the distribution of r affected children in a family of size s is binomial,

$$(1) \qquad P(r;s,p) = \binom{s}{r} p^r (1-p)^{s-r} \qquad 0<p<1, \quad 0\le r\le s$$

This and the other distributions encountered in segregation analysis are most easily and efficiently studied by maximum likelihood scores, which provide both iterative estimates and tests of hypotheses (Morton, 1958).

ADMIXTURE OF FAMILIES THAT CANNOT SEGREGATE (COMPLETE SELECTION)

Suppose that the segregation frequency is uniform in families able to produce affected children, but in the mating type under study there is admixed a proportion h of families unable to segregate, and indistinguishable from the potentially segregating families which with probability $(1 - p)^s$ happen to produce no affected children. Then the distribution of r affected children in a family of size s is an *augmented binomial,*

$$(2) \qquad P(r=0;s,p,h) = h + (1-h)(1-p)^s \qquad 0<p,h<1,$$
$$P(r>0;\, s,p,h) = (1-h)\binom{s}{r} p^r (1-p)^{s-r} \qquad 0\le r\le s$$

There are three principal reasons for families unable to segregate (i.e., $h>0$): homozygosity, phenocopies, and bivalent alleles (Morton et al., 1966). The probability of homozygosity for a dominant allele can be calculated from the genotype frequencies in the population (Morton, 1958).

A second reason for parents who cannot segregate is *phenocopies* (rare nontransmissible phenotypes that simulate affected phenotypes). For example, retinoblastoma is sometimes due to a rare dominant gene with incomplete penetrance and sometimes to a phenocopy that may be a somatic mutation. Then h is the probability that an affected parent in a family under complete selection be a phenocopy, in which case the probability of a child developing retinoblastoma is virtually nil. In the remaining proportion $1 - h$ of families the affected parent can transmit a gene for retinoblastoma, and the probability that a child will have the disease is p (nearly 1/2).

Analysis of Equation (2) permits estimation of h and p from families of any size, whereas other methods use only exceptionally large families which can be classified as low or high risk by inspection (Steinberg, 1962). Genetic counselors usually confound h and p by stating that the risk for an affected child from an affected parent is $(1 - h)p$, thereby underestimating the disease risk p in families with one or more affected children and exaggerating the risk in families with only normal children. If there are s children beyond the age of manifestation, all normal, the posterior probability that the affected parent is a phenocopy is

$$(3) \qquad \theta = \frac{h}{h + (1-h)(1-p)^s}$$

and the mean risk for a subsequent child is only $(1 - \theta)p$.

Some phenocopies are technical errors. For example, in the mating MN x M, the probability of all MN children may be taken to be $h + (1 - h)(1 - p)^s$, where h is the probability that a parent of phenotype MN be actually of genotype NN but give a false positive reaction with anti-M serum. This distribution has been used to argue that technical errors are not frequent enough to explain the excess of MN children observed in published family studies, which therefore may be due to preferential survival of MN fetuses (Morton and Chung, 1959), or even meiotic drive (Hiraizumi, 1964). However, this argument is not decisive.

A third possibility for $h>0$ is that some alleles in the MN system react consistently with both anti-M and anti-N. Such rare *bivalent alleles,* transmitting in coupling two factors which ordinarily occur only in repulsion, produce families that cannot segregate. Morton et al. (1966) used segregation analysis to search for rare bivalent alleles.

Equation (2) may be generalized to double segregation. Let p and h be defined as above, and y be the probability that the mating be unable to segregate normal children. Then

$$(4) \quad P(r = 0;\ s, p, h, y) = h + (1 - h - y)\ (1 - p)^s \qquad 0<p, h, y<1$$

$$P(0<r<s;\ s, p, h, y) = (1 - h - y)\ \binom{s}{r} p^r (1 - p)^{s - r} \qquad 0<r<s$$

$$P(r = s; s, p, h, y) = y + (1 - h - y)\ (1 - p)^s$$

In different applications y may represent matings of allelic homozygotes (Chung et al., 1958), technical errors (Morton and Chung, 1959) and bivalent alleles (Morton et al., 1966). Clearly Equations (1) and (2) are special cases of (4).

BIMODAL SEGREGATION FREQUENCY (COMPLETE SELECTION)

The previous model dealt with a mixture of null-risk and high-risk groups. Sometimes all families are exposed to a small risk which, however, is not negligible. This is equivalent to a mixture of two uniform segregation frequencies. Let the low risk be m and the high risk be p, and let the proportion of low-risk families be w. The segregation frequency in high-risk families is $p + m - pm$, and the probability of r affected in a family of size s is

$$(5) \quad P(r; s, m, w, p) = \binom{s}{r} \{ w m^r (1 - m)^{s - r} + (1 - w)\ (p + m - pm)^r (1 - p - m + pm)^{s - r} \} \qquad 0<p, m, w<1,\ 0\leq r\leq s$$

The previous model was the special case $m = 0$, $w = h$. Severe mental defect and many other diseases of complex etiology conform approximately to this model, which is therefore useful in genetic counseling as well as for population genetics. The high-risk group corresponds to major genes (especially recessive idiomorphs), chromosomal anomalies, and familial nongenetic factors which can usually be recognized, whereas the low-risk group is due to complications of pregnancy and labor, postnatal injury and infection, complex genetic causes, and other mechanisms which do not often recur in families. While the parameters of this model can be estimated in any large body of data, proof that the segregation frequency is bimodal requires either $m \ll p$, large families, or evidence that the two groups are of different etiology (Dewey et al., (1965). Even if critical evidence for bimodality is lacking, the reduction in χ^2 due to simultaneous estimation of m and w is a powerful test for heterogeneity of p.

Extramarital parentage is one source of low-risk families. In the MNS system, matings S x s can give S+s- progeny either by extramarital parentage or segregation of an S-s- allele, which is very rare in Caucasians and Orientals but not uncommon in Negroes. By equating transmission of an S-s- allele to the high-risk group it is possible to infer the frequency of extramarital parentage not otherwise detected, to determine that most S-s- alleles inferred from segregation in Negroes are not due to extramarital parentage, to establish the gene frequency of S-s- alleles, and to demonstrate that they are part of the MN system, rather than an unlinked suppressor (Morton et al., 1966).

CONTINUOUSLY VARIABLE SEGREGATION FREQUENCY (COMPLETE SELECTION)

A reasonable generalization of Equation (1) is the distribution named after Skellam (1948), which assumes that p is not constant among families, but takes a value v according to the beta density,

$$(6) \quad f(v) = \frac{(z + \xi + 1)!}{z!\,\xi!}\ v^z (1 - v)^\xi \qquad z, \xi > -1,\quad 0<v<1$$

Then the Skellam distribution of r affected in families of size s is

$$(7) \quad P(r; s, z, \xi) = \int_0^1 f(v)\ \binom{s}{r} v^r (1 - v)^{s - r} dv = \frac{\binom{r + z}{r} \binom{s - r + \xi}{s - r}}{\binom{s + z + \xi + 1}{s}}$$

The mean segregation frequency is

$$p = \int_0^1 v f(v) dv = \frac{z + 1}{z + \xi + 2}$$

It is convenient to estimate p and one nuisance parameter, z, using the ML score

$$u_p = \frac{u_\xi}{\partial p/\partial \xi} + \frac{u_z}{\partial p/\partial z}$$

This model appears promising for quasi-continuous variation (Edwards, 1967) to estimate the risk p_{sr} for affection of child following s sibs of whom r are affected, when there is no birth order effect,

$$(8) \quad p_{sr} = \frac{\int_0^1 v f(v)\ \binom{s}{r} v^r (1 - v)^{s - r} dv}{\int_0^1 f(v)\ \binom{s}{r} v^r (1 - v)^{s - r} dv} = \frac{z + r + 1}{z + \xi + s + 2} = \frac{(z + r + 1) p}{z + sp + 1}$$

INCOMPLETE SELECTION

Idiomorphs not recognizable in heterozygotes and rare traits of uncertain etiology are usually studied only in segregating families. Even polymorphs are sometimes analyzed inefficiently in this way, to avoid gene frequency estimation in mixtures of families that cannot segregate. Such selection of families through the children, with exclusion of nonsegregating families, is called incomplete selection. This may involve both sampling of families and later exclusion of sampled families with certain segregation patterns.

PROBANDS AND THE ASCERTAINMENT PROBABILITY

Analysis of incomplete selection is based on the concept of a proband, defined as an affected person who at any time was detected independently of the other members of the family, and who would therefore be sufficient to assure selection of the family in the absence of other probands. Probands may be ascertained through hospital records, death certificates, inquiries to physicians, examination of a population sample, or other direct means. Affected individuals not represented in these primary sources and detected only through family study of probands are called secondary cases. In a family of s classified children with r affected we let a be the number of probands, so that $r - a$ is the number of secondary cases ($a \leq r$).

The concept of a proband leads directly to the ascertainment probability. Suppose that in a given finite population there are R affected individuals, of whom A are detected as probands. Then the ascertainment probability is defined as

$$(9) \quad \pi = \frac{A}{R},$$

the probability that an affected member of the population be detected as a proband. Under incomplete selection an estimate of π is essential for a valid analysis of segregation and determination of the number of affected individuals in the population,

$$R = \frac{A}{\pi} \tag{10}$$

In the general case of multiple selection there may be from 1 to r probands in an ascertained family with r affected, and each proband may have $t \geq 1$ ascertainments, with $0 < \pi \leq 1$. Two limiting cases are of particular interest:

Truncate selection ($\pi = 1$). Segregating families are selected at random, so that families with many affected children are no more likely to be selected than families with only one affected child. If the segregation frequency is uniform, then in families of a given size s the distribution of the number of affected children is a truncated binomial, lacking only the first term corresponding to nonsegregating families ($r = 0$). Since the a priori probability of this excluded event is $(1 - p)^s$,

$$P(r \mid r > 0, \pi = 1; s, p) = \frac{\binom{s}{r} p^r (1 - p)^{s - r}}{\sum_{r=1}^{s} \binom{s}{r} p^r (1 - p)^{s - r}} = \frac{\binom{s}{r} p^r (1 - p)^{s - r}}{1 - (1 - p)^s} \tag{11}$$

Single selection ($\pi \to 0$). The probability of ascertainment is so small that there is virtually no chance of having two probands in the same family, and the probability that a family be ascertained is proportional to the number of affected children,

$$P(r \mid r > 0, \pi \to 0; s, p) = \frac{r\binom{s}{r} p^r (1 - p)^{s - r}}{\sum_{r=1}^{s} r\binom{s}{r} p^r (1 - p)^{s - r}} = \frac{s\binom{s-1}{r-1} p^r (1 - p)^{s-r}}{sp \sum_{r=1}^{s} \binom{s-1}{r-1} p^{r-1} (1 - p)^{s - r}} = \binom{s-1}{r-1} p^{r-1} (1 - p)^{s-r} \tag{12}$$

Thus single selection is equivalent to complete selection of the sibs of the proband.

Despite the attractive simplicity of this result, single selection should be avoided whenever possible for two reasons:

1. When $\pi \to 0$, it is impossible to estimate the number of affected individuals in the population by Equation (10).
2. Single selection gives a poor representation of families with isolated cases ($r = 1$), which may depend on interesting genetic or nongenetic mechanisms different from familial cases ($r > 1$).

SPORADIC CASES

Even if many affected individuals are due to simple genetic mechanisms and therefore occur in high-risk families, it is the rule rather than the exception for some cases to be sporadic due to mutations, phenocopies, technical errors, extramarital conceptions, rare instances of heterozygous expression of a recessive gene, chromosomal nondisjunction, or complex genetic mechanisms. These sporadics, of different origin from the high-risk cases, must be distinguished from chance-isolated cases whose siblings, although normal, have the same a priori risk p of being affected. Sometimes the distinction between sporadic and chance-isolated cases can be made phenotypically or from the history of the isolated case. Often, however, such a distinction is difficult or impractical, but the proportion of sporadic cases can be determined. An isolated case should never be termed sporadic unless there is reason to believe that it is not chance-isolated, i. e., that it is of different origin from familial cases. A family with an isolated case is called simplex, and a family with more than one affected child is said to be multiplex.

The concept of sporadic cases may be derived from the bimodal segregation frequency of Equation (5) under incomplete selection. If a proportion w of the families in the population have low-risk m and the remainder have high-risk $p+m - pm$, then the frequency of affected persons is

$$wm + (1 - w)\ (p+m - pm), \tag{13}$$

of which the first term is due to affection in low-risk families. The probability that a randomly chosen individual come from a low-risk family is therefore

$$x = \frac{wm}{wm + (1 - w)\ (p+m - pm)} \tag{14}$$

The probability that a random affected individual come from a low-risk family and have at least one of $s - 1$ sibs affected is

$$x\{1-(1-m)^{s-1}\}$$

The probability that a random affected individual come from a high-risk family and have at least one of $s - 1$ sibs affected is

$$(1 - x)\{1 - (1 - p - m+pm)^{s - 1}\}.$$

We conclude that the probability that a multiplex family be low-risk is

$$\xi_s = \frac{x\{1 - (1 - m)^{s - 1}\}}{x\{1 - (1 - m)^{s - 1}\}+(1 - x)\{1 - (1 - p - m+pm)^{s - 1}\}} \tag{15}$$

If this is negligibly small, low-risk mechanisms make no significant contribution to multiplex families. Stated otherwise, almost all cases in low-risk families are sporadic, and we are justified in calling x the proportion of sporadic cases among affected individuals in the population. Clearly $\xi_s \to 0$ under two different conditions:

1. The proportion of low-risk families in the population is small ($x \to 0$), implying from Equation (14) either $w \to 0$ or $m \to 0$.
2. There is a large difference between low and high risks, the latter group exists in appreciable frequency, and families are not extremely large ($m \ll p$, $ms \ll 1$).

For the rare traits which are efficiently analyzed under incomplete selection, one or both of these conditions is usually met. Traits which may violate these conditions should be studied under complete selection. The proportion x of sporadic cases among affected individuals is not the same as the proportion ϵ_s of families with sporadic cases among all segregating families of size s. Excluding sporadic cases, the mean number of affected children in a segregating sibship of size s is

$$\bar{r} = E(r \mid r > 0) = \frac{\sum_{r=1}^{s} r\binom{s}{r} p^r (1 - p)^{s - r}}{\sum_{r=1}^{s} \binom{s}{r} p^r (1 - p)^{s - r}} = \frac{sp}{1 - (1 - p)^s} \tag{16}$$

Each family with a sporadic case contributes one affected, while the average family with a nonsporadic case contributes $\bar{r}$ affected. By definition, the proportion of sporadic cases is

$$x = \frac{\epsilon_s(1)}{\epsilon_s(1) + (1 - \epsilon_s)\bar{r}}$$

Substituting for $\bar{r}$ and rearranging,

(17) $$\epsilon_s = \frac{xsp}{xsp + (1 - x)[1 - (1-p)^s]}$$

Since $1 - (1 - p)^s < sp$, it follows that $x < \epsilon_s$. If sporadic cases are not related to parity or parental age, it may be shown that x is independent of s. For then the expected number of sporadic cases in a family of size s is sm, and the expected number of nonsporadic cases is $s(p + m - pm)$. Therefore the frequency of sporadic cases in families of size s is

$$\frac{w(sm)}{w(sm) + (1 - w)\{s(p + m - pm)\}}$$

which is independent of s. Since ϵ_s increases with s, it is a less useful parameter than x.

Additional evidence of sporadic cases is given by a variable with mean μ_I for n_I probands in simplex families and μ_F for n_F probands in multiplex families, where $\mu_I \neq \mu_F$. Then we may estimate x if we know the mean μ_0 for sporadic cases. We argue as follows: probands are assumed to be drawn at random from isolated and familial cases. Then $n_I/(n_I + n_F)$ is an unbiased estimate of the proportion of isolated cases in the population. We assume that isolated probands are made up of a mixture of sporadic cases in proportion y and chance-isolated cases having the same mean as familial cases. Then the expected value of μ_I is

$$\mu_I = y\mu_0 + (1 - y)\mu_F, \text{ or}$$

$$y = \frac{\mu_F - \mu_I}{\mu_F - \mu_0}$$

and

(18) $$x = \left(\frac{n_I}{n_I + n_F}\right)\left(\frac{\mu_F - \mu_I}{\mu_F - \mu_0}\right)$$

In the most common special case, a rare recessive trait, μ stands for the inbreeding coefficient.

MULTIPLE SELECTION WITH A MIXTURE OF SPORADIC CASES AND A UNIFORM SEGREGATION FREQUENCY (ASCERTAINMENT PROBABILITY UNIFORM)

If the ascertainment probability π is constant for affected individuals among and within families, and if probands are independently ascertained, the probability that none of r affected sibs be ascertained is $(1 - \pi)^r$, and therefore the probability of ascertaining such a family is $1 - (1 - \pi)^r$. Note that if $\pi = 1$ (truncate selection), the probability of ascertaining a segregating family ($r>0$) is 1, independent of r, whereas the probability of ascertaining a nonsegregating family ($r = 0$) is 0. However, if π approaches 0 the probability of ascertaining a segregating family is

$$1 - (1 - \pi)^r \rightarrow 1 - \{1 - r\pi + O(\pi)^2\}.$$

For small π, terms in π^2, π^3, etc. can be neglected, and so $1 - (1 - \pi)^r \rightarrow r\pi$, justifying our earlier assertion (12) that under single selection the probability of ascertaining a family with r affected is proportional to r.

For the general case of multiple selection ($0 < \pi \leq 1$), the distribution of r affected among ascertained families of size s, excluding sporadic cases, is

(19) $$P(r \mid a>0;\ s,p,\pi) = \frac{\binom{s}{r}p^r(1-p)^{s-r}\{1 - (1-\pi)^r\}}{\sum_{r=0}^{s}\binom{s}{r}p^r(1-p)^{s-r}\{1-(1-\pi)^r\}}$$

$$= \frac{\binom{s}{r}p^r(1-p)^r\{1 - (1-\pi)^r\}}{1 - (1 - p\pi)^s}$$

Note that $(1 - p\pi)^s$ is the probability of having no proband in a family of size s.

Now we introduce sporadic cases. Among segregating families in the population a proportion ϵ_s have sporadic cases, with probability π for ascertainment, and $1 - \epsilon_s$ have nonsporadic cases. Among the latter, the probability of an ascertained family among all segregating families is the ratio

$$\frac{1 - (1 - p\pi)^s}{1 - (1 - p)^s}$$

We see that the probability of an ascertained family of size s having either a sporadic or nonsporadic case is

$$P(a>0;\ s,p,\pi,\epsilon_s) = \epsilon_s\pi + \frac{(1 - \epsilon_s)\{1 - (1 - p\pi)^s\}}{1 - (1 - p)^s}$$

Substituting for ϵ_s from (17), the probability of an ascertained family of size s among segregating families becomes

$$P(a>0 \mid r>0;s,p,\pi,x) = \frac{xsp\pi + (1 - x)\{1 - (1 - p\pi)^s\}}{xsp + (1 - x)\{1 - (1 - p)^s\}}$$

The first term in the numerator comes from sporadic cases, and the second term from nonsporadic ones, including chance-isolated cases. We saw in Equation (19) that

$$1 - (1 - p\pi)^s = \sum_{r=0}^{s}\binom{s}{r}p^r(1 - p)^{s-r}\{1 - (1 - \pi)^r\}.$$

The first term in this sum is $sp\pi(1 - p)^{s-1}$, corresponding to ascertained chance-isolated cases. Thus the distribution of r affected in ascertained families of size s is

(20) $$P(r = 1 \mid a>0;s,p,\pi,x) = \frac{sp\pi\{x + (1 - x)(1 - p)^{s-1}\}}{xsp\pi + (1 - x)\{1 - (1 - p\pi)^s\}}$$

$$P(r>1 \mid a>0;s,p,\pi,x) = \frac{(1 - x)\binom{s}{r}p^r(1 - p)^{s-r}\{1 - (1 - \pi)^r\}}{xsp\pi + (1 - x)\{1 - (1 - p\pi)^s\}}$$

Note from Equation (20) that the frequency of families with sporadic cases among ascertained families of size s is

(21) $$\zeta_s = \frac{xsp\pi}{xsp\pi + (1 - x)\{1 - (1 - p\pi)^s\}}$$

By comparison with (17) we see that for $\pi = 1$, $\zeta_s = \epsilon_s$, the proportion of families in the population with sporadic cases among all segregating families of size s, but $\zeta_s < \epsilon_s$ for $\pi < 1$. In particular, $\zeta_s = x$ for single selection, so that

$$x \leq \zeta_s \leq \epsilon_s$$

Equation (20) is the most general formulation for segregation under incomplete selection, admitting a mixture of sporadic cases with a uniform segregation frequency. Methods of analysis for continuously variable or bimodal segregation patterns corresponding to Equations (5) and (7) have not yet been developed for incomplete selection, where the sample size is usually too small to justify this elaboration.

DISTRIBUTION OF PROBANDS AMONG AFFECTED SIBS

If the ascertainment probability π is uniform and ascertainments are independent, the distribution of a probands among r affected sibs is a truncated binomial,

$$(22) \qquad P(a \mid a>0;r,\pi) = \frac{\binom{r}{a}\pi^{a}(1-\pi)^{r-a}}{\sum_{a=1}^{r}\binom{r}{a}\pi^{a}(1-\pi)^{r-a}} = \frac{\binom{r}{a}\pi^{a}(1-\pi)^{r-a}}{1-(1-\pi)^{r}}$$

The definition of probands may be difficult under certain methods of ascertainment, for example, through membership in a society of families with affected children. Since the sibship is the unit of analysis, a sibship with no proband, but one or more secondary cases ascertained through related probands, is obtained by truncate selection, as selection is assured in such a sibship by presence of a single affected child. Such secondary sibships should be scored for $\pi = 1$ and appropriate values of p and x, but they give no information about the ascertainment probability. Unrelated affected individuals ascertained in the course of a family study are probands in their sibship, and such fortuitous detection is one mode of ascertainment. The investigator should remember that ascertainment is defined not by words but by conformity to a probability model.

Sometimes it is impractical to define probands so that they are independent. For example, a severe disease like hemophilia may have a high level of ascertainment in a given population, but ascertainment of cases who died in the previous generation may be poor. Then we are tempted to define a proband as a living ascertained case. The material for analysis will be a mixture of young sibships with most of the affected members living and old sibships with few living affected.

For such ascertainment we may suppose that the ascertainment probability is constant within families but variable among families. Let the ascertainment probability for the ith family with r affected be π_i, so that the probability of a probands in that family is

$$P(a;r,\pi_i) = \binom{r}{a}\pi_i^{a}(1-\pi_i)^{r-a}$$

Our objective is to obtain an accurate estimate of $\pi = 1 - P(0)$, given a good representation of P(a) for a = 1, . , r, the goodness of fit being tested by χ^2. It seems reasonable to assume that π_i varies among families in accordance with the beta distribution,

$$f(\pi_i) = \frac{(\mu+b+1)!}{\mu!\ b!}\pi_i^{\mu}(1-\pi_i)^{b} \qquad \mu, b > -1, \quad 0 < \pi_i < 1$$

On this hypothesis the distribution of a probands among all families with r affected is a Skellam distribution like that given in Equation (7).

$$(23) \qquad P(a;r) = \int_0^1 f(\pi_i)P(a;r,\pi_i)d\pi_i = \frac{\binom{a+\mu}{a}\binom{r-a+b}{r-a}}{\binom{r+\mu+b+1}{r}}$$

Skellam (1948) noted, "Since this form of distribution is capable of a variety of shapes, no severe limitation is imposed on the way the probability fluctuates. In practice we could, at least in most cases, take this form of distribution as a convenient approximation."

In the population the ascertainment probability is

$$(24) \qquad \pi = 1 - P(0;1) = \frac{\mu+1}{\mu+b+2},$$

which is the mean value of π_i.

The probability of ascertaining a family with r affected is

$$1 - P(0,1) = 1 - \frac{\binom{r+b}{r}}{\binom{r+\mu+b+1}{r}}$$

Ascertained families follow the truncated Skellam distribution

$$(25) \qquad P(a \mid a>0;r) = \frac{\binom{a+\mu}{a}\binom{r-a+b}{r-a}}{\binom{r+\mu+b+1}{r}\{1-P(0;r)\}}$$

We may eliminate the nuisance parameter μ to give the ML score for π

$$u_\pi = \frac{u_\mu}{\partial\pi/\partial\mu} + \frac{u_b}{\partial\pi/\partial b}$$

To iterate on π and b we need to know the corresponding value of μ at each iteration; by (24) this is

$$(26) \qquad \mu = \frac{\pi(b+2)-1}{1-\pi}$$

It has not so far seemed necessary to modify Equation (20) by introducing a variable ascertainment probability. Consideration of a number of published studies which used quite different definitions of probands indicates that segregation analysis is reasonably robust, with a sufficiently small covariance between π and the other parameters so as not to be disturbed appreciably by small departures from the ascertainment model (Morton, 1962). Since even the incomplete and crude data provided by published genetic studies can apparently be tolerated, an enlightened effort to define probands appropriately is all that is usually required to assure a valid segregation analysis by (20), using (25) where (22) is demonstrably invalid.

DISTRIBUTION OF ASCERTAINMENT PER PROBANDS

The distribution of probands among r affected sibs gives information about π only for multiplex families ($r>1$), since isolated cases are necessarily probands. If most cases are sporadic, or if p or s is small, even a large sample will give little information about the ascertainment probability by this method. We are led to look for a more powerful technique that will yield information even for isolated cases.

Suppose a proband can be ascertained in an indefinitely large number of ways. If ρ_i is the probability of ascertainment from the ith source and there are T independent sources, with

$$\Psi = \sum_{i=1}^{T}\rho_i,$$

the probability of ascertainment by t of these is shown in elementary statistics to approach the Poisson distribution

$$P(t) = \frac{\Psi^{t}e^{-\Psi}}{t!}$$

as $T \to \infty$. The ascertainment probability under these conditions is

$$\pi = 1 - P(0) = 1 - e^{-\Psi}$$

so that $\Psi = -\ln(1-\pi)$. The truncated Poisson distribution may therefore be written

$$(27) \qquad P(t \mid t>0;\pi) = \frac{\{-\ln(1-\pi)\}^{t}(1-\pi)}{t!\,\pi}$$

The distribution of t gives most of the information about π, since both isolated and familial cases can be used. In two test cases, the distribution of probands among affected gave less than 1/4 as much information as the distribution of ascertainments per proband (Morton and Chung, 1959). However, for limb-girdle muscular dystrophy the estimates on the hypothesis of independent ascertainment and uniform ascertainment probability were significantly heterogeneous. It was suggested that the ascertainment distribution tends to underestimate π because ascertainment by one source tends to make another less likely, as one physician's referral tends to preclude another. On the other hand, the distribution of probands on the assumption of a uniform ascertainment probability may overestimate π because two or more siblings may be examined and counted as probands, even though only one of them would have submitted to examination independently. Pooling the two independent sources of information gives an intermediate value that is closer to the lower estimate. Further experience with the same population suggested that the ascertainment distribution was more reliable than the proband distribution and that the pooled estimate was substantially correct.

Unless the ascertainment model is clearly understood, ascertainments will not be defined appropriately. Multiple referrals by the same physician are usually not independent. For example, if a proband was ascertained through a hospital record, the referring and consulting physician should not be counted as separate sources of ascertainment unless they also report the proband independently of the hospital record.

The distribution of ascertainments per proband can be freed from the hypothesis of uniform ascertainment probability, just as the proband distribution was. Suppose that the expected mean number of ascertainments Ψ_i is different for each proband, and has a gamma distribution ($0 < \Psi_i < \infty$). The purpose of this assumption is merely to obtain an accurate estimate of $\pi = 1 - P(0)$, given a good representation of P(t) for t = 1, 2, ..., the goodness of fit being tested by χ^2. Fisher (1941) showed that our hypothesis gives a truncated negative binomial distribution,

$$(28)\quad P(t \mid t > 0; z, u, \pi) = \binom{z}{t}\mu^t(1-\mu)^{z-t}/\pi \qquad \mu, z < o$$

The ascertainment probability is

$$\pi = 1 - P(0) = 1 - (1-\mu)^z,$$

and so we can eliminate μ by setting

$$(29)\quad u_\pi = \frac{u_\mu}{\partial\pi/\partial\mu} + \frac{u_z}{\partial\pi/\partial z}$$

$$\mu = 1 - (1-\pi)^{1/z}$$

Although the negative binomial distribution is sufficiently flexible to describe most kinds of variable and dependent ascertainment, it assumes that the expected mean number of ascertainments varies continuously around a single mode. Sometimes affection is a mixture of mild and severe forms. If these are sharply discontinuous, the expected number of ascertainments may be bimodal. In this case, reliable genetic analysis requires separation of the different entities.

MULTIPLEX FAMILIES

If affection is a mixture of sporadic and high-risk cases, data on the latter component may be collected by omitting isolated cases. From (20) the distribution of r affected in sibships of size s under a uniform ascertainment probability is

$$(30)\quad P(r \mid r>1, a>0; s, p, \pi) = \frac{\binom{s}{r}p^r(1-p)^{s-r}\{1-(1-\pi)^r\}}{\sum_{r=0}^{s}\binom{s}{r}p^r(1-p)^{s-r}\{1-(1-\pi)^r\}}$$

$$= \frac{\binom{s}{r}p^r(1-p)^{s-r}\{1-(1-\pi)^r\}}{1-(1-p\pi)^s - sp\pi(1-p)^{s-1}}$$

MULTIPLEX FAMILIES WITH TWO OR MORE PROBANDS

Although the preceding distribution allows us to concentrate on high-risk mechanisms by exclusion of sporadic cases, it assumes that simplex families are detected but excluded from further study. It is easier to select families with two or more probands. Then simplex families will not enter the sample and the investigator focuses at once on multiplex families. The distribution of r affected in sibships of size s is

$$(31)\quad P(r \mid a>1; s, p, \pi)$$

$$= \frac{\binom{s}{r}p^r(1-p)^{s-r}\{1-(1-\pi)^r - r\pi(1-\pi)^{r-1}.\}}{1-(1-p\pi)^s - sp\pi(1-p\pi)^{s-1}}$$

The distribution of a probands among r affected is

$$(32)\quad P(a \mid a>1; r, \pi) = \frac{\binom{r}{a}\pi^a(1-\pi)^{r-a}}{1-(1-\pi)^r - r\pi(1-\pi)^{r-1}}$$

POISSON FAMILY SIZE DISTRIBUTION

Under complete selection the distribution of family size contributes no genetic information, except with regard to possible fertility differentials among mating types. Under incomplete selection, however, family distributions are useful to determine relations between population and sample frequencies. With selection of multiplex families the distribution of their size is essential to determine the number of affected individuals in the population.

Two distributions have been used for family size: the 1-parameter Poisson distribution and the 2-parameter negative binomial. We are only concerned with fertile families (s>0), since childless families do not contribute to segregation analysis, but it is convenient to consider the complete distribution, s = 0, 1, 2, ... For the Poisson this is

$$P(s) = \frac{m^s e^{-m}}{s!} \qquad m>0$$

Under multiple selection the probability from (20) that a family of size s have at least one proband, among all families segregating or not, is

$$P(a>0; s, p, \pi) = xsp\pi + (1-x)\{1-(1-p\pi)^s\}$$

Then the conditional probability of a family of size s among ascertained families is

$$(33)\quad P(s \mid a>0) = \frac{P(s)P(a>0;s)}{\sum_{s=0}^{\infty} P(s)P(a>0;s)}$$

The denominator is

$$e^{-m}\sum_{s=0}^{\infty}\frac{m^s\{xsp\pi + (1-x)[1-(1-p\pi)^s]\}}{s!}$$

$$= xmp\pi + (1-x)(1-e^{-mp\pi})$$

Therefore

$$(34)\quad P(s \mid a>0) = \frac{m^s e^{-m}\{xsp\pi + (1-x)[1-(1-p\pi)^s]\}}{s!\{xmp\pi + (1-x)(1-e^{-mp\pi})\}}$$

It has been found that the distribution of family size contributes no significant information about the segregation parameters p, π, and x (Barrai et al., 1965). The value of (34) is twofold: it gives the estimate

$$E(s) = m$$

for the fertility of families able to produce affected children, and it compounds with ϵ_s (17) and ζ_s (21) to give

$$(35) \quad \epsilon = \sum_{s=0}^{\infty} \epsilon_s P(s \mid a>0, \pi = 1) = \frac{xmp}{xmp + (1-x)(1-e^{-mp})}$$

as the proportion of families with sporadic cases among all segregating families in the population, and

$$(36) \quad \zeta = \sum_{s=0}^{\infty} \zeta_s P(s \mid a>0)$$

$$= \frac{xmp\pi}{xmp\pi + (1-x)(1-e^{-mp\pi})}$$

as the proportion of families with sporadic cases among ascertained families, on the hypothesis of a truncated Poisson distribution of s among fertile families.

Under selection of multiplex families the probability that a family of size s have at least one proband, among all families segregating or not, is from (30)

$$P(a>0 \mid r>1; s) = 1 - (1-p\pi)^s - sp\pi(1-p)^{s-1}$$

Therefore the conditional probability of a family of size s among ascertained families is

$$(37) \quad P(s \mid a>0, r>1) = \frac{P(s)P(a>0 \mid r>1; s)}{\sum_{s=0}^{\infty} P(s)P(a>0 \mid r>1; s)}$$

$$= \frac{m^s e^{-m}\{1 - (1-p\pi)^s - sp\pi(1-p)^{s-1}\}}{s!\,(1 - e^{-mp\pi} - mp\pi e^{-mp})}$$

This mode of ascertainment omits probands among chance-isolated cases. To estimate the frequency of high-risk cases in the population we need to know θ, the frequency of probands in selected families among all probands, including chance-isolated cases.

The probability that a proband have $s - 1$ sibs is

$$\frac{sP(s)}{\sum_{s=0}^{\infty} sP(s)}$$

The probability that all $s - 1$ sibs be unaffected is $(1-p)^{s-1}$, and the mean for all sibship sizes is

$$\frac{\Sigma sP(s)(1-p)^{s-1}}{\sum_{s=0}^{\infty} sP(s)}$$

The frequency of probands in selected families among all probands is the complement of this, or

$$\theta = 1 - \frac{\sum_{s=0}^{\infty} sP(s)(1-p)^{s-1}}{\sum_{s=0}^{\infty} sP(s)}$$

For a Poisson distribution of family size and selection of multiplex families,

$$(38) \quad \theta = 1 - \frac{\Sigma sm^s(1-p)^s/s!}{\Sigma sm^s/s!}$$

$$= 1 - e^{-mp}$$

Since θ is a more important parameter in segregation analysis than m, we may set

$$(39) \quad u_\theta = \frac{u_m}{\partial\theta/\partial m} + \frac{u_p}{\partial\theta/\partial p}$$

$$m = \frac{-\ln(1-\theta)}{p},$$

and iterate θ simultaneously with p and π.

Under selection of multiplex families with two or more probands the probability of ascertaining a family of size s is from (31)

$$P(a>1; s) = 1 - (1-p\pi)^s - sp\pi(1-p\pi)^{s-1}$$

Therefore the conditional probability of a family of size s among ascertained families is

$$(40) \quad P(s \mid a>1; s) = \frac{P(s)P(a>1; s)}{\sum_{s=0}^{\infty} P(s)P(a>1; s)}$$

$$= \frac{m^s e^{-m}\{1 - (1-p\pi)^s - sp\pi(1-p\pi)^{s-1}\}}{s!\,(1 - e^{-mp\pi} - mp\pi e^{-mp\pi})}$$

The frequency of probands in selected families is the probability that a proband have at least one proband sib, or

$$(41) \quad \theta = 1 - \frac{\sum_{s=0}^{\infty} sP(s)(1-p\pi)^{s-1}}{\sum_{s=0}^{\infty} sP(s)}$$

$$= 1 - e^{-mp\pi}$$

We may score for θ by

$$(42) \quad u_\theta = \frac{u_m}{\partial\theta/\partial m} + \frac{u_p}{\partial\theta/\partial p} + \frac{u_\pi}{\partial\theta/\partial \pi}$$

$$m = \frac{-\ln(1-\theta)}{p\pi}$$

NEGATIVE BINOMIAL FAMILY SIZE DISTRIBUTION

Often the Poisson distribution of family size is found by χ^2 or the likelihood ratio test to give a significantly poor fit. A more flexible distribution is the negative binomial

$$(43) \quad P(s) = \binom{z}{s} m^s (1-m)^{z-s} \qquad s = 0, 1, 2, \ldots \quad m, z<0$$

with $E(s) = mz$, which has been used to describe events with varying probabilities such as accidents (Greenwood and Yule, 1920), abortions (James, 1963), and family size (Kojima and Kelleher, 1962). This gives a geometric distribution when $z = -1$ and approaches a Poisson distribution as $z \to \infty$, $m \to 0$, and mz remains constant. The truncated negative binomial approaches a logarithmic distribution when $z \to 0$ (Kendall and Stuart, 1958). Thus Equation (43) is rather general. We are of course concerned only with its truncated form for $s = 1, 2, \ldots$

The conditional probability of a family of size s among families with at least one proband corresponds to (34) and is

$$(44) \quad P(s \mid a>0) = \frac{\binom{z}{s} m^s (1-m)^{z-s}\{xsp\pi + (1-x)[1 - (1-p\pi)^s]\}}{xmzp\pi + (1-x)\{1 - (1-mp\pi)^z\}}$$

The proportion of families with sporadic cases among all segregating families in the population corresponds to (35) and is

$$(45) \quad \epsilon = \frac{xmzp}{xmzp + (1-x)\{1 - (1-mp)^z\}}$$

The proportion of families with sporadic cases among ascertained families corresponds to (36) and is

(46) $$\zeta = \frac{xmzp\pi}{xmzp\pi + (1 - x)\{1 - (1 - mp\pi)^z\}}$$

Under selection of multiplex families the conditional probability of a family of size $\underline{s}$ among ascertained families corresponds to (37) and is

(47) $$P(s \mid a>0, r>1) = \frac{\binom{z}{s} m^s (1 - m)^{z - s}\{1 - (1 - p\pi)^s - sp\pi(1 - p)^{s - 1}\}}{1 - (1 - mp\pi)^z - mp\pi z(1 - mp)^{z - 1}}$$

The frequency of probands in selected families is

(48) $$\theta = 1 - (1 - mp)^{z - 1}$$

We may eliminate the nuisance parameter $\underline{m}$ by the transformation

(49) $$u_\theta = \frac{u_m}{\partial\theta/\partial m} + \frac{u_p}{\partial\theta/\partial p} + \frac{u_z}{\partial\theta/\partial z}$$

$$m = \frac{1 - (1 - \theta)^{1/(z - 1)}}{p}$$

Finally, under selection of multiplex families with two or more probands the conditional probability of a family of size $\underline{s}$ among ascertained families is

(50) $$P(s \mid a>1; s) = \frac{\binom{z}{s} m^s (1 - m)^{z - s}\{1 - (1 - p\pi)^s - sp\pi(1 - p\pi)^{s - 1}\}}{1 - (1 - p\pi)^z - mp\pi z(1 - mp\pi)^{z - 1}}$$

The frequency of probands in selected families is

(51) $$\theta = 1 - (1 - mp\pi)^{z - 1}$$

We may eliminate the nuisance parameter $\underline{m}$ by the transformation

(52) $$u_\theta = \frac{u_m}{\partial\theta/\partial m} + \frac{u_p}{\partial\theta/\partial p} + \frac{u_\pi}{\partial\theta/\partial \pi} + \frac{u_z}{\partial\theta/\partial z}$$

$$m = \frac{1 - (1 - \theta)^{1/(z - 1)}}{p\pi}$$

PENETRANCE AND PREMATURE MORTALITY

The mendelian segregation frequencies 1/4, 1/2, etc. are altered by incomplete penetrance. If ω is the average penetrance in a sample of families and p_o is the theoretical segregation frequency, the effective segregation frequency is

(53) $$p = \omega p_o$$

Delayed onset constitutes an important special case of incomplete penetrance. Let f(z) be the frequency of age $\underline{z}$ at death or last examination among normal and affected siblings, $f_1(z)$ be this frequency with the index cases excluded, and G(z) be the probability of onset before age $\underline{z}$ among affected cases. Then if incomplete penetrance is entirely due to delayed onset, the estimate of the average penetrance in the sample is

(54) $$\omega = \sum_z f(z)G(z) \text{ for complete selection}$$

$$= \sum_z f_1(z)G(z) \text{ for single selection}$$

Since these are the two limiting cases, the best estimate of ω should lie between these values.

If the age distribution in the population is h(z), the corresponding penetrance is

(55) $$\gamma = \sum_z h(z)G(z)$$

The assumption that p is constant neglects inter-family heterogeneity in penetrance and age at onset. However, significant variation due to this has never been detected.

The calculations are simplified if the distribution of age is normal in the population,

(56) $$h(x) = \frac{e^{-x^2/2}}{\sqrt{2\pi}}$$

$$\text{where } x = \frac{z - \mu}{\sigma}$$

and μ, σ are the mean and standard deviation of z. We suppose that the probability of manifestation before age $\underline{z}$ is the smoothly increasing function

(57) $$G(x) = e^{a + bx}$$

which has been introduced for quasi-continuous variation (Edwards, 1967). The population penetrance is

(58) $$\gamma = \int_{-\infty}^{\infty} h(x)G(x)dx = e^{a + b^2/2}$$

Similarly, if the probability of mortality due to affection before age $\underline{z}$ is the smoothly increasing function

(59) $$M(x) = e^{c + dx},$$

the probability that an affected person be living at the time of the investigation is

(60) $$\Lambda = \gamma - \int_{-\infty}^{\infty} h(x)M(x)dx = e^{a + b^2/2} - e^{c + d^2/2}$$

where the constants a, b, c, d may be estimated by tabulating the cumulative penetrance (57) and mortality (59).

If viability up to the age of diagnosis is reduced to a fraction $\underline{f}$ of normal, the number of survivors in families with theoretical segregation frequency p_o is reduced by a fraction $p_o(1 - f)$, and the effective segregation frequency becomes

(61) $$p = \frac{fp_o}{1 - p_o(1 - f)}$$

This fact has been useful in testing by segregation analysis for effects of genetic markers on mortality before the age of examination (Morton et al., 1966).

PREVALENCE AND INCIDENCE

We define prevalence P as the frequency of a phenotype in a given population at a given time. The incidence I is the frequency at birth of trait-bearers, who with probability equal to the penetrance will develop the trait during their lifetime. Estimation of these two frequencies is one of the objectives of segregation analysis under incomplete ascertainment. For a common trait it is feasible and clearly desirable to estimate prevalence from a random sample of the general population (Lilienfeld, 1962). However, this is often prohibitively expensive for the rare traits with which a geneticist often has to deal, since reliable estimates would require examination of a large fraction of the population (e.g., more than 100,000 individuals for a trait with a prevalence of 1/10,000). The estimate of prevalence leads to an estimate of incidence, from which, if the genetic basis is simple, we can proceed to a calculation of gene frequency and (under certain conditions) the mutation rate. We saw in (10) that

(62) $$P = \frac{A}{\pi N}$$

where A is the number of probands, π is the ascertainment probability, and N is the population size. Under selection of multiplex families

$$A = \frac{A^*}{\theta}$$

(63) so $$P = \frac{A^*}{\pi \theta N}$$

where A* is the number of probands in selected families and θ is the proportion of probands in selected families.

The standard error of prevalence involves some intricate problems of sampling from a finite population. Assume as a convenient approximation that the method of sampling is to collect a fixed number A or A* of probands and then study their relatives. Since the population is finite, the chance of detecting a family with $\underline{r}$ affected when there are $\underline{R}$ affected individuals in the population is

$$1 - \frac{\binom{A}{0}\binom{R-A}{r}}{\binom{R}{r}} = 1 - \prod_{i=1}^{r} \left(1 - \frac{A}{R - i + 1}\right)$$

If $r << R$, the effect of sampling without replacement is negligible and the chance of detecting the family approaches $1 - (1 - \pi)^r$, where $\pi = A/R$ is the ascertainment probability. Since A (or A*), R, and N are all fixed numbers, the variance of (62) is

(64) $$\sigma_P^2 = \left(\frac{\partial P}{\partial \pi}\right)^2 K^{\pi\pi}$$

$$= \frac{A^2 K^{\pi\pi}}{N^2 \pi^4}$$

$$= \left(\frac{P}{\pi}\right)^2 K^{\pi\pi}$$

Similarly the variance of (63) is

(65) $$\sigma_P^2 = \left(\frac{\partial P}{\partial \pi}\right)^2 K^{\pi\pi} + \left(\frac{\partial P}{\partial \theta}\right)^2 K^{\theta\theta} + 2\left(\frac{\partial P}{\partial \pi}\right)\left(\frac{\partial P}{\partial \theta}\right) K^{\pi\theta}$$

$$= P^2 \left(\frac{K^{\pi\pi}}{\pi^2} + \frac{K^{\theta\theta}}{\theta^2} + \frac{2K^{\pi\theta}}{\pi\theta}\right)$$

where the K's are elements of the covariance matrix.

Estimation of incidence from prevalence depends on how probands and population are defined. If the population consists of individuals born during a certain interval, whether living or dead at the time of investigation, and if the ascertainment probability is equal for living and dead trait-bearers in this population, the prevalence estimate will include dead members of the population if dead probands are counted in A or A*. Then the estimate of prevalence must be corrected only for penetrance γ (58) and

(66) $$I = \frac{P}{\gamma}$$

with variance

(67) $$\sigma_I^2 = \sigma_P^2 / \gamma^2$$

If an affected individual must be living to be counted as a proband, the values of A, A* will be reduced. The ascertainment probability will diminish correspondingly and be variable so that the distribution of $\underline{r}$ should approximate (25), but (62) and (63) are unchanged.

Finally, if both living and dead probands are used to estimate π, but only living probands are counted in (62) and (63), then the prevalence estimate will include only living members of the population. Both delayed onset and premature mortality must be considered in estimating incidence. Let Λ be the probability that an affected person be living at the time of the investigation (60). Then

(68) $$I = \frac{P}{\Lambda}$$

with variance $$\sigma_I^2 = \sigma_P^2 / \Lambda^2$$

This will be a poorer estimate than the preceding if the ascertainment probability is different for dead and living members of the population.

SEX LINKAGE

Sex-linked genes present interesting problems, such as that of distinguishing the mutation rates in the two sexes. For a deleterious sex-linked recessive trait, the probability that an affected male be sporadic is

(69) $$x = \frac{(1 - f)u}{2u + v}$$

where f is the relative fitness of affected males, u and v are the mutation rates in egg and sperm, respectively, and it is assumed that carrier females have normal fertility and that all cases are sex-linked (Morton and Chung, 1959). The test of the hypothesis that u = v reduces to testing whether $x = \frac{1 - f}{3}$.

Corroboration of this test may be sought in the distribution of affected maternal uncles, assuming ascertainment through nephews. If two or more nephews are affected, the probability that at least one of $\underline{s}$ maternal uncles be affected is

(70) $$P(r>0; s, p, x') = (1 - x')\{1 - (1 - p)^s\}$$

The expected value of x′ is 1/2 on the hypothesis of sex linkage (Morton and Chung, 1959). Similarly, if only one nephew is affected in a sibship of $\underline{n}$ nephews, the probability that at least one maternal uncle be affected, if ascertainment is through the nephew, is

(71) $$\{P(r>0; s, p, x, x')\} = \frac{(1 - x)(1 - x')(1 - p)^{n-1}\{1 - (1 - p)^s\}}{x + (1 - x)(1 - p)^{n-1}}$$

These distributions provide a test of homogeneity of x and x′, of the deviation x′ from 1/2, and of x from (1-f)/3, the last deviation being accepted as significant only if the others are nonsignificant.

These ancillary tests are particularly important if sex-linked and autosomal cases may be confused, in which event families with autosomal or sporadic cases may be recognized if they contain at least one affected girl. The probabilities for isolated and familial cases under this condition (Morton, 1958) are

(72) $P(r = 1 \mid a>0)$

$$= \frac{sp\pi[x + (1 - x)(1 - p)^{s-1}]}{xsp\pi + 2(1 - x)[1 - (1 - p\pi)^s - (1 - p/2)^s + (1 - p/2 - p\pi/2)^s]}$$

$P(r>1 \mid a>0)$

$$= \frac{(1 - x)\binom{s}{r} p^r (1 - p)^{s-r} [1 - (1 - \pi)^r][1 - (\frac{1}{2})^r]}{xsp\pi + 2(1 - x)[1 - (1 - p\pi)^s - (1 - p/2)^s + (1 - p/2 - p\pi/2)^s]}$$

SEGRAN: A GENERAL PROGRAM FOR SEGREGATION ANALYSIS

The first version of this program was written in 1959 by R. A. Hedberg and Nancy Jones at the University of Wisconsin under the author's direction. The second edition was written by P. Juetten and V. Pratt in 1961 for the CDC 1604 computer at the University of Wisconsin. Simplified versions were produced in 1963 by I. Barrai for an Olivetti computer at the University of Pavia and by C. S. Chung for the Honeywell 800 at the National Institutes of Health. The program for the CDC 1604 was adapted to the IBM 7040 computer at the University of Hawaii in 1964 by M. P. Mi, who developed a new program in Fortran IV language for the CDC 3100 computer in 1967. The latest version of SEGRAN has been written under my direction to cover all the distributions discussed above.

A segregation assembly program (SAP) reduces individual records (one for father, mother, and each observed child) into family records, giving the mating type of father and mother and a summary of the sibship. SEGRAN generates six kinds of summary tables from family records:

- SR tables of the number r of affected among s sibs
- RA tables of the number a of probands among r affected sibs
- T tables of the number of ascertainments per proband
- S tables of the number of s sibs under incomplete selection
- UI tables of the number of segregating and nonsegregating sets of maternal uncles for an isolated affected nephew under sex linkage
- UF tables of the number of segregating and nonsegregating sets of maternal uncles for familial affected nephews under sex linkage

Summary tables give the number of sibships observed for each count of a probands, r affected, r-s normal sibs, ru affected maternal uncles, nu normal maternal uncles, and t ascertainments per proband. Families with s>15 are scored and then pooled on output.

Maximum likelihood iteration uses the expected number of observations m_k in the kth class to compute

$$K_{ij} = \sum_k m_k \, u_{ik} \, u_{jk}$$

as the amount of information about the ith and j^{th} parameters. The parameters available for iteration and tests of significance are p, m, w, h, y, and z for complete selection and p, x, n, b, θ, and z for incomplete selection. The parameter x' for sex linkage is represented by xp in SEGRAN, and m in Equations (34) and (44) by TH. If an S table is analyzed, the ancillary parameters ϵ, ζ, and NP are estimated with standard error

$$\sqrt{\partial' K^{-1} \partial} ,$$

where ∂ is the vector of partial derivatives with respect to the iterated parameters and K is the corresponding information matrix. Various segregation models are given by equation number in Table 1. For any given set of sibships with specified affected and normal phenotypes only one table of a given type should be analyzed simultaneously.

Sometimes iteration fails to converge even after several good guesses of trial values. Then regula falsi interpolation may work. If this does not succeed, it is necessary to reduce the rank of the hypothesis, for example, by assuming a Poisson distribution for numbers of ascertainments when the 2-parameter negative binomial fails to converge. The most common convergence problem is with the nuisance parameters b and z under incomplete selection. Then any value in the neighborhood of the root may be assumed, continuing iteration on the more important parameter π. While inelegant, these solutions are satisfactory in practice.

TABLE 1. Segregation models by equation number

Mode of selection	Hypothesis about p and π	Hypothesis about s	Tables: SR	RA	T	S	UI	UF
Complete $(0 \le r \le s)$	p constant	-	4	-	-	-	-	-
	p bimodal	-	5	-	-	-	-	-
	p unimodal	-	7	-	-	-	-	-
Incomplete, simplex families included (a>0)	π constant	Poisson	20	22	27	34	-	-
		neg. binomial	20	22	27	44	-	-
	π variable	Poisson	20	25	28	34	-	-
		neg. binomial	20	25	28	44	-	-
Incomplete, multiplex families (a>0, r>1)	π constant	Poisson	30	22	27	37	-	-
		neg. binomial	30	22	27	47	-	-
	π variable	Poisson	30	25	28	37	-	-
		neg. binomial	30	25	28	47	-	-
Incomplete, two or more probands (a>1)	π constant	Poisson	31	32	27	40	-	-
		neg. binomial	31	32	27	50	-	-
	π variable	Poisson	31	-	28	40	-	-
		neg. binomial	31	-	28	50	-	-
Incomplete, at least one affected girl	π constant	-	72	22	27	-	-	-
	π variable	-	72	25	28	-	-	-
Incomplete, uncles of isolated nephews under sex linkage	-	-	-	-	-	-	71	-
Incomplete, uncles of familial nephews under sex linkage	-	-	-	-	-	-	-	70

SUMMARY

The theory behind the latest version of the general SEGRAN program for segregation analysis is described.

ACKNOWLEDGEMENT

PGL Paper No. 26. This work was supported by U.S. Public Health Service grants GM 10424 and GM 15421 from the National Institutes of Health.

REFERENCES

Barrai, I., Mi, M. P., Morton, N. E., and Yasuda, N. 1965. Estimation of prevalence under incomplete selection. Amer. J. Human Genet. 17: 221-236.

Chung, C. S., Robison, O. W., and Morton, N. E. 1959. A note on deaf mutism. Ann. Hum. Genet. 23: 357-366.

Dewey, W. J., Barrai, I., Morton, N. E., and Mi, M. P. 1965. Recessive genes in severe mental defect. Amer. J. Hum. Genet. 17: 237-256.

Edwards, J. H. 1967. Linkage studies of whole populations. Proc. Third Int. Cong. Human Genet., ed. J. F. Crow and J. V. Neel, pp. 483-489. Baltimore: Johns Hopkins Press.

Fisher, R. A. 1941. The negative binomial distribution. Ann. Eugen. (Lond.) 11: 182-187.

Greenwood, M., and Yule, G. Y. 1920. An inquiry into the nature of frequency distributions representative of multiple happenings with particular reference to the occurrence of multiple attacks of disease or of repeated accidents. J. Roy. Stat. Soc. 83: 255.

Hiraizumi, Y. 1964. Prezygotic selection as a factor in the maintenance of variability. Cold Sp. Harbor Sym. Quant. Biol. 29: 51-60.

James, W. H. 1963. Notes towards an epidemiology of spontaneous abortion. Amer. J. Hum. Genet. 15: 223-240.

Kendall, M. G., and Stuart, A. 1958. The Advanced Theory of Statistics. New York: Hafner Publ. Co.

Kojima, K., and Kelleher, T. M. 1962. Survival of mutant genes. Amer. Naturalist 96: 329-346.

Lilienfeld, A. M. 1962. Sampling techniques and significance tests. In Methodology in Human Genetics, W. J. Burdette (ed.). San Francisco: Holden-Day, pp. 3-16.

Morton, N. E. 1958. Segregation analysis in human genetics. Science 127: 79-80.

Morton, N. E. 1959. Genetic tests under incomplete ascertainment. Amer. J. Hum. Genet. 11: 1-16.

Morton, N. E. and Chung, C. S. 1959. Are the MN blood groups maintained by selection? Am. J. Hum. Genet. 11: 237-251.

Morton, N. E. and Chung, C. S. 1959. Formal genetics of muscular dystrophy. Amer. J. Hum. Genet. 11: 360-379.

Morton, N. E. 1962. Segregation and linkage. In Methodology in Human Genetics, W. J. Burdette (ed.). San Francisco: Holden-Day, pp. 17-52.

Morton, N. E., Mi, M. P., and Yasuda, N. 1966. A study of the S^u alleles in northeastern Brazil. Vox Sang. 11: 194-208.

Morton, N. E., Krieger, H., and Mi, M. P. 1966. Natural selection on polymorphisms in northeastern Brazil. Amer. J. Hum. Genet. 18: 153-171.

Morton, N. E., Mi, M. P., and Yasuda, N. 1966. Bivalent alleles. Amer. J. Hum. Genet. 18: 233-242.

Skellam, J. G. 1948. A probability distribution derived from the binomial distribution by regarding the probability of success as variable between the sets of trials. J. Roy. Stat. Soc. B 10: 257-261.

GENETIC TAXONOMY

A. W. F. Edwards
Gonville and Caius College
Cambridge

Taxonomy is traditionally concerned with the interpretation of morphological differences for purposes of classification or the construction of phylogenetic trees. In recent years more diverse sources of information have been used, many of them chemical in origin, such as the results of biochemical tests on bacteria, or the structures of homologous proteins. At the same time the traditional methods have been augmented by computer techniques under the title 'numerical taxonomy'. One consequence of this double revolution has been an increased amount of discussion on the basic principles of taxonomy, much of which has been aimed at classification rather than the construction of phylogenies. If science involves the building and testing of models, then classification is indeed an art; by contrast, the drawing of phylogenetic trees is based on a model for the origin of the observed diversity, namely evolution by natural selection and associated forces. When that model is made specific we may adopt the term 'genetic taxonomy', for only then are the phylogenetic conclusions based on a detailed consideration of the genetical processes which are fundamental to the evolution of diversity.

The basic event of evolution, as represented in classical population genetics, is the substitution of one gene for another in a population. In the light of our growing knowledge about the fine structure of the genetic material, and of the progress that has been made towards an understanding of the population genetics of linked complexes of genes, this view may seem excessively naive. But one must start somewhere, and in genetic taxonomy even the simplest model turns out, as we shall see, to be difficult to handle.

At the present stage of our knowledge we cannot easily, however, base our phylogenies on complete gene substitutions, first because of the lack of information, and secondly because each gene substitution may not be of equal evolutionary importance.

Over the past five years Cavalli-Sforza and I, in a series of papers listed below, have attempted to overcome these difficulties by concentrating on a consideration of the partial substitution of genes of no apparent selective value. We have emphasized the importance of adopting a specific model, and have considered the ways in which various factors can upset our assumptions; and we have tried to apply our model to data on the distribution of the human blood groups.

I would like to review our efforts briefly, to give an account of the present state of the work, and to indicate likely lines of development in the future.

THE PHYLOGENETIC MODEL

Our phylogenetic model is exactly appropriate (except in that it is a continuous approximation to a discrete process) when the evolution of a group of populations consists of the random genetic drifting of selectively-neutral unlinked genes, accompanied by the random splitting of populations into two, each population having identical properties. As the derived populations split in turn, so an evolutionary tree develops. Although we have argued that this model may in fact be rather robust with respect to some departures from it, we agree with those critics who find it a naive mirror of evolution. We never supposed otherwise. But we do insist that conclusions based upon a specific model, whose imperfections may be discussed, have a greater scientific validity than conclusions based upon concepts of evolution so general that they are beyond criticism. If we can learn to handle the simple model adequately, more complex assumptions may be incorporated later.

It must not be supposed that a model of evolution based essentially upon the random walk of gene frequencies will necessarily prove uninformative. Boyd (1963) assumed this when he wrote of human evolution that "unless the blood groups are adaptive, they are not going to be very useful in classification." For any variable which shows a persistently high correlation in successive generations should be valuable; far from selection increasing the utility of a character, it may be expected to lessen it, either by reducing the informative variability, if it is stabilizing selection, increasing the variability, if it is disruptive selection, or simply confusing the picture, if it is directional selection. Indeed, any pattern of variability amongst populations can be accommodated by any phylogenetic tree if directional selection is invoked. Intuitively there is, in fact, a temptation to seek solutions which involve the minimum possible amount of evolution, a method I will refer to later.

Before we can apply a random-walk model to gene-frequency data, it is necessary to stabilize the variances and to combine the information from several loci. Cavalli-Sforza and I have described the necessary manipulations, involving the multi-dimensional angular transformation, in our papers. The corresponding program is COOR2.

MAXIMUM LIKELIHOOD ESTIMATION

The statistical model appropriate to the data thus transformed is a combined Brownian-motion (for the random walk of the gene frequencies) and Yule (for the random splitting of the populations) process. It is not a difficult matter to set up the equations for maximum likelihood estimation on this model. There is no need to give them here, or to indulge in a detailed discussion of the difficulties encountered in trying to solve them, for I have recently submitted an account of the problem for publication, and duplicated copies of the paper, "Likelihood Estimation of the Branch Points of a Brownian-Motion/Yule Process," are available.

The conclusion reads "It is shown that estimation of the branch points of a Brownian-motion/Yule process given the positions of the particles [i.e., populations] at a certain time is not, in many cases, possible: the positions may not be sufficiently informative about the history of the process.

"The equations of estimation by maximum likelihood are given for the case where the tree form and node times are known. They provide an explicit solution for the node coordinates, the rate of diffusion of the Brownian-motion process, and the rate of splitting of the Yule process."

In other words, having attempted to jump in at the shallow end by adopting a very simple model for the evolutionary process, we have nevertheless ended up in deep water by finding ourselves unable to conclude very much if we use this model. Could it be that Boyd put his finger on the weak spot after all? Before discussing this point, this is a convenient moment to record that listings of the program EVOTREE for performing these calculations are available, together with those of a small simulation program GROW2 which produces examples of the Brownian-motion/Yule process in two dimensions.

In starting a discussion of the genetical implications of the near-impossibility of drawing phylogenetic conclusions on the basis of the current model, I shall assume that most people are agreed that the phylogenetic tree for 15 populations derived by Cavalli-Sforza and myself from data on five blood-group systems, and published in several of our papers, is not unreasonable. It was not, however, derived by using the Brownian-motion/Yule process as a model; indeed, the use of EVOTREE on these data shows that there is, almost certainly, no non-singular maximum to the likelihood. We must therefore conclude that either the estimation procedure or the model is inappropriate.

Felsenstein (1968) has argued that the estimation difficulties can be overcome by omitting to estimate the coordinates of the nodes, and concentrating on their times alone. I find this unacceptable, for reasons which are given in the paper "Likelihood Estimation ......" Somewhat reluctantly I am forced to the other conclusion, that the model is inappropriate for the particular problem. At this stage it is only fair to point out that Cavalli-Sforza and I have repeatedly emphasized that one of the anticipated benefits of arguing from a specific model is that the results are likely to be informative about the relevance of that model. Perhaps I should also add that the views I am now airing are mine alone, and are not necessarily shared by Cavalli-Sforza.

In the case of the human blood-group gene frequencies the most obvious way in which the model may be departing from reality is in the presumed absence of directional selection. Is it possible to allow for directional selection in the model, given that nothing specific is known about selection amongst the populations during the period under consideration?

If selection is slow we may presume, in order to sketch out a rough argument, that it would be represented by a constant-velocity shift of the population in the transformed gene-frequency space. In fact to accommodate selection the transformation should be the logit rather than the angular, but both cannot, of course, be performed simultaneously. Let us suppose that the rate of change of transformed gene frequency is a constant for any single branch of the evolutionary tree, but that this rate is itself drawn from a normal distribution with mean zero and variance ρ^2.

Under random genetic drift alone the probability density a distance $\underline{d}$ from an original population after a time $\underline{t}$ has elapsed is, in the one-dimensional case, $N(O, t\sigma^2)$, where σ^2 is the variance per unit time. Under selection alone the density would be $N(O, t^2\rho^2)$, for the distribution of selection rates is $N(O, \rho^2)$, and the distance travelled is the rate multiplied by the time.

Assuming drift and selection to be independent, the new model incorporating both would have a probability density of $N(O, t\sigma^2 + t^2\rho^2)$. The modification of the program EVOTREE to include this change is difficult unless $\sigma^2 = O$ and the variation in gene frequencies is due to selection alone. Experience with this version of the program has been no more encouraging than with the original version; nor is this surprising, because the difference in using t and t^2 is not very great, and the singularities in the likelihood surface appear in the same way. Besides, it is very artificial to suppose that selection is constant until a population splits, and then acts upon the daughter populations quite independently.

Such, then, is the current state of the problem. Specific models that are tractable seem, in general, to yield no results. We could set up a model that included selection, but allowed it to change only slowly with time, so that the evolutionary tree as estimated would have no sharp bends in its branches, but this model is likely to be intractable. Or we can return to the method which has at least provided some acceptable results: the method of minimum evolution.

METHOD OF MINIMUM EVOLUTION

When I first propounded this method in 1963 it was partly because I was impressed with its intuitive appeal and partly because I was fairly sure that it would correspond to the maximum-likelihood solution based on a proper probability model, when that was obtained. This was not to be, however.

The intuitive appeal remains. Estimation on the Brownian-motion/Yule process unites two adjacent nodes of a tree by a straight line, and the method of minimum evolution extends this shortest-distance approach to the whole tree. As mentioned earlier, the desire to explain the observed variability by invoking the minimum amount of directional selection is strong. I shall not attempt to justify the method further here, although I would be very glad to have some discussion of it, but it is important to note that it is a method of estimation, and not a principle under which evolution is supposed to act. In order to get some idea of the validity of the solutions it gives, data should be divided at random into several sets, and the method applied to each set separately. In addition, it should be borne in mind that the ideal representation of the populations is in a non-Euclidean space, but the prospect of looking for minimum networks in such a space is daunting, even if we can define what we mean by "minimum."

The program EVOMIN is available, and I have just completed a program-complex TREE which links COOR2 and EVOMIN by a sum-of-squares clustering program KNAUT thus giving fully-automated tree-building.

Because of the effort that has been put into elucidating the difficulties encountered in the maximum likelihood approach, there is, as yet, little experience of the practical application of EVOMIN. Recently, however, I have completed the investigation of the 15-populations data, the results of which will be published elsewhere, and as part of this I divided the data into two groups of loci (ABO, MNS, Di, and Rh, Fy) and ran them through the program-complex TREE. It was encouraging to note that, even with such limited information, the similarities between the two independent trees were much too great to attribute to chance.

An alternative approach to constructing phylogenies, using the method of least squares, was advocated by Cavalli-Sforza and me in some of the earlier papers, but I have not made any further use of this method.

CONTOUR MAP REPRESENTATION

Even if based on a satisfactory genetic model, it must always be remembered that phylogenies estimated from gene frequencies do not, in general, make use of all the available information about the relationships amongst populations. With human material the linguistic, cultural and geographic information may yet be more important than the genetic, and the genetical phylogeny must be thought of as a succinct way of summarizing the gene-frequency information, rather than as a definitive estimate. It is thus to be compared with the maps of gene frequencies, which carry much the same information in a different way.

The advantage of drawing gene-frequency contours is that geographical information is incorporated in the presentation, but the disadvantage is that each gene requires a separate map, so that with many genes and loci an overall view is difficult. One might manage to place the information from a triallelic locus on a single map by printing it in the three primary colours, but in general a new approach is needed which allows the geographical information of the map and the general genetic distance of the tree to be presented simultaneously.

I suggest that, having calculated the pair-wise genetic distances between populations from all the data, and having thus derived population coordinates, a Principal Component Analysis is undertaken, and the first principal component used as a general measure of divergence which can be plotted on a contour map. Indeed, the first three principal components could be printed simultaneously using the primary colours, or individually in black and white, and these may be expected to account for the major part of the information.

Acknowledgements

I am indebted to the World Health Organisation and the Medical Research Council (London) for the provision of, and support for, computer facilities, and to Dr. L. L. Cavalli-Sforza for his continued encouragement.

REFERENCES

Boyd, W. C. 1963. Genetics and the human race. Science 140: 1057-1064.

Cavalli-Sforza, L. L., Barrai, I., and Edwards, A. W. F. 1964. Analysis of human evolution under random genetic drift. Cold Spring Harbor Symp. Quant. Biol. 29:9-20 .

Cavalli-Sforza, L. L. and Edwards, A. W. F. 1964. Analysis of human evolution. Proc. 11th Int. Congr. of Genet. 3: 923-933.

Cavalli-Sforza, L. L. and Edwards, A. W. F. 1966. Estimation procedures for evolutionary branching processes. Bull. Internat. Stat. Inst. 41:803-808.

Cavalli-Sforza, L. L. and Edwards, A. W. F. 1967. Phylogenetic analysis. Evolution, 21: 550-570; Am. J. Hum. Genet. Suppl. 19: 233-257.

Edwards, A. W. F. 1966. Studying human evolution by computer. New Scientist 30: 438-440.

Edwards, A. W. F. 1968. Likelihood estimation of the branch points of a Brownian-motion/Yule process. Submitted to J. Roy. Stat. Soc.

Edwards, A. W. F. and Cavalli-Sforza, L. L. 1963. The reconstruction of evolution (summary) Ann. Hum. Genet. London 27:104-105; Heredity 18:553.

Edwards, A. W. F. and Cavalli-Sforza, L. L. 1964. Reconstruction of evolutionary trees. Systematics Assoc. Publ. No. 6, 67-76. Phenetic and Phylogenetic Classification.

Felsenstein, J. 1968. Statistical inference and the estimation of phylogenies. Ph.D. dissertation, University of Chicago.

DISCUSSION ON GENETIC TAXONOMY

V. Balakrishnan, T. W. Kurczynski,
J. H. Edwards, and N. E. Morton

BALAKRISHNAN: During the last 4 or 5 years, there have been a series of papers by Dr. Edwards and Dr. Cavalli-Sforza on the subject of the analysis of evolution. These papers broke new ground and stimulated thinking on the subject.

In the paper under discussion, Dr. Edwards has presented some of the mathematical difficulties encountered by him in the estimation of the parameters in his model and has come to the conclusion that, in general, there is insufficient information for the estimation to be successful.

Apart from these mathematical hurdles, we had entertained some doubts regarding his model. His model is based essentially on the assumption that random genetic drift plays the leading role in evolution. As far as selection is concerned, directional selection, of course, is not important in the study of evolution if the direction is the same in all the populations. However, this becomes important when the selection acts in different ways in different populations. It is hard to believe that such selection has not been the cause of most of the adaptive characters which help to distinguish the major races of mankind.

Cavalli-Sforza et al. (1964) have themselves come to the conclusion that if, in recent human evolution, drift has played a role unmitigated by stabilizing selection, then effective breeding size must have been large, say of the order of 10,000. This is unrealistically high and hence most of the polymorphisms must have been buffered by some type of stabilizing selection. The reduction in the speed of evolution caused by such stabilising selection remains unestimable without a knowledge of the selection differentials.

It is true, as pointed out by Cavalli-Sforza and Edwards (1967) that any observed diversity can be explained by any evolutionary tree, provided we are willing to postulate the necessary selection. However, this does not seem to be a sufficient argument for neglecting this important factor.

Of course, selection trends will be detectable only if data from the past are available. But non-availability of data is not sufficient reason to draw conclusions from a simple but unrealistic model. We can only try to fill in the gaps by the study of present-day populations of various degrees of primitiveness and hope that this legacy of ours will help to formulate realistic models in the future.

Another unrealistic assumption is the constancy of the speed of evolution over space and time. Data from present-day populations will be of help in elucidating their interrelationships and in classifying them into groups. This in turn will enable us to study how various forces of evolution act on the various loci in different population groups.

With this limited but realistic goal we, in our laboratory, have tried to evolve suitable measures to study the interrelationships between populations using genetic data. For this purpose, Sanghvi (1952) proposed a measure of distance, G^2_s, based on the chi-square statistic.

In 1965 we started using another measure, B^2, which is analogous to Mahalanobis's D^2. We still entertain some theoretical doubts about this measure and have published it only recently (Balakrishnan and Sanghvi, 1968). This is defined as:

$$B^2_{mn} = \sum_{j=1}^{r} \sum_{k=1}^{s_j} \sum_{l=1}^{s_j} C_j^{kl} \quad d_{jk} \quad d_{jl}$$

where $d_{jk} = p_{mjk} - p_{njk}$ and $(C_j^{kl}) = (C_j)^{-1}$, p_{mjk} = frequency of the kth allele at the jth locus in the mth group, $s_j + 1$ = number of alleles at the jth locus,

r = number of loci, and (C_j) = the common dispersion matrix of the variables X_{jk} whose means are estimated by p_{mjk} and which take only the values O and 1, subject to the condition that one, and only one, of the X_{jkl}, for each j, can be unity on any individual. The common dispersion matrix is obtained as weighed mean over all the populations under study, with sample sizes as weights. In each population, the dispersion matrix of the X variables is derived from the dispersion matrix of the allelic frequencies obtained during the maximum likelihood estimation.

This extension of Mahalanobis's D^2 differs in some crucial respects from the extension as originally proposed by Steinberg et al. (1966). I would like here to thank Dr. Steinberg and Dr. Kurczynski for the discussions we had in 1966 on our separate measures. Their measure is, in fact, a multiple of Sanghvi's G^2 divested of some non-essential factors.

The theoretical framework for G^2_s worked out by Steinberg et al. (1966) enabled us to define yet another measure, G^2_c, which differs from B^2 only in the manner in which the common dispersion matrix is obtained. In G^2_c, it is taken to be the dispersion matrix in a population showing allelic frequencies equal to the weighted means of the allelic frequencies in each population. In terms of measurable characters, this corresponds to the total dispersion matrix whereas that used in B^2 corresponds to the within dispersion matrix.

Edwards and Cavalli-Sforza (1964) have approached this problem in a quite different manner. Their distance is based on the square-root transformation of the allelic frequencies and neglects the correlations between them, considering them as negligible. However, the square of the correlation between 2 frequencies p and q is $pq/(1-p)(1-q)$ which will be 0.67 with $p = 0.5$ and $q = 0.4$. Moreover, this method is not really an extension of the arcsine transformation to multiallelic loci since they do not explicitly get the transformations and treat them as normal variables with constant variance.

Coming to the theoretical difficulty mentioned above in respect of B^2, it may be pointed out that it is common to B, G_c, and G_s (and hence to Kurczynski's D^2_k). The fundamental assumption of a common dispersion matrix is really not tenable with multinomial distributions, since the variances and covariances are dependent on the means. This makes it questionable to use these indices when the populations are obviously far apart.

However, this does not appear to cause any real trouble in practice. We have compared the four indices using the allelic frequencies on the 15 populations analysed by Cavalli-Sforza and Edwards (1963). Clustering was done by the technique of Edwards and Cavalli-Sforza (1965) modified to reduce the number of splits to be considered substantially. The process is made to start with a set of two populations showing the smallest distance and the group is built up by the addition of one population at a time giving the highest between-groups sum of squares. With this restriction the number of splits to be considered at the first stage is reduced from $2^{q-1}-1$ to $(q - 1)(q - 2)/2$, that is from 16383 to 91 for 15 populations.

All the four indices gave a broadly similar picture showing that the above defect is not really serious in practice. It was, however, found that B^2 separated clearly the Negroes of Africa from the Europeans, which remained together as one primary group on the basis of the other 3 indices. The rudimentary knowledge which we possess of early human migrations and admixture supports such an isolation of sub-Sahara Africans rather than a primary split which separates the Europeans from the Asians. "This genetical uniqueness of Africans is in itself an interesting problem. It would appear to result from a long genetical isolation of Africans from the rest of mankind----an isolation by no means complete but sufficient to allow the production of mutations and their subjection to natural selection almost entirely separately from the corresponding process going on in the rest of the human race." (Mourant, 1961.)

Sneath and Sokal (1962) have defined numerical taxonomy as "the numerical evaluation of the affinity or similarity between taxonomic units and the ordering of these units into taxa on the basis of their affinities." The affinity coefficients or distances are subjected to cluster analysis and the operational taxonomic units are arranged into a reasonable hierarchic system or dendrogram showing the degree of relationship with the nearest stem. Such diagrams, though not phylogenetic trees when used with present day populations alone, will be of great help in elucidating their interrelationships. They can also be used to study the resemblances and differences of the populations within a group in respect of the various characters used and also to see how the groups differ in respect of these characters. If information on the past is available, such diagrams may even help in phylogenetic ordering, as has been done by Camin and Sokal (1965) with data on fossil horses.

Sneath and Sokal (1962) have very correctly stressed that the purpose of numerical taxonomy is not directly related to phylogenetic speculations. The objections of Edwards and Cavalli-Sforza (1964) against a 'general' purpose classification appear to be based on some misunderstanding. Such classification with as many randomly selected characters as possible will definitely be of help in clarifying the overall relationships. No classification, as such, will be of help in studying evolution unless other assumptions are made.

It cannot be said that the transformation used by Edwards and Cavalli-Sforza (1964) is equivalent to character-weighting. Stabilization of variance is, of course, necessary, but this does not give weights to the various loci in accordance with their evolutionary importance.

Regarding the points raised with respect to likelihood estimation, I would think that more than the singularities of the likelihood surface, which generally occur at some extreme points, the flatness of the likelihood surface as revealed by Fig. 2 in Edwards (1968) is more disturbing. In this situation, a wide range of values for the parameter gives more or less the same likelihood. Also, even when there is a maximum, it differs very little from the general value of likelihood. More thought would have to be given to devising a suitable method of estimation under such circumstances.

My thanks to Dr. L. D. Sanghvi for the many valuable discussions we had on the points presented in this paper and to Dr. A. W. F. Edwards for supplying the allelic frequencies of the 15 populations.

Balakrishnan, V., and Sanghvi, L. D. 1968. Distance between populations on the basis of attribute data. Biometrics, 24:859-865.

Cavalli-Sforza, L. L., Barrai, I., and Edwards, A. W. F. 1964. Analysis of human evolution under random genetic drift. Cold Spring Harbour Symp. Quant. Biol., Vol. XXIX, 9-20.

Cavalli-Sforza, L. L. and Edwards, A. W. F. 1963. Analysis of human evolution. Genetics today. Proc. XI Int. Congr. Genet., The Hague, The Netherlands, 923-933, Pergamon Press, Oxford.

Cavalli-Sforza, L. L. 1967. Phylogenetic analysis. Models and estimation procedures. Am. J. Hum. Genet., Supplement, 19:233-257.

Edwards, A. W. F., and Cavalli-Sforza, L. L. 1964. Reconstruction of evolutionary trees. Phenetic and Phylogenetic Classification., 67-76. The Systematics Assoc. Pub. No. 6, The Systematics Assoc., London.

Edwards, A. W. F. 1965. A method for cluster analysis. Biometrics, 21:362-375.

Mourant, A. E. 1961. Blood groups. Symp. Soc. for study of Hum. Biol. Vol. IV, Genetic Variation in Human Population. Pp. 1-15, Pergamon Press, Oxford.

Sanghvi, L. D. 1952. Biological studies on some endogamous groups in Bombay, India. Thesis submitted for Ph.D., Columbia University, New York.

Steinberg, A. G., Bleibtreu, H. K., Kurczynski, T. W., Martin, A. O., and Kurczynski, E. M. 1966. Genetic studies on an inbred human isolate. Proc. 3rd Int. Congr. Hum. Genet., Chicago, 267-289. John Hopkins Press, Baltimore.

Camin, Joseph E., and Sokal, Robert R. 1965. A method for deducing branching sequences in phylogeny. Evolution, 19:311-326.

Sneath, P. H. A., and Sokal, Robert R. 1962. Numerical taxonomy. Nature, London, 193:855-860.

KURCZYNSKI: In this discussion of Dr. Edward's paper we would like to consider two general areas: first, the model of human evolution based on a Brownian motion/ Yule process, second, some of the methodological problems involved in applying the model to man. Any mathematical model of some real process must attempt to satisfy two often opposing conditions. On the one hand it must be a reasonable representation of the process, on the other hand, it must be tractable enough to provide answers to the questions being asked. Since the model developed by Edwards and Cavalli-Sforza has been applied to the evolution of human populations, it is appropriate to consider how the model may deviate from the pattern of human evolution since the mid-Pleistocene. Approximately 500,000 years ago, at the time of the mid-Pleistocene, man in the form of *Homo erectus* was already living in diverse geographical areas. Sinanthropus was in Asia, Pithecanthropus in Indonesia, Ternefine man in Africa, and Heidelberg man in Europe. It should be pointed out that these fossil forms, previously believed in some cases to represent former genera, are now considered subspecies or races of a single polytypic species, *Homo erectus* (Coon, 1963; Dobzhansky, 1963; Mayr, 1963; Simpson, 1963). *Homo erectus* already possessed many distinctively human characteristics. He walked erect, he was a tool user and tool maker, he probably used fire, and he killed and ate members of his own species. His post-cranial skeleton was essentially modern but the skull had a number of distinguishing features including a cranial capacity on the low side, about 1000 cc (LeGros Clark, 1955; Coon, 1963). As the polytypic species *erectus* evolved into the polytypic species *sapiens* there was an increase in cultural adaptation as evidenced by the complexity and diversity of tool traditions characterizing the Middle and Upper Paleolithic (Clark, 1965). In view of the long standing polytypic nature of the human species the general pattern of human evolution is characterized by predominantly anagenesis, i. e., change without diversification as opposed to cladogenesis, or change by branching and splitting of ancestral forms (Dobzhansky, 1963). Thus, the evolutionary pattern of man is more like a bush than a tree. As a result of past migrations some degree of gene flow probably occurred between many populations. More importantly, selection must have been important in the changes in physical form, for example the increase in cranial capacity by 50 per cent (to 1500 cc) as well as in cultural adaptation. Culture is man's most important means of adaptation, and it is reasonable to think it was strongly selected for. Moreover, if races are considered as adaptive in any sense, then selective differences must have existed at least some time in the past. It is difficult otherwise to explain the past or present diversity of human populations. This is not to deny the importance of random processes but merely to suggest that selective factors cannot be ignored. The effects of such deviations from the proposed random walk model do not necessarily invalidate it in general since there are undoubtedly many situations in which such a model could be realistically applied. Two such cases are considered later.

As Dr. Edwards has pointed out, in applying the model to human data difficulties were encountered in solving the maximum likelihood equations. In fact he concluded that estimation of the branch points given the positions of the populations at some point in time was not possible in many cases. However, these essentially statistical problems may occur regardless of whether the model is appropriate in a given instance and thus do not invalidate the model to principle. In addition, these methodological problems have led to other approaches such as least squares estimation which may yet prove valuable. In this regard some comments may be made about the method of minimum evolution. Although its mathematical basis seems reasonable, the biological implications remain to be clarified, for example, the meaning or reality of some minimum or efficient principle in evolution.

Cavalli-Sforza, Barrai, and Edwards (1964) applied the method of minimum evolution to the construction of a phylogenetic tree for fifteen diverse human populations. The view that the tree from such an evolutionary scheme is likely to be an over-simplification has already been presented by Dr. Edwards, and also the argument against such an evolutionary pattern was already discussed. For this and reasons given below it is not unexpected that a number of relationships suggested by the phylogenetic tree are open to question. For example, the implication of a Maori origin from the New World has already been criticized by Simpson (1965). Archaelogical, ethnological, and linguistic data overwhelmingly support the theory of an Old World origin for Polynesians (Clark, 1965; Suggs, 1960). Blood group data, however, suggests *some* similarities to the New World (Simmons et al., 1955), and based on these data alone it is not surprising that New World affinities were deduced for the Maori. Furthermore, the Maori, American Indians, and Eskimos are grouped as descendants from a common ancestral group. This is also unlikely since Polynesians are traced ultimately to Southeast Asia (Suggs, 1960) and American Indians and Eskimos migrated to the New World at different times and are only distantly related (Laughlin, 1963).

The preceeding discussion of the Maori and their affinities illustrate an established principle of modern taxonomy, namely that in constructing a phylogenetic classification all the available information on the groups being classified must be utilized. To the extent that a taxonomy fails to include or be consistent with other relevant information it is bound to be deficient. As Dr. Edwards has stated, the abundance of anthropological and genetic data on human populations provides the genetic taxonomist with the opportunity (1) of applying particular models to appropriate situations and (2) of testing the results of predictions based on the model. As illustration of a model of genetic drift consider Polynesian populations. The migration of Austronesian speakers into Polynesia began about 3,000 years ago (Suggs, 1960). Successive island groups were colonized by small numbers of people. Because of isolation and small effective sizes the opportunities for genetic drift were probably great. Since new islands were settled by groups from islands previously inhabited, a type of Brownian motion/Yule process model would seem appropriate. Predictions from this model could be compared with the known movements of people across the Pacific.

Another population suitable for the study of genetic drift is the Hutterites. These people migrated to the United States in the 1870's and formed three original colonies, designated the S-leut, L-leut, and D-leut. Some genetic and anthropological studies on these people have already been reported (Mange, 1964; Bleibtreu, 1964; Steinberg et al., 1967). Each of the original colonies has subsequently undergone successive splits so that three colony lines may be distinguished. Furthermore, there has been essentially no gene exchange between the colony lines since the early 1900's. The data available on the population permit a detailed analysis of the effects of genetic drift. One simple approach was to compare the gene frequencies of the S-leut and L-leut. An extension of the Mahalanobis generalized distance was developed for gene frequency data to provide an overall measure of divergence based on any number of multiallelic systems. The distance measure is defined as $D^2 = (P_1 - P_2)\, S^{-1}\, (P_1 - P_2)$, where P_1 and P_2 are the vectors of gene frequencies in two populations and S^{-1} is the inverse of the joint covariance matrix. Using gene frequencies for the ABO, Rh, MN, and Kell systems, the distance between the S-leut and L-leut was found to be 0.406. This value is larger than the distance between Poles from Cracow and Englishmen (.341) or the distance between Austrians from Vienna and Italians from Ferrara (.308). Thus, the extent of genetic drift between primary subdivisions of the Hutterites is considerable compared to the variation found in European populations. Perhaps Dr. Edward's program could be applied to this population to test whether it can predict a known pattern accurately.

The use of models in studying evolution is only beginning. With their increasing refinement and greater application our knowledge of the relative importance of various evolutionary factors may be extended and made more precise. Indeed successful models may be used to predict evolutionary changes for a species or a portion of a species. If the species is a rapidly breeding one, the accuracy of the prediction may be evaluated.

Bleibtreu, H. K. 1964. Ph.D. thesis, Harvard University.

Cavalli-Sforza, L. L., Barrai, I. and Edwards, A. W. F. 1964. Analysis of human evolution under random genetic drift. Cold Spring Harbor Symp. Quant. Biol. 29:9-20.

Clark, G. 1965. World Prehistory, Cambridge Univ. Press.

Coon, C. S. 1963. The Origin of Races, Alfred A. Knopf.

Dobzhansky, T. 1963. Genetic Entities in Hominid Evolution, in Classification and Human Evolution, (ed. S. L. Washburn), Aldine.

Laughlin, W. S. 1963. Eskimos and Aleuts: their origins and evolution, Science 142:633-645.

LeGros Clark, W. E. 1955. The fossil evidence for human evolution. Univ. of Chicago Press.

Mange, A. P. 1964. Growth and inbreeding of a human isolate, Hum. Biol. 36:104-133.

Mayr, E. 1963. The taxonomic evaluation of fossil hominids, in Classification and Human Evolution, (ed. S. L. Washburn), Aldine.

Simpson, G. G. 1965. Current issues in taxonomic theory, Science 148:1078.

Steinberg, A. G., Bleibtreu, H. K., Kurczynski, T. W., Martin, A., Kurczynski, E. M. 1967. Genetic studies on an inbred human isolate. Proc. Third Int. Cong. Hum. Genet., ed. J.F. Crow and J.V. Neel.

Suggs, R. C. 1960. The Island Civilizations of Polynesia, Mentor Books.

Note: Supported by funds from PHS Training Grant No. GM226 from the National Institute of General Medical Sciences.

* * *

J. H. EDWARDS: Most solutions to the problem of relating differences in allele proportions, or gene frequency, at several loci to distance (by which is usually meant time) assume some function which is monotonic over all loci in the sense that every increment in any distance increases the estimate of total time. In fact, under many conditions, this is not so as there will be plenty of time for the small changes within the span required for the large changes, so that it may be preferable to consider only the most divergent loci.

A very simple and computationally tractable model is to consider the inverse of an explosion in n-dimensional Euclidean space in which gravitation defines direction of motion but speed is constant and mass is proportional to "weight" in the statistical sense.

MORTON: I am puzzled by the way in which gene frequencies are transformed and genetic systems are combined in the studies on genetic taxonomy. It is well known that the coefficient of kinship appropriate to a cross between two populations is estimated as

$$\varphi_{ijk} = \frac{(q_{ik} - \bar{q}_k)\ (q_{jk} - \bar{q}_k)}{\bar{q}_k\,[1 - \bar{q}_k]}$$

where q_{ik} q_{jk} are the gene frequencies of allele q_k in the ith and j^{th} populations respectively and $\bar{q}_k$ is the mean gene frequency. Such estimates may be averaged over alleles by the Yasuda equation

$$\varphi = \sum_k \bar{q}_k \phi_{ijk},$$

A preferable computation scheme estimates φ from pairs of phenotypes (Yasuda, 1968). The estimates may be averaged over different loci simply by weighting with the amounts of information.

The advantage of this formulation is that the measure of genetic distance is itself a genetic parameter (the coefficient of kinship) and has a simple interpretation in terms of the expected phenotype frequencies in a cross between populations. None of the other measure of genetic distance has a simple interpretation. No transformation of gene frequencies is necessary or desirable, and there is no need to estimate the latent roots of a matrix. The estimates of φ are of course specific for the alleles and loci used in estimation, but that is true of any taxonomic procedure.

More attention should be paid to relating genetic taxonomy to genetics. In view of the considerations indicated by Kurczynski and Steinberg and the usual experience of taxonomists, I wonder whether taxonomy can reasonably hope to do more than recognize phenotypic similarity without being able to impose on plausible assumptions any simple branching process. In any case, the coefficient of kinship would seem to contain all the taxonomic information.

Yasuda, N. 1968. An extension of Wahlund's principle to evaluate mating type frequency. Amer. J. Hum. Genet. 20:1-23.

DETERMINISTIC SIMULATION OF EVOLUTIONARY CHANGES IN THREE-LOCUS GENETIC SYSTEMS

Ken-ichi Kojima and Andrea Klekar
Department of Zoology
University of Texas at Austin

During the past decade, there has been a continuous accumulation of theoretical information on the evolutionary process of multilocus systems in mendelian populations. Kimura (1956), Lewontin and Kojima (1960), Bodmer and Parsons (1962), Kojima (1965) and Bodmer and Felsenstein (1967) investigated evolutionary behavior of various two-locus systems and the conditions under which non-trivial stable equilibria might be maintained with respect to both loci. The majority of these activities were inspired by an earlier work by Fisher (1930) in which he stated that a closer linkage might result between a pair of loci on the same chromosome when equilibrium was maintained by a certain type of epistasis between the loci.

One of the major problems in the evolution of multilocus systems is the existence of linkage disequilibrium, which refers to the deviations of actual gametic frequencies from the values computed by multiplying the frequencies of alleles in a population. Because of such deviations, the frequency changes in multilocus systems must be formulated in terms of gametic, rather than allelic, frequencies. Thus, the frequency changes in an n-locus system with two alleles per locus will be expressed by 2^n simultaneous equations, each representing a differential or difference equation of a gamete frequency change with respect to time. Theoretically, these equations will provide all the information on possible genetic equilibria when they are set to zero and the solutions are sought.

However, in practice analytical solutions for an equilibrium turn out to be impossible to obtain even for the simple case of two loci when the values of fitness for the genotypes involved are not restricted. Analytical solutions were successfully obtained only when the simultaneous equations were reducible to a polynomial equation of one variable by virtue of symmetry in the model of fitness values.

For a general fitness model the only approach used so far has been the method of numerical solutions, as in the case of Kojima (1965). This approach consisted of (a) programming the simultaneous equations of gametic frequency changes on a computer, (b) seeking numerical solutions of a set of gametic frequencies with which no change was to take place in the gametic frequencies, and (c) determining the stability of such equilibrium solutions. Such an approach can be applied to a genetic model of any number of loci. Lewontin (1964a,b) used a similar approach for the investigation of epistatic linked gene complexes made up with five loci and twenty loci. The main modes of gene action used by Lewontin were a general heterotic model and Wright's optimum model.

The objective of this paper is to investigate some aspects of evolutionary behavior of multilocus systems with special emphasis on the effects of changing linkage intensity among loci with various modes of non-allelic gene interactions. The latter emphasis was prompted by a recent paper by Turner (1967), who argued that there should be an optimum value of recombination in a multilocus system where the population fitness is maximized. Turner's view differs somewhat from the earlier conclusion which predicted ever-diminishing recombination values based upon Fisher's model.

GAMETE FREQUENCY CHANGES

Let A vs. a, B vs. b and C vs. c be the alternative alleles at the three loci. When they are on the same chromosome, their positions are assumed to be in the order of A-B-C. Without considering linkage phases, there are 27 distinct genotypes in a diploid population. The genotype containing i A's, j B's and k C's is denoted as (i, j, k) where each of i, j and k takes the value of either 0, 1 or 2. The fitness values of genotype (i, j, k) is designated as W_{ijk}. Possible gametes in such a population are ABC, ABc, AbC, aBC, Abc, aBc, abC and abc and their frequencies are represented by X_8, X_7, ..., X_1, respectively. The type of any particular gamete may be identified by the subscript of X's (e.g., ABC is gamete 8, ABc gamete 7, etc.).

Under the assumption of random union of gametes, the genotypes and their frequencies at the time of fertilization may be arrayed as in Table 1. The eight cells on the main diagonal represent eight homozygotes, and the eight cells on the other diagonal contain the triple heterozygotes in four different linkage phases. The rest are six different double heterozygotes in 24 cells, and 12 different single heterozygotes in another 24 cells. The margins of Table 1 contain the eight marginal means of fitness values, W_8, W_7, ..., W_1, and the population fitness is shown by $\overline{W}$. The marginal means are computed by multiplying the eight W's in a given row or column by the corresponding gametic frequencies and adding the eight products. For example,

$$W_8 = X_8 W_{222} + X_7 W_{221} + X_6 W_{212} + X_5 W_{122} + X_4 W_{211} + X_3 W_{121} + X_2 W_{112} + X_1 W_{111} \tag{1}$$

The (mean) population fitness is given by

$$\overline{W} = \sum_{i=1}^{8} X_i W_i \tag{2}$$

The recombination fractions between loci A and B and between loci B and C are given as r_1 and r_2, respectively. In the absence of crossover interference, an appropriate combination of r_1 and r_2 will specify all possible recombinational events. They are

i) No recombination $\qquad R_{00} = (1-r_1)(1-r_2)$

ii) Recombination in A-B but no recombination in B-C $\qquad R_{10} = r_1(1-r_2)$

iii) No recombination in A-B but recombination in B-C $\qquad R_{01} = (1-r_1)r_2$

iv) Recombination in A-B and in B-C $\qquad R_{11} = r_1 r_2$

v) Recombination in A-C $\qquad R_{10} + R_{01}$

It should be noted that $R_{00} + R_{10} + R_{01} + R_{11} = 1$.

The derivation of the change in a gametic frequency, ΔX_i, may be demonstrated by using ΔX_8 as an example. Suppose that there is no recombination; that is to say that the eight gametic types in Table 1 represent eight multiple alleles. Then, the change in X_8 is

$$X_8 (W_8 - \overline{W}) / \overline{W} \tag{3}$$

which is the average excess of gamete ABC over the population mean, multiplied by X_8. However, the actual contribution from the column or row of X_8 to ΔX_8 is reduced by the amount of recombination in the double and triple heterozygotes in the column or row. The reduction from ABC/abc is

$$W_{111} X_1 X_8 \{R_{10} + R_{01} + R_{11}\} / \overline{W} \tag{4}$$

since any one of the three recombinational events in the curly bracket prohibits the production of gamete ABC. The reduction from the double heterozygotes are

$$[W_{211}X_8X_4r_2 + W_{121}X_8X_3(R_{10} + R_{01}) + W_{112}X_8X_2r_1] \;/\; \overline{W} \qquad (5)$$

On the other hand, there are contributions of ABC gametes due to recombination in double and triple heterozygotes which are not in the X_8 column or row. There are three other triple heterozygotes and they contribute to ΔX_8 by

$$W_{111}\{X_2X_7R_{01} + X_6X_3R_{11} + X_5X_4R_{10}\} \;/\; \overline{W} \qquad (6)$$

Three double heterozygotes, (2, 1, 1), (1, 2, 1) and (1, 1, 2), in the columns or rows other than those of X_8 also produce ABC gametes through recombination. The contribution from these double heterozygotes is

$$[W_{211}X_7X_6r_2 + W_{121}X_7X_5(R_{10} + R_{01}) + W_{112}X_6X_5r_1] \;/\; \overline{W} \qquad (7)$$

There is no other genotype which can produce ABC gametes with or without recombination. Thus, adding (3), (4) and (5) and subtracting (6) and (7), the total change in X_8 is

$$\Delta X_8 = [X_8(W_8 - \overline{W}) - W_{112}(X_8X_2 - X_6X_5)r_1 - W_{211}(X_8X_4 - X_7X_6)r_2 - W_{121}(X_8X_3 - X_7X_5)(R_{10} + R_{01}) - W_{111}\{R_{10}(X_1X_8 - X_5X_4) + R_{01}(X_1X_8 - X_2X_7) + R_{11}(X_1X_8 - X_6X_3)\}] \;/\; \overline{W} \qquad (8)$$

In the above expression it should be noted that all contributions through recombination in the double heterozygotes are the products of double heterozygote fitness values, linkage disequilibria and recombination fractions corresponding to the respective double heterozygotes. The contributions from the triple heterozygotes are also the products of their fitness values, the excesses of X_1X_8 over the frequencies of the other triple heterozygotes, and relevant recombination values for such excesses. These ground rules apply to all ΔX's. Thus, in general

$$\Delta X_i = [X_i(W_i - \overline{W}) - W_{11t}(X_iX_a - X_dX_e)\,r_1 - W_{u11}(X_iX_b - X_fX_g)r_2 - W_{1v1}(X_iX_c - X_hX_j)(R_{10} + R_{01}) - W_{111}\{(X_iX_k - X_lX_m)R_{10} + (X_iX_k - X_nX_o)R_{01} + (X_iX_k - X_pX_q)R_{11}\}] / \overline{W} \qquad (9)$$

where

i) each of t, u and v takes either value 0 or 2, depending upon X_i. For example, t=u=v=2 for ΔX_8, and t=0 and u=v=2 for ΔX_7.

ii) X_a, X_b and X_c must form the cell frequencies of genotypes (1, 1, t), (u, 1, 1) and (1, v, 1) when multiplied by X_i.

iii) X_dX_e, X_fX_g and X_hX_j are the cell frequencies of genotypes (1, 1, t), (u, 1, 1) and (1, v, 1) in the opposite linkage phase from those specified in (ii).

iv) subscript k specifies the gamete which forms a triple heterozygote with gamete i.

v) X_lX_m, X_nX_o and X_pX_q are the cell frequencies of the triple heterozygotes in three different linkage phases excepting X_i X_k, and these triple heterozygotes should produce gamete i with recombination events of R_{10}, R_{01} and R_{11}, respectively.

A COMPUTER PROGRAM FOR STUDYING DETERMINISTIC EVOLUTIONARY CHANGES

Using the equations for the eight gametic frequency changes as in (9), it was possible to derive the formula for the rate of change in $\overline{W}$ as in Kojima and Kelleher (1961). Theoretically, the behavior of such a formula in the neighborhood of the point of equilibrium should tell whether a tighter linkage is always advantageous or not to a population of multilocus system. However, as pointed out before for a general two-locus system, it is not always

Table 1. Zygotic assay produced by random union of gametes. Each cell contains a genotype specification in parenthesis and its frequency.

Table 1 -- continued

	X_8 ABC	X_7 ABc	X_6 AbC	X_5 aBC	X_4 Abc	X_3 aBc	X_2 abC	X_1 abc	
X_8 ABC	X_8^2 (2, 2, 2)	X_8X_7 (2, 2, 1)	X_8X_6 (2, 1, 2)	X_8X_5 (1, 2, 2)	X_8X_4 (2, 1, 1)	X_8X_3 (1, 2, 1)	X_8X_2 (1, 1, 2)	X_8X_1 (1, 1, 1)	W_8
X_7 ABc	X_8X_7 (2, 2, 1)	X_7^2 (2, 2, 0)	X_7X_6 (2, 1, 1)	X_7X_5 (1, 2, 1)	X_7X_4 (2, 1, 0)	X_7X_3 (1, 2, 0)	X_7X_2 (1, 1, 1)	X_7X_1 (1, 1, 0)	W_7
X_6 AbC	X_8X_6 (2, 1, 2)	X_7X_6 (2, 1, 1)	X_6^2 (2, 0, 2)	X_6X_5 (1, 1, 2)	X_6X_4 (2, 0, 1)	X_6X_3 (1, 1, 1)	X_6X_2 (1, 0, 2)	X_6X_1 (1, 0, 1)	W_6
X_5 aBC	X_8X_5 (1, 2, 2)	X_7X_5 (1, 2, 1)	X_6X_5 (1, 1, 2)	X_5^2 (0, 2, 2)	X_5X_4 (1, 1, 1)	X_5X_3 (0, 2, 1)	X_5X_2 (0, 1, 2)	X_5X_1 (0, 1, 1)	W_5
X_4 Abc	X_8X_4 (2, 1, 1)	X_7X_4 (2, 1, 0)	X_6X_4 (2, 0, 1)	X_5X_4 (1, 1, 1)	X_4^2 (2, 0, 0)	X_4X_3 (1, 1, 0)	X_4X_2 (1, 0, 1)	X_4X_1 (1, 0, 0)	W_4
X_3 aBc	X_8X_3 (1, 2, 1)	X_7X_3 (1, 2, 0)	X_6X_3 (1, 1, 1)	X_5X_3 (0, 2, 1)	X_4X_3 (1, 1, 0)	X_3^2 (0, 2, 0)	X_3X_2 (0, 1, 1)	X_3X_1 (0, 1, 1)	W_3
X_2 abC	X_8X_2 (1, 1, 2)	X_7X_2 (1, 1, 1)	X_6X_2 (1, 0, 2)	X_5X_2 (0, 1, 2)	X_4X_2 (1, 0, 1)	X_3X_2 (0, 1, 1)	X_2^2 (0, 0, 2)	X_2X_1 (0, 0, 1)	W_2
X_1 abc	X_8X_1 (1, 1, 1)	X_7X_1 (1, 1, 0)	X_6X_1 (1, 0, 1)	X_5X_1 (0, 1, 1)	X_4X_1 (1, 0, 0)	X_3X_1 (0, 1, 1)	X_2X_1 (0, 0, 1)	X_1^2 (0, 0, 0)	W_1
	W_8	W_7	W_6	W_5	W_4	W_3	W_2	W_1	$\overline{W}$

possible to obtain the gametic frequencies at a point of genetic equilibrium. Consequently, a meaningful examination of the formula for $\overline{W}$ for the present purpose cannot be carried out analytically at this stage.

An alternative approach is a numerical analysis of such a system by the use of a computer program which simulates deterministically the evolutionary behavior of a given multilocus system. Such a program using a three-locus system was written for

i) Computing the eight ΔX's every generation, starting from a given initial gametic frequency, iteratively until either the largest of $|\Delta X|$'s became less than 10^{-5} or 1,000 generations were completed;

ii) Incrementing X's in the same direction as the population was moving during stage (i) until every X value was within the distance of 10^{-5} from its equilibrium value;

iii) Testing the stability of the equilibrium obtained by examining characteristic roots of the matrix containing as the element the values of $\partial(\Delta X_i)/\partial X_j$ evaluated at the point of equilibrium (i-1, 2, ---, 8 and j-1, 2, ---, 8);

iv) Computing ΔX's and $\overline{W}$ after recombination fractions were changed, namely after reproportioning of linkage distance (see section on Linkage Parameterization) to see whether a new linkage situation would lead to a new point of equilibrium, and if so;

v) Testing the stability of the new equilibrium.

Stages (i) and (iv) are self-explanatory, but some comments are in order for explaining stages (ii), (iii) and (v). The purpose of stage (ii) is to speed up the approach to the point of equilibrium. For this, the last set of ΔX's from stage (i) was multiplied by 10^3 and was used as the increments to individual X's. The new set of ΔX's was computed using the incremented X values. The last ΔX's from stage (i) and the new ΔX's were compared with respect to the signs of individual ΔX-pairs. If none of the ΔX's changed the sign, the incrementation successfully pushed the population closer to the point of equilibrium. If one or more ΔX-pairs changed the sign, the incrementation had over-shot the point of equilibrium to which the population was heading during stage (i). In the latter case, all the increments were halved, and the same testing process as for the initial increments was started. When the largest increment in absolute value became less than 10^{-5}, stage (ii) stopped and stage (iii) began.

Stages (iii) and (v) started computing the elements of an 8 x 8 matrix. The element in the ith row and the jth column was the value of $\partial(\Delta X_i)/\partial X_j$ evaluated at the point of equilibrium reached after stage (ii) or (iv). Thus, the element (i, j) represents the linear tendency of ΔX_i in the direction X_j. This is reduced to the 7 x 7 matrix using the constraint $\Sigma \Delta X_i = 0$. The characteristic roots of this matrix were computed, and the signs and magnitudes of individual roots examined for the test of stability of the systems. The necessary and sufficient condition for stable equilibrium in difference equations after linearization is that all the roots are inside a circle with radius 1, centered at the real axis -1 and imaginary axis 0. (See Kojima, 1965)

The output of the program consisted of gametic frequencies $\overline{W}$ and $\Delta\overline{W}$ at every generation during stage (i); the increments used and $\overline{W}$ during stage (ii); the characteristic matrix and its roots in stage (iii); gametic frequencies $\overline{W}$ and $\Delta\overline{W}$ at every generation during stage (iv); the characteristic matrix and its roots in stage (v).

LINKAGE PARAMETERIZATION

The linkage relation among the three loci, A, B and C, was determined according to following manner. First, the recombination fraction between the A and C loci was set at either loose linkage (L) of 0.50, medium linkage (M) of 0.125 or tight linkage (T) of 0.01. Then, the position of the B locus was varied - in general, at 1/6, 1/5, 1/4, 1/3, 1/2, 2/3, 3/4, 4/5 and 5/6 of the total distance between the A and C loci. Thus, there were nine linkage parameterizations for given initial conditions (initial gamete frequency, W values, and the total A-C distance). However, there are cases where only a few of the above nine conditions were run.

INITIAL GAMETE FREQUENCY

Except for some special cases, the initial gamete frequencies were set at either linkage equilibrium (E) of $X_1 = X_2 = ---- = X_8 = .125$, extreme coupling (C) of $X_1 = X_8 = 0.5$ or extreme repulsion (R) of $X_3 = X_6 = 0.5$. In all these cases, the allele frequency was 0.5 at any given locus. Special cases were those in which perturbation was made in the gamete frequency near a point of unstable equilibrium.

FITNESS MODELS

All of 27 fitness values in various models studied were assumed to be constant over time. These values were chosen to fit several concepts concerning the determination of selective values.

OPTIMUM MODELS

This model was adapted from Wright's concept (1935) which related the fitness value of a genotype to the primary value of that genotype. In this model, the fitness of a genotype decreased according to a squared deviation of additive genotype value from a given optimum value on the scale of genotype value. Letting α_i, β_j and γ_k be the values of the genotypes at the loci A, B and C, respectively, the fitness value of zygote (i, j, k) was defined as

$$W_{ijk} = 1 - K(\alpha_i + \beta_j + \gamma_k - \overline{0})^2 \tag{10}$$

where i, j and k take the values as defined before, $\overline{0}$ was the optimum value, and K was a constant for adjusting all W values to be within the range of 0 and 1.0. Of course, there existed an infinite number of possible W sets under the model in (10). For the present study, the following restrictions were made:

$$\alpha_2 = \beta_2 = \gamma_2 = 1.0$$

$$\alpha_1 = \beta_1 = \gamma_1 = 0.7$$

$$\alpha_o = \beta_o = \gamma_o = -1.0$$

According to Kojima (1959) and Lewontin (1964), the value of $\overline{0}$ between approximately 1.7 and 2.7 would provide a stable equilibrium for the above model when linkages are loose. Thus, 1.4, 1.7, 1.8, 2.0 and 2.5 were chosen as the value of $\overline{0}$ in this study.

SYMMETRIC MODELS (S-MODELS)

The word "symmetric" refers to the condition of all double heterozygotes having equal W values, all single heterozygotes having equal W values, all homozygotes having equal W values and the triple heterozygote having its own W value.

(i) S-Model 1 (optimum heterozygosity model)

This model represented the situation where the double heterozygotes were the best in fitness (W=1.00), the triple heterozygote was the 2nd (W=0.800), the single heterozygote was the 3rd (W=0.500) and the homozygote was the poorest (W=0.10). In other words, heterozygosity increased the value of fitness up to a certain degree, but decreased it thereafter.

(ii) S-Model 2

In this model the W values of all homozygotes and the triple heterozygote were set to 1.00, those of all double heterozygotes to 0.50, and those of all single heterozygotes to 0.10.

(iii) S-Model 3

In this model the triple heterozygote had W = .10, all double heterozygotes W = 0.50, all single heterozygotes W = 1.00, and all homozygotes W = 0.80.

ASYMMETRIC MODELS

Some degree of asymmetry was introduced into the above three symmetric models.

(i) AS-Model 1 A

Asymmetry was introduced by changing W_{221} from 0.50 to 1.00, W_{011} from 1.00 to 0.10 and W_{021}, W_{012}, W_{010} and W_{001} from 0.50 to 0.10 in the S-model 1. These changes were intended to create selective differences among genotypes 2, 1, and 0 at locus A.

(ii) AS-Model 1 B

To create selective differences among genotypes 2, 1, and 0 at locus B, asymmetry was introduced by changing W_{101} to 0.1, W_{212} to 1.0, and W_{201}, W_{102}, W_{100}, W_{001} to 0.1 in the S-Model 1.

(iii) AS-Model 1 C

In the S-Model 1, W_{110} was changed to 0.1, W_{122} to 1.0, and W_{210}, W_{120}, W_{100}, W_{010} to 0.1. This model is symmetric to AS-Model 1 A and gives a selective advantage to allele C as opposed to allele c.

(iv) AS-Model 2

In the S-Model 2, W_{000} was changed from 1.00 to 0.10 to create AS-Model 2. This change made aa, bb, and cc genotypes less favorable in selective values than those in the S-Model 2.

(v) AS-Model 3 A

In the S-Model 3, W_{112} was changed from 0.50 to 1.00, giving an advantage to the double heterozygote AaBbCC.

(vi) AS-Model 3B

In addition to the AS-Model 3A, W_{110} was changed from 0.50 to 0.80.

(vii) AS-Model 3 C

In addition to the AS-Model 3 A, W_{110} was changed from 0.50 to 1.00 and W_{200} from 0.8 to 1.00.

RESULTS

To describe the entire frequency change results of individual runs is unprofitable and troublesome. For this reason the results (Table 2-16) are summarized with respect to the types of equilibrium reached, the final $\overline{W}$ and the changes in $\overline{W}$ over generations. These results are given for each initial gamete frequency under a specified A-C linkage distance for a given fitness model.

The procedure for linkage proportioning (1/6, 1/5, ---, 1/2, 2/3, ---, 5/6) was to run the case of equi-distance between A-B and B-C (i.e., the case of 1/2) for the first time, then when the population reached a point of equilibrium, other proportioning runs were carried out. If, however, the population reached a point of fixation at the 1/2 linkage case, no further reproportioned case was run.

Each cell in the table contains three types of information; stability (S = stable equilibrium, U = unstable equilibrium, F = fixation of at least one locus), the final value of $\overline{W}$, and the mode of change in $\overline{W}$ over generations. The last information was summarized by symbols:

a: continuous increase

Table 2. Optimum Model with $\overline{O}$ = 2.5

Position of B from A

		1/6	1/5	1/4	1/3	1/2	2/3	3/4	4/5	5/6
C	L	S .99304 d	S .99304 d	S .99304 d	S .99304 d	S .99304 a	S .99304 d	S .99304 d	S .99304 d	S .99304 d
	M	S .99306 d	S .99306 d	S .99305 d	S .99305 d	S .99305 a	S .99305 d	S .99305 d	S .99306 d	S .99306 d
	T	S .99317 a	S .99317 a	S .99317 a	S .99317 a	S .99317 a	S .99317 a	S .99317 a	S .99317 a	S .99317 a
	L	S .99304 d	S .99304 d	S .99304 d	S .99304 d	S .99304 a	S .99304 d	S .99304 d	S .99304 d	S .99304 d

Table 2 -- continued

		1/6	1/5	1/4	1/3	1/2	2/3	3/4	4/5	5/6
E	M	S .99306 d	S .99306 d	S .99305 d	S .99305 b	S · .99305 a	S .99305 b	S .93305 d	S .99306 d	S .99306 d
	T	S .99317 d	S .99317 b	S .99317 b	S .99317 b	S .99317 c	S .99317 b	S .99317 b	S .99317 b	S .99317 d
R	L	S .99304 d	S .99304 d	S .99304 d	S .99304 d	S .99304 a	S .99304 d	S .99304 d	S .99304 d	S .99304 d
	M	S .99306 d	S .99306 d	S .99306 d	S .99306 b	S .99306 a	S .99306 b	S .99306 d	S .99306 d	S .99306 d
	T	S .99317 d	S .99317 b	S .99317 b	S .99317 b	S .99317 a	S .99317 b	S .99317 b	S .99317 b	S .99317 d

Table 3. Optimum Model with $\bar{0}$ = 2.0

Position of B from A

		1/6	1/5	1/4	1/3	1/2	2/3	3/4	4/5	5/6
	L	S .97283 d	S .97278 d	S .97274 d	S .97270 d	S .97268 e	S .97270 d	S .97274 d	S .97278 d	S .97283 d
C	M	S .97383 d			S .97340 d	S .97332 a	S .97340 d			S .97383 d
	T	S .97813 d			S .97799 d	S .97794 a	S .97799 d			S .97813 d
	L	S .97283 d	S .97278 d	S .97274 d	S .97270 d	S .97268 c	S .97270 d	S .97274 d	S .97278 d	S .97283 d

Table 4. Optimum Model with $\bar{0}$ = 1.8

Position of B from A

		1/6	1/5	1/4	1/3	1/2	2/3	3/4	4/5	5/6
	L	U .96470 d			U .96418 d	S .96409 e	U .96418 d			U .96470 d
C	M	U .96891 d			U .96648 a	U .96576 a	U .96648 a			U .96891 d
	T	S .97822 a	S .97786 a	S .97732 a	S .97650 a	S .97571 a	S .97650 a	S .97732 a	S .97786 a	S .97822 a
	L	U .96471 d			U .96418 d	S .96409 e	U .96418 d			U .96471 d

Table 3 -- continued

		1/6	1/5	1/4	1/3	1/2	2/3	3/4	4/5	5/6
E	M	S .97383 d			S .97340 d	S .97332 c	S .97340 d			S .97383 d
	T	S .97813 d			S .97799 d	S .97794 a	S .97799 d			S .97813 d
	L	S .97283 d	S .97278 d	S .97274 d	S .97270 d	S .97268 a	S .97270 d	S .97274 d	S .97278 d	S .97283 d
R	M	S .97383 d			S .97340 d	S .97332 c	S .97340 d			S .97383 d
	T	S .97813 d			S .97799 d	S .97794 c	S .97799 d			S .97813 d

Table 4 -- continued

		1/6	1/5	1/4	1/3	1/2	2/3	3/4	4/5	4/6
E	M	U .96894 d			U .96686 a	U .96580 e	U .96686 a			U .96894 d
	T	S .97822 d			S .97651 d	S .97577 a	S .97651 d			S .97822 d
	L	F .97438 a			F .97438 a	S .96985 a	F .97438 a			F .97438 a
R	M					F .97429 a				
	T	S .97823 d			S .97651 d	S .97582 c	S .97651 d			S .97823 d

Table 5. Optimum Model with $\bar{0}$ = 1.7

Position of B from A

		1/6	1/5	1/4	1/3	1/2	2/3	3/4	4/5	5/6
	L	S .98038 d			S .98038 a	U .95992 e	S .98038 a			S .98038 d
C	M	S .98039 a			S .98039 a	S .97041 a	S .98039 a			S .98039 a
	T	U .97914 a	U .97899 a	S .97887 a	S .97888 a	S .97462 a	S .97888 a	S .97887 a	U .97889 a	U .97914 a
	L	S .98038 a			S .98038 a	U .96053 e	S .98038 a			S .98038 a

Table 6. Optimum Model with $\bar{0}$ = 1.4

Position of B from A

		1/6	1/5	1/4	1/3	1/2	2/3	3/4	5/6	5/6
	L					F .99358 a				
C	M					F .99358 a				
	T	F .99359 a	F .99359 a	F .99359 a	F .99359 a	S .99359 a	F .99359 a	F .99359 a	F .99359 a	F .99359 a
	L					F .99358 a				

Table 5 -- continued

		1/6	1/5	1/4	1/3	1/2	2/3	3/4	4/5	5/6
E	M	S .98039 a			S .98039 a	S .98017 e	S .98039 a			S .98039 a
	T	S .98003 a			S .98002 a	S .97769 a	S .98002 a			S .98003 a
	L	S .98039 a			S .98039 a	S .98038 a	S .98039 a			S .98039 a
R	M	S .98039 a			S .98039 a	S .98039 a	S .98039 a			S 98039 a
	T	S .98013 c			S .98008 a	S .98003 a	S .98008 a			S .98013 c

Table 6 -- continued

		1/6	1/5	1/4	1/3	1/2	2/3	3/4	4/5	5/6
E	M					F .99358 a				
	T					F .99358 a				
	L					F .99358 e				
R	M					F .99358 a				
	T	F .99359 a	F .99359 a	F .99359 a	F .99359 a	S .99359 a	F .99359 a	F .99359 a	F .99359 a	F .99359 a

Table 7. Symmetric Model 1

Position of B from A

		1/6	1/5	1/4	1/3	1/2	2/3	3/4	4/5	5/6
	L	S .67500 a	S .67500 a	S .67500 a	S .67500 a	S .67500 a	S .67500 a	S .67500 a	S .67500 a	S .67500 a
C	M	S .67500 a	S .67600 a	S .67500 a	S .67500 a	S .67500 a	S .67500 a	S .67500 a	S .67500 a	S .67500 a
	T	U .67500 a	U .67500 a	U .67500 a	U .67500 a	U .67500 a	U .67500 a	U .67500 a	U .67500 a	U .67500 a
	L	S .67500 a	S .67500 a	S .67500 a	S .67500 a	S .67500 h	S .67500 a	S .67500 a	S .67500 a	S .67500 a
E	M	S .67500 a	S .67500 a	S .67500 a	S .67500 a	S .67500 h	S .67500 a	S .67500 a	S .67500 a	S .67500 a

Table 8. Asymmetric Model 1 A

Position of B from A

		1/6	1/5	1/4	1/3	1/2	2/3	3/4	4/5	5/6	5/6
	L	S .65984 a	S .65930 a	S .65878 a	S .65829 a	S .65784 c	S .65762 b	S .65755 b	S .65752 b	S .65750 b	U 96470 d
C	M	S .66705 a	S .66609 a	S .66479 a	S .66308 a	S .66096 c	S .65984 b	S .65948 b	S .65930 b	S .95919 b	U 96891 d
	T	S .70159 a	S .70151 a	S .70138 a	S .70116 a	S .70073 a	S .70031 b	S .70010 b	S .69997 b	S .69989 b	S .97822 a
	L	S .65984 a	S .65930 a	S .65878 a	S .65829 a	S .65784 a	S .65762 b	S .65755 b	S .65752 b	S .65750 b	U .96471 d

Table 7 -- continued

		1/6	1/5	1/4	1/3	1/2	2/3	3/4	4/5	5/6
	T	U .67500 a	U .65700 a	U .67500 a	U .67500 a	U .67500 a	U .67500 a	U .67500 a	U .67500 a	U .67500 a
	L	S .67500 a	S .67500 a	S .67500 a	S .67500 a	S .67500 a	S .67500 a	S .67500 a	S .67500 a	S .67500 a
R	M	S .67500 a	S .67500 a	S .67500 a	S .67500 a	S .67500 a	S .67500 a	S .67500 a	S .67500 a	S .67500 a
	T	U .67500 a	U .67500 a	U .67500 a	U .67500 a	U .67500 a	U .67500 a	U .67500 a	U .67500 a	U .67500 a
S*	T	S .76501 a			S .76502 a	S .76501 a	S .76502 a			S .76501 a

S*: $X_8 = X_2 = 0.1900$ $X_7 = X_1 = 0.1850$ $X_6 = X_5 = X_4 = X_3 = .0625$

Table 8 -- continued

		1/6	1/5	1/4	1/3	1/2	2/3	3/4	4/5	5/6	'6
E	M	S .66705 a	S .66609 a	S .66479 a	S .66308 a	S .66095 a	S .65984 b	S .65948 b	S .65930 b	S .65919 b	394
	T	U .67216 a	U .67207 a	U .67193 a	U .67171 a	U .67125 a	U .67079 b	U .67056 b	U .67043 b	U .67034 b	822
	L	S .65984 a	S .65930 a	S .65878 a	S .65829 a	S .65784 c	S .65762 b	S .65755 b	S .65752 b	S .65750 b	F 438
R	M	S .66705 a	S .66609 a	S .66479 a	S .66308 a	S .66095 a	S .65984 b	S .65948 b	S .65930 b	S .65919 b	
	T	S .70159 a	S .70151 a	S .70138 a	S .70116 a	S .70073 a	S .70031 b	S .70010 b	S .69997 b	S .69989 b	S 7823 d

Table 9. Asymmetric Model 1 B
Position of B from A

		1/6	1/5	1/4	1/3	1/2	2/3	3/4	4/5	5/6
	L	S .64008 d	S .63928 d	S .63840 d	S .63752 d	S .63696 a	S .63752 d	S .63840 d	S .63928 d	S .64008 d
C	M	S .65335 d	S .65297 d	S .65249 d	S .65189 d	S .65141 a	S .65189 d	S .65249 d	S .65297 d	S .65335 d
	T	S .69184 d	S .69184 d	S .69184 d	S .69184 d	S .69184 g	S .69184 d	S .69184 d	S .69184 d	S .69184 d
	L	S .64008 d	S .63928 d	S .63840 d	S .63752 d	S .63696 a	S .63752 d	S .63840 d	S .63928 d	S 64008 d

Table 9 -- continued

		1/6	1/5	1/4	1/3	1/2	2/3	3/4	4/5	5/6
E	M	S .65335 d	S .65297 d	S .65249 d	S .65189 d	S .65142 a	S .65189 d	S .65249 d	S .65297 d	S .65335 d
	T	S .69184 d	S .69184 d	S .69184 d	S .69184 d	S .69184 a	S .69184 d	S .69184 d	S .69184 d	S .69184 d
	L	S .64008 d	S .63928 d	S .63840 d	S .63752 d	S .63696 a	S .63752 d	S .63840 d	S .63928 d	S .64008 d
R	M	S .65335 d	S .65297 d	S .65249 d	S .65189 d	S .65142 a	S .65189 d	S .65249 d	S .65297 d	S .65335 d
	T	S .69184 d	S .69184 d	S .69184 d	S .69184 a	S .69184 a	S .69184 a	S .69184 d	S .69184 d	S .69184 d

Table 10. Asymmetric Model 1 C
Position of B from A

		1/6	1/5	1/4	1/3	1/2	2/3	3/4	4/5	5/6
	L	S .65750 b	S .65752 b	S .65755 b	S .65762 b	S .65784 c	S .65829 a	S .65878 a	S .65930 a	S .65984 a
C	M	S .65919 b	S .65930 b	S .65948 b	S .65984 b	S .66096 c	S .66308 a	S .66479 a	S .66609 a	S .66705 a
	T	S .69989 b	S .69997 b	S .70010 b	S .70031 b	S .70073 a	S .70116 a	S .70138 a	S .70151 a	S .70159 a
	L	S .65750 b	S .65752 b	S .65755 b	S .65762 b	S .65784 a	S .65829 a	S .65878 a	S .65930 a	S .65984 a
E	M	S .65919 b	S .65930 b	S .65948 b	S .65984 b	S .66095 a	S .66308 a	S .66479 a	S .66609 a	S .66705 a

Table 10 -- continued

		1/6	1/5	1/4	1/3	1/2	2/3	3/4	4/5	5/6
	T	U .67034 b	U .67043 b	U .67056 b	U .67079 b	U .67125 d	U .67171 a	U .67193 a	U .67207 a	U .67216 a
	L	S .65750 b	S .65752 b	S .65755 b	S .65762 b	S .65784 c	S .65829 a	S .65878 a	S .65930 a	S .65984 a
R	M	S .65919 b	S .65930 b	S .65948 b	S .65984 b	S .66095 a	S .66308 a	S .66479 a	S .66609 a	S .66705 a
	T	S .69989 b	S .69997 b	S .70010 b	S .70031 b	S .70073 a	S .70116 a	S .70138 a	S .70151 a	S .70159 a

Table 11. Symmetric Model 2

Position of B from A

		1/6	1/5	1/4	1/3	1/2	2/3	3/4	4/5	5/6
C	L	U .58747 a	U .57496 a	U .55624 d	U .52500 d	S .47500 b	U .52500 d	U .55624 d	U .57496 a	U .58747 a
	M	U .87709 f	U .87741 f	U .87782 f	U .87834 f	U .87876 b	U .87834 f	U .87782 f	U .87741 f	U .87709 f
	T	U .99001 b	U .99002 b	U .99002 b	U .99002 b	U .99003 b	U .99002 b	U .99002 b	U .99002 b	U .99001 b
	L	U .47500 a	U .47500 a	U .47500 a	U .47500 a	S .47500 a	U .47500 a	U .47500 a	U .47500 a	U .47500 a

Table 11 -- continued

		1/6	1/5	1/4	1/3	1/2	2/3	3/4	4/5	5/6
E	M	U .47500 a	U .47500 a	U .47500 a	U .47500 a	U .47500 a	U .47500 a	U .47500 a	U .47500 a	U .47500 a
	T	U .47500 a	U .47500 a	U .47500 a	U .47500 a	U .47500 a	U .47500 a	U .47500 a	U .47500 a	U .47500 a
	L	U .58747 a	U .57496 a	U .55624 d	U .52500 d	S .47500 b	U .52500 d	U .55624 d	U .57496 a	U .58747 a
R	M	U .87709 f	U .87741 f	U .87782 f	U .87834 f	U .87876 b	U .87834 f	U .87782 f	U .87741 f	U .87709 f
	T	U .99001 b	U .99002 b	U .99002 b	U .99002 b	U .99003 b	U .99002 b	U .99002 b	U .99002 b	U .99001 b

Table 12. Asymmetric Model 2

Position of B from A

		1/6	1/5	1/4	1/3	1/2	2/3	3/4	4/5	5/6
C	L					F .99999 d				
	M	F 1.0000 a	F 1.0000 a	F 1.0000 a	F 1.0000 a	S .99999 a	F 1.0000 a	F 1.0000 a	F 1.0000 a	F 1.0000 a
	T	S 1.0000* a	S 1.0000* a	S 1.0000* a	S 1.0000* a	S 1.0000* a	S 1.0000* a	S 1.0000* a	S 1.0000* a	S 1.0000* a
	L					F .99999 a				

Table 12 -- continued

		1/6	1/5	1/4	1/3	1/2	2/3	3/4	4/5	5/6
E	M	F 1.0000 a	F 1.0000 a	F 1.0000 a	F 1.0000 a	S .99999 a	F 1.0000 a	F 1.0000 a	F 1.0000 a	F 1.0000 a
	T	S 1.0000 a	S 1.0000 a	S 1.0000 a	S 1.0000 a	S 1.0000 a	S 1.0000 a	S 1.0000 a	S 1.0000 a	S 1.0000 a
						F .99999 d				
R	M	U .87709 f	U .87742 f	U .87782 f	U .87834 f	U .87877 b	U .87834 f	U .87782 f	U .87742 f	U .87709 f
	T	U .99001 b	U .99002 b	U .99002 b	U .99002 b	U .99003 b	U .99002 b	U .99002 b	U .99002 b	U .99001 b

Table 13. Symmetric Model 3

Position of B from A

		1/6	1/5	1/4	1/3	1/2	2/3	3/4	4/5	5/6
	L	U .67500 a	U .67500 a	U .67500 a	U .67500 a	U .67500 a	U .67500 a	U .67500 a	U .67500 a	U .67500 a
C	M	U .67500 a	U .67500 a	U .67500 a	U .67500 a	U .67500 a	U .67500 a	U .67500 a	U .67500 a	U .67500 a
	T	F .90000 a	F .90000 a	F .90000 a	F .90000 a	U .67500 a	F .90000 a	F .90000 a	F .90000 a	F .90000 a
	L	U .67500 a	U .67500 a	U .67500 a	U .67500 a	U .67500 h	U .67500 a	U .67500 a	U .67500 a	U .67500 a

Table 14. Asymmetric Model 3 A
Position of B from A

		1/6	1/5	1/4	1/3	1/2	2/3	3/4	4/5	5/6
	L					F .95000 a				
C	M					F .95000 a				
	T					F .95000 a				
	L	S .94998 a	S .94998 a	S .94998 a	S .94998 a	S .94997 a	S .94998 a	S .94998 a	S .94998 a	S .94998 a

156

Table 13 -- continued

		1/6	1/5	1/4	1/3	1/2	2/3	3/4	4/5	5/6
E	M	U .67500 a	U .67500 a	U .67500 a	U .67500 a	U .67500 h	U .67500 a	U .67500 a	U .67500 a	U .67500 a
	T	U .67500 a	U .67500 a	U .67500 a	U .67500 a	U .67500 h	U .67500 a	U .67500 a	U .67500 a	U .67500 a
	L	U .67500 a	U .67500 a	U .67500 a	U .67500 a	U .67500 a	U .67500 a	U .67500 a	U .67500 a	U .67500 a
R	M	U .67500 a	U .67500 a	U .67500 a	U .67500 a	U .67500 a	U .67500 a	U .67500 a	U .67500 a	U .67500 a
	T	U .67500 a	U .67500 a	U .67500 a	U .67500 a	U .67500 a	U .67500 a	U .67500 a	U .67500 a	U .67500 a

Table 14 -- continued

		1/6	1/5	1/4	1/3	1/2	2/3	3/4	4/5	5/6
E	M	S .94998 a	S .94998 a	S .94998 a	S .94998 a	S .94997 a	S .94998 a	S .94998 a	S .94998 a	S 94998 a
	T	S .94998 a	S .94998 a	S .94998 a	S .94998 a	S .94997 a	S .94998 a	S .94998 a	S .94998 a	S .94998 a
	L					F .95000 a				
R	M					F .95000 a				
	T					F .95000 a				

Table 15. Asymmetric Model 3 B

		1/6	1/5	1/4	1/3	1/2	2/3	3/4	4/5	5/6
C	L					F .95000 a				
	M					F .95000 a				
	T					F .95000 a				
	L	S .94998 a	S .94998 a	S .94998 a	S .94998 a	S .94996 a	S .94998 a	S .98998 a	S .94998 a	S .94998 a

Table 15 -- continued

		1/6	1/5	1/4	1/3	1/2	2/3	3/4	4/5	5/6
E	M	S .94998 a	S .94998 a	S .94998 a	S .94998 a	S .94996 a	S .94998 a	S .94998 a	S .94998 a	S 94998 a
	T	S .94998 a	S .94998 a	S .94998 a	S .94998 a	S .94996 a	S .94998 a	S .94998 a	S .94998 a	S .94998 a
	L					F .95000 a				
R	M					F .95000 a				
	T					F .95000 a				

Table 16. Asymmetric Model 3 C

		1/6	1/5	1/4	1/3	1/2	2/3	3/4	4/5	5/6
	L					F .99999 a				
C	M					F .99999 a				
	T					F .99998 a				
	L					F .99999 a				

Table 16 -- continued

		1/6	1/5	1/4	1/3	1/2	2/3	3/4	4/5	5/6
E	M					F .99999 a				
	T					F .99998 a				
	L					F .99999 a				
R	M					F .99999 a				
	T					F .99998 a				

b: continuous decrease

c: decrease followed by increase

d: decrease followed by increase

e: increase followed by decrease followed by increase

f: decrease followed by increase followed by decrease

g: increase followed by decrease followed by increase followed by decrease

h: no change.

Because of the symmetry of the model and initial gamete frequency, the equilibrium points obtained were symmetric for reproportioning: 1/3 vs. 2/3, 1/4 vs. 3/4, 1/5 vs. 4/5, 1/6 vs. 5/6. When the model was asymmetric, such as asymmetric models 1 A and 1 C, the points of equilibrium were different for these reproportionings.

The Optimum Models

When linkage was loose in $\bar{O}$ = 2.5 and $\bar{O}$ = 2.0, regardless of initial gamete frequencies (coupling, equal, repulsion), the population tended toward a stable equilibrium with a high frequency of gamete ABC. As linkage tightened, M and T conditions, the frequency of gamete ABC started declining and those of gametes ABc, AbC, aBC increased, with AbC the more frequent gamete. In this range of $\bar{O}$ values, the degree of heterozygosity was higher with a smaller $\bar{O}$ value and tighter linkage.

As the value of $\bar{O}$ became 1.8, the system, under loose linkage in conditions C and E, behaved in a manner similar to that of $\bar{O}$ = 2.5 and $\bar{O}$ = 2.0. Under these linkage conditions, with the B locus at the middle, the population reached a stable equilibrium. However, when linkages were reproportioned in such equilibrium populations, the stability disappeared and the populations moved toward unstable equilibria with a high frequency of gamete AbC. With loose linkage and the repulsion condition of initial gamete frequencies, the population started moving toward a stable equilibrium with a high frequency of gamete AbC. However, any reproportioning increased the frequency of this gamete and fixation occurred for alleles A and C.

Under the condition of medium linkage for initial gamete frequencies C and E, the population attained an unstable equilibrium with a high frequency of gametes ABC and AbC. When the initial gamete condition was R, the frequency of AbC became extremely high and fixation occurred for alleles A and C.

With the tight linkage other stable equilibria were reached under any condition of initial gamete frequency and linkage reproportioning. In general, the frequency of gamete AbC was high except in cases where linkage reproportioning was made under the initial conditions of E and R. Respectively in the latter cases, the frequency of ABc tended to build up when the B locus was moved closer to locus C; the frequency of aBC increased when B was moved closer to locus A.

When $\bar{O}$ assumed the value of 1.7, the population was unstable for all initial gamete frequency and linkage distance conditions when the frequency of AbC was less than about 0.70, but stable when the frequency exceeded this point. These stable equilibria, however, were those with near fixation of AbC.

With the optimum value of 1.4, under loose and medium linkage and the coupling initial condition (C), the population approached near fixation of ABc, with only the C locus segregating. Under equal and repulsion initial conditions (E and R), the populations went to near fixation of AbC, with only the B locus segregating. With tight linkage two cases, namely coupling and repulsion initial conditions with the B locus at the middle, resulted in stable equilibria, where ABc and AbC respectively, were high in frequency. All other reproportioning in these two cases resulted in fixation of either ABc in the coupling initial condition or AbC in the repulsion initial condition. Under conditions of equal initial gamete frequencies, the frequency of AbC became high and eventually A and b were fixed with only locus C segregating.

The S-Model 1

For loose and intermediate linkage in all initial gamete conditions, stable equilibrium emerged at the point where all gamete frequencies were equal. This equilibrium became unstable with tight linkage. Under tight linkage, another point of stable equilibrium emerged at relatively high frequencies of gametes ABC, Abc, aBc, abC. This point resulted when perturbations were made away from the unstable equilibrium found with tight linkage and was represented by S* in Table 7. The unstable equilibrium represented a point of linkage equilibrium, while stable equilibrium points were out of linkage equilibrium.

The AS-Model 1 A

Stable equilibrium was reached with a high frequency of ABC and ABc for loose and medium linkage conditions in all initial gamete frequency inputs. Reproportioning of linkage had little effect on this situation. With tight linkage, gametes ABc and abc attained a high frequency and $\overline{W}$ improved considerably for C and R initial gamete frequencies. With initial gamete frequency E and tight linkage, the population reached a state of unstable equilibrium with gametes ABC and ABc in high frequency. Undoubtedly a perturbation away from this unstable point would lead to a stable equilibrium similar to that attained in the cases of C and R initial conditions.

The AS-Model 1 B

In all the initial conditions associated with loose linkage, the frequency of ABC became high at the point of stable equilibrium. As the linkage distance was tightened (M and T conditions), the gamete AbC tended to increase in frequency at the expense of gamete ABC and became the most frequent gamete at the tight linkage condition. Reproportioning of the given A-C linkage distance in any initial gamete condition increased the frequency of AbC. All the equilibria reached in this model were stable.

The AS-Model 1 C

The results of this model were a mirror image of AS-Model 1 A. This can be observed by comparing Tables 8 and 10.

The S-Model 2

In C, E, and R initial conditions with loose linkage and the locus B at the middle, a stable equilibrium emerged at the point of all gametic frequencies being equal. In the coupling situation, reproportioning of 1/3, 1/4, 1/5, 1/6 tended to build up the frequencies of gametes ABC, ABc, abC, abc, while reproportioning of 2/3, 3/4, 4/5; 5/6 increased the frequencies of ABC, aBC, Abc, abc. Reproportioning in the equal initial condition sent the population to an unstable equilibrium of equal gamete frequencies. In the initial condition of repulsion, reproportioning of 1/3, 1/4, 1/5, 1/6 caused an increase in the frequencies of AbC, aBC, Abc, aBc. Reproportioning of 2/3, 3/4, 4/5, 5/6 increased the frequencies of ABc, AbC, ABc, abC. All the points of equilibrium attained with reproportioning were unstable.

In medium and tight linkage associated with the equal initial gamete frequency condition, an unstable equilibrium was reached with the B locus. This unstable equilibrium was at the point of equal gamete frequencies. Unstable equilibria were quickly reached with a high frequency of gametes ABC and abc in the coupling condition and a high frequency of AbC and aBc in the repulsion condition. Any perturbation of these unstable points would result in the fixation of any one of the gametes, depending upon which gamete was increased in the perturbation.

The AS-Model 2

For the initial conditions of C, E, and R associated with loose linkage, the fixation of gamete types ABC, AbC, and aBc occurred, respectively.

The medium linkage condition with the locus B at the middle was stable for the initial conditions C and E and unstable for R. The stable equilibria occurred near the fixation of ABC in the case of coupling and of aBc in the case of equal initial gamete frequencies. These fixation points were evidenced in the reproportioning situations of these two cases. The reproportionings in the case of repulsion tended to break down the high frequency of AbC and aBc, but the equilibrium remained unstable. A perturbation of the gamete frequencies at this unstable point would effect the fixation of any one of the gamete types other than gamete abc because the survival value of $W_{000} = 0.10$ while all other homozygotes W value were equal to 1.00.

With tight linkage, stable points of equilibrium existed with near fixation of ABC for the coupling initial condition and aBc for the equal frequency initial condition. Any reproportioning did not alter this situation. In the repulsion case, a high level of gametes AbC and aBc was maintained at an unstable equilibrium. Reproportioning had no appreciable effect on this picture. However, perturbation of the gamete frequencies at this point would probably effect the fixation of any one of the gamete types except abc.

The S-Model 3

There was an unstable center point to which the populations converged for all the initial gamete frequencies and given linkage distance with B at the middle. In general, any reproportioning resulted in the convergence of the population to an unstable central point, except for the case of tight linkage in the coupling initial condition where the fixation of gametes abC and abc was evidenced.

It is most likely that some perturbation from the center unstable point would lead to the fixation of alleles at a pair of loci.

The AS-Model 3 A

When the population was started with equal initial gamete frequencies, conditions L, M, and T and associated reproportionings lead the population to stable equilibria which were located very close to the fixation of the C allele. At this point, the gametes which did not contain the c allele were equally represented with a frequency slightly less than 0.25. With coupling and repulsion as initials, the population moved to the fixation of the C allele, keeping A and B loci segregating. The fixation of the C allele occurred because preference was given to its survival value. In this model, the fixation of a certain allele depends on which allele is given a survival advantage.

The AS-Model 3 B

In this model the situation was the same as in the AS-Model 3 A.

The AS Model 3 C

In all the cases, the fixation of c occurred with gamete Abc in high frequency. An advantage for the survival of c resulted in the loss of the C allele and the continued segregation of the A and B loci.

CONCLUDING REMARKS

Several observations of a general nature were made in this study:

(1) A three-locus genetic population was of extreme complexity even under fairly simplified conditions, and this made analysis extremely difficult.

(2) A tighter linkage resulted in a higher $\overline{W}$ than the looser linkage.

(3) The existence of stable equilibrium was dictated to a great extent by the fitness models and to a lesser extent by linkage relations among loci.

(4) The mean fitness value of a given population usually moved up to a certain level, and then might go through various combinations of ups and downs under a given set of conditions, indicating that the mean fitness is only a gross measure of determining population adaptability.

(5) The mean fitness changes associated with linkage reproportionings were very slight, and for this reason optimization of linkage relations does not seem to be affected by the mean fitness value changes. Any other first order effect such as the existence of linkage modifying factors must be of primary importance for the adjustment of linkage relations.

Acknowledgement

A part of this research was financed by Public Health Service Grant GM-15769.

REFERENCES

Bodmer, W. F. and Felsenstein, J. 1967. Linkage and selection: theoretical analysis of the deterministic two locus random mating model. Genetics 57:237-265.

Fisher, R. A. 1930. The genetical theory of natural selection. Oxford University Press, 2nd ed. 1958. Dover.

Kimura, M. 1956. A model of a genetic system which leads to closer linkage by natural selection. Evolution 10:278-287.

Kojima, K. 1959. Stable equilibria for the optimum model. Proc. Nat. Acad. Sci. 45:989-993.

_________ 1965. The evolutionary dynamics of two-gene systems. In Computers in Biomedical Research, vol. 1: 197-220. Academic Press, N. Y.

Kojima, K. and Kelleher, T. M. 1961. Changes of mean fitness in random mating population when epistasis and linkage are present. Genetics 46:527-540.

Lewontin, R. C. 1964a. The interaction of selection and linkage I. General considerations: heterotic models. Genetics 49:49-67.

_________ 1967b. The interaction of selection and linkage II. Optimum models. Genetics 50:757-782.

Lewontin, R. C. and Kojima, K. 1960. The evolutionary dynamics of complex polymorphisms. Evolution 14:457-472.

Turner, J. R. G. 1967. Why does the genotype not congeal? Evolution 21: 645-656.

Wright, S. 1935. Evolution in populations in approximate equilibrium. J. Genet. 30:257-266.

DISCUSSION ON DETERMINISTIC EQUILIBRIUM

A. W. F. Edwards

EDWARDS: In supposing that because a tightening of linkage is often accompanied by a raising of the mean fitness, natural selection would in fact lead to tighter linkage, Kojima is using a weak argument. I presume he has in mind some modifier loci capable of influencing recombination, but it is far from clear that genes favoring closer recombination would in fact be selected for simply because of a higher mean fitness. The selective pressures for this are second-order, and even with first-order situations the mean fitness can decrease under selection, as noted by Kojima with one of the three-locus models, and as pointed out by Moran for a particular two-locus model. Mean fitness arguments, though appropriate for interpopulation competition and for intrapopulation competition under very restricted conditions, should not be relied upon in general.

KOJIMA: I do agree with Dr. Edwards' comment.

REMARKS ON EQUILIBRIUM CONDITIONS IN CERTAIN TRIMORPHIC, SELF-INCOMPATIBLE SYSTEMS

P. T. Spieth and E. Novitski
Biology Department, University of Oregon

Self-incompatibility in mating systems is an extremely widespread phenomenon, the most familiar example of which is dimorphic sex with its two phenotypic classes, males and females. When sexual dimorphism is genetically determined, the two sexes will tend to be equally frequent (Fisher, 1930). This numerical equality of self-incompatible phenotypes has been termed "isoplethy" (Finney, 1952). In the simplest case, where dimorphic sex is genetically based on heteromorphic sex chromosomes or allelic heterozygosity, it can readily be shown that a state of stable equilibrium exists when the two sexes are equally frequent.

When there are more than two self-incompatible types, however, an analysis of the mechanisms and conditions governing the phenotypic and genotypic frequencies presents unexpected complications. The trimorphic case, superficially only slightly more involved than the dimorphic, is more than an interesting extension of the latter. Indeed, it may present intractable analytic problems (Fisher, 1941; Moran, 1962). Examples of this type include *Oxalis rosea* and *Lythrum salicaria*.

The experimental determination of the genetic basis for trimorphism has been accomplished in only a few instances. This paper takes the theoretical approach suggested initially by Fabergé and developed by Finney (1952), of postulating various related genetic formulations and determining their mathematical consequences. An essential feature of such models is that gene frequencies usually do not remain fixed from one generation to the next. In a trimorphic system the effective contribution that a given type makes to the next generation may depend not on its own frequency alone, but on its relative value with respect to the second type in mating with the third. Thus if the frequencies of types A, B and C are .01, .04 and .95, respectively, then twenty per cent of the matings with C will be by A (and eighty per cent by B). Clearly this will magnify the genetic contribution of A, and the genes characterizing A will increase in frequency. On the other hand, if the respective frequencies are .01, .49, and .50, then only two per cent of the matings with C will be by A, and there will not be such a significant magnification of A's contribution.

These changes in gene frequencies may very likely cease after an indefinite number of generations; at this time the population is considered to be in equilibrium. It is the relation of the genetic formulation hypothesized to the equilibrium values eventually attained (if any) that occupies our attention in this paper. For example, one very common result at equilibrium is equality of the three phenotypes (isoplethy). Such a numerically simple result suggests that an extremely simple relation must exist within the genetic formulation. Naively one might suspect that isoplethy (1:1:1) could be explained as simply as Mendel's segregation ratios.

Despite its commonness and apparent simplicity there is as yet no way to predict from the genetic formulation whether or not a particular system will be isoplethic. Consider the following two hypothetical systems which are both based upon genetic formulations involving two loci with two alleles at each locus:

SYSTEM 1

Phenotype 1	Phenotype 2	Phenotype 3
aabb	aab'b'	a'a'bb
aabb'	aa'b'b'	a'a'bb'
aa'bb		a'a'b'b'
aa'bb'		

SYSTEM 2

Phenotype 1	Phenotype 2	Phenotype 3
aabb	aab'b'	a'a'bb
aabb'	a'a'bb'	aa'b'b'
aa'bb		a'a'b'b'
aa'bb'		

These two systems are identical except for the underlined genotypes which are reversed in System 2 from the way they are assigned in System 1. Nevertheless System 1 is isoplethic, whereas System 2 has an equilibrium in which the phenotypic frequencies are .332, .296 and .373, respectively. Contrarily totally dissimilar genetic formulations may share the common attribute of isoplethy at equilibrium.

Determining necessary conditions for isoplethy is one of the most intriguing aspects of the self-incompatibility problem. Finney (1952) considered the general case of all polymorphic self-incompatibility systems based upon formulations involving one locus with multiple alleles. In that context he presented a limited theorem specifying sufficient conditions for isoplethy. However, Cotterman (personal communication) reports having found counter-examples. In any case, the mechanisms controlling isoplethy still remain an unanswered challenge.

It is evident that there are several levels of complexity involved in the total problem, of which isoplethy is only one interesting aspect. Self-incompatibility is a property of phenotypes; the ability of two individuals to mate and produce viable offspring depends upon whether or not they have the same phenotype. In contrast, genotypes determine phenotypes and determine what the genotypes (and phenotypes) of the next generation will be. Therefore, if one is to formulate an incompatibility system meaningfully, one must specify: (1) the number of different, self-incompatible phenotypes; (2) the genetic basis for the system in terms of loci, alleles and chromosomes involved in determining the phenotypes; and (3) a scheme or list indicating which phenotype is produced by each of the possible genotypes. For example, saying that man has two sexes specifies a dimorphic incompatibility, but genetic predictions cannot be made without some knowledge of the basis for sex determination. More typically, a system is specified by postulating the number of loci and alleles involved and then depicting the genetic formulation in a chart such as the one used above.

One of the things that makes the question of trimorphic incompatibility so interesting is that point one (above) is known in a number of natural populations, but points two and three are only vaguely known or guessed at. A principal question underlying any study of self-incompatibility is whether or not an equilibrium state exists, and, if so, what the genotypic and phenotypic frequencies will be at equilibrium. These all depend upon the genetic formulation underlying the particular system being considered. From the above discussion of what defines a system it is clear that for a purely theoretical approach the number and kinds of hypothetical formulations available are limited only by the imagination of the investigator.

Our approach involves a comparative study of a class of related hypothetical systems. By observing correlations between systems and their respective equilibrium points, possibly some generalizations can be made relating a system to its equilibrium point. In particular we have considered the class of all trimorphic systems determined by two unlinked loci with two alleles at each locus. In other words, points one and two (above) have been fixed and are identical for all systems under consideration. The only variate among members of the class is point three. Table 1 lists the nine possible genotypes. In this context a "system" consists of any particular way of assigning one of the phenotypes to each of the genotypes. Systems 1 and 2 described earlier are both examples of this class.

TABLE 1

Initial Genotypic Frequencies

	Genotype	First Run	Second Run
1.	aabb	.1	.005
2.	aabb'	.1	.200
3.	aab'b'	.1	.475
4.	aa'bb	.1	.005
5.	aa'bb'	.2	.010
6.	aa'b'b'	.1	.295
7.	a'a'bb	.1	.001
8.	a'a'bb'	.1	.003
9.	a'a'b'b'	.1	.006

From a practical point of view, the first consideration in making such a comparative study is determining the equilibrium values for a given system. In principle this is no problem, but in actual practice obtaining explicit numerical equilibrium values becomes difficult. Consider the very simple system based upon one locus with two alleles, a and a'. Suppose there are three phenotypes, A, B, and C, that are produced by the genotypes aa, aa', and a'a', respectively. There is no direct method available for obtaining the equilibrium values. However, from mendelian principles it is clear that in a cross between A (aa) and B (aa') half the progeny will be type A and half will be type B. Likewise, the results of the other possible crosses can be expressed mathematically in terms of what types of progeny will be produced. From the earlier example it was shown that the relative frequency of viable crosses can be computed from the relative frequencies of the three phenotypes. So it is clearly possible to write an equation for each genotype that will mathematically express its relative frequency in the next generation in terms of the relative frequencies of all the genotypes in the current generation.

In general this sort of procedure yields a set of equations, one for each genotype in the system. Such equations do not give direct information about equilibrium values, but they do provide mathematical expressions for the amount of change in the genotypic frequencies from one generation to the next. If equilibrium is defined as that state where the genotypic frequencies do not change from generation to generation, then such a set of equations provides an indirect method for determining equilibrium values.

There are primarily two ways of utilizing such equations. The first way is to handle them with direct mathematical analysis. If the equations are written in a form that expresses the change in genotypic frequencies, then they constitute a set of simultaneous equations which must then be solved algebraically for the values of the genotypic frequencies which will result in all the changes being zero.

In simple systems this method is ideal. For instance Finney used it to show that the system in the last example is at equilibrium when A, B, and C have the frequencies 0.219, 0.562, and 0.219, respectively. Two loci with two alleles at each locus systems are only slightly more complicated than one locus multiple allele systems; however, the added complication is enough that in general it becomes very difficult if not impossible to solve the simultaneous equations.

For specific systems the method has been used successfully. Fisher (1941) used it to find equilibrium values for a particular model for *Lythrum salicaria*. However, this model was simpler than general systems since three genotypes were always zero. Furthermore, Fisher solved it in an indirect manner by making the a priori assumption that the system was isoplethic and then by making an appropriate substitution to simplify the equations.

Since there are nine genotypes there are usually nine simultaneous equations to be solved. Each equation is a sum of terms that are the product of the frequency of the cross with the product of two gametic frequencies. Consequently each of the nine equations is cubic or quadratic. (See the next section for an example of such an equation.) Finding a general solution to a set of so many non-linear simultaneous equations is usually impossible. Even the assumption of isoplethy does not necessarily simplify the equations enough to make the system solvable (Moran, 1962). As a rule of thumb, solutions can only be found when there is some simplifying factor in the system such as symmetry or the total lack of some genotypes as in Fisher's system.

The second method for finding equilibrium values of a particular system is by computer simulation. Since the equations for the system give the values of the genotypes in terms of the values in the preceding generation, a computer is ideally suited to calculate the frequencies in generation after generation. The only requirement is that some initial values for the genotypes must be supplied to begin the process. After computing a number of such simulated generations, if the genotypic frequencies cease changing, then it can be assumed that an equilibrium has been reached.

This method is less general than the first in that there is no guarantee that the simulation will ever reach an equilibrium state, even if one exists for the system. Or if it does reach one, there is no assurance that there are not other equilibria. Perhaps the greatest limitation to computer simulation is that it provides answers without generating any mathematical insights. For instance, the knowledge that a particular system is isoplethic does not by itself provide any further understanding of the conditions necessary for isoplethy.

Despite all its drawbacks, computer simulation has several overwhelming advantages. The most important is that it will generally provide equilibrium values, whereas direct algebraic solutions often are not possible. Furthermore, it is practical. The equations for computing one generation from the previous one can usually be programmed in sufficiently broad scope so as to allow one program to be utilized for a large class of hypothetical systems, such as all two loci, two alleles per locus systems. Finally, with the aid of devices to speed up the amount of time needed to reach equilibrium, equilibrium values for any given system can usually be obtained in an amount of time that is reasonable by modern high-speed computer standards. For these reasons computer simulation was our method of choice.

Before getting into our specific study, there is one further introductory point that should be mentioned. A fundamental distinction must be made between two alternative mechanisms of self-incompatibility. In the case of a species with two sexes it is clearly impossible for two sperm or two eggs to unite to form a zygote. In other words, "wrong" crosses never occur. The analogous situation in plants is the case where pollen produced by a plant of phenotype X fails to develop a pollen tube on the style of any plant with phenotype X. Such incompatibility mechanisms are known as gamete competition systems, the "wrong" gametes always losing in the competition for fertilization. Finney (1952) and Moran (1962) have referred to this method as pollen elimination, since they were thinking in terms of models for specific organisms. We prefer the term gamete competition because it is more general. It is just as reasonable to suppose that all types of gamete unions occur, regardless of the phenotypes of the individuals producing the gametes, but that zygotes resulting from incompatible crosses abort. This alternative situation is known as zygote elimination. The distinction between these two alternative mechanisms is necessary because they require different mathematical formulae to express the relative frequencies of each of the viable crosses. Moreover, it is only in isoplethic systems that the two formulae give the same numerical results. The derivations for each type of formula were first given by Finney (1952).

METHOD OF ATTACK

As already mentioned, our method of attacking the problem has been to make a comparative study of a single class of systems by using computer simulation to find equilibrium values for every distinct hypothetical trimorphic system based on two unlinked loci with two alleles at each locus.

The following is a specific example of one possible system:

Phenotype 1	Phenotype 2	Phenotype 3
aabb - p	aab'b' - t	a'a'bb - v
aabb' - q	aa'b'b' - u	a'a'bb' - w
aa'bb - r		a'a'b'b' - x
aa'bb' - s		

where p, q, r, etc. denote the various genotypic frequencies.

The recursion formulae for gamete competition were used since they are probably more typical of the actual situation in nature. If the three phenotypic frequencies are given by:

$$P_1 = p + q + r + s$$

$$P_2 = t + u$$

$$P_3 = v + w + x$$

then the weight factor for a cross between members of phenotypes i and j is giver by:

$$F_{ij} = \frac{1}{1 - P_i} + \frac{1}{1 - P_j} \quad \text{for } i \neq j$$

$$= 0 \quad \text{for } i = j$$

For a complete derivation see Finney, 1952. Letting a prime denote a succeeding value, the recursion formulae for this particular system become:

$p' = 0$ (Note that the gamete ab is produced only by Phenotype 1.)

$q' = (p + q/2 + r/2 + s/4)(t + u/2)F_{12}$

$r' = (p + q/2 + r/2 + s/4)(v + w/2)F_{13}$

$s' = (p + q/2 + r/2 + s/4)(u/2)F_{12} + (p + q/2 + r/2 + s/4)(x + w/2)F_{13} + (q/2 + s/4)(v + w/2)F_{13} + (t + u/2)(r/2 + s/4)F_{12} + (t + u/2)(v + w/2)F_{23}$

$t' = (q/2 + s/4)(t + u/2)F_{12}$

$u' = (q/2 + s/4)(u/2)F_{12} + (q/2 + s/4)(x + w/2)F_{13} + (t + u/2)(s/4)F_{12} + (t + u/2)(x + w/2)F_{23}$

$v' = (r/2 + s/4)(v + w/2)F_{13}$

$w' = (r/2 + s/4)(u/2)F_{12} + (r/2 + s/4)(x + w/2)F_{13} + (v + w/2)(s/4)F_{13} + (v + w/2)(u/2)F_{23}$

$x' = (s/4)(u/2)F_{12} + (s/4)(x + w/2)F_{13} + (u/2)(x + w/2)F_{23}$

These formulae are obtained by simply using mendelian ratios to calculate the frequencies of the various different gametes in each phenotypic class, and then enumerating all the different gamete combinations that will result from all the different phenotype pairings.

Taking the frequencies of the genotypes to be those given by the column titled "First Run" in Table 1, it is then relatively easy to compute the new frequencies of the population after one generation. First calculating the phenotypic frequencies and the weight factors gives:

$P_1 = .5 \qquad P_2 = .2 \qquad P_3 = .3$

$F_{12} = 3.25 \qquad F_{13} = 3.43 \qquad F_{23} = 2.68$

Therefore direct calculations may be made by the above recursion formulae to get:

$p' = 0$

$q' = (.25)(.15)(3.25) = .122$

$r' = (.25)(.15)(3.43) = .129$

$s' = (.25)(.05)(3.25) + (.25)(.15)(3.43) + (.1)(.15)(3.43) + (.15)(.1)(3.25) + (.15)(.15)(2.68) = .330$

$t' = (.1)(.15)(3.25) = .049$

$u' = (.1)(.05)(3.25) + (.1)(.15)(3.43) + (.15)(.05)(3.25) + (.15)(.15)(2.68) = .152$

$v' = (.1)(.15)(3.43) = .051$

$w' = (.1)(.05)(3.25) + (.1)(.15)(3.43) + (.15)(.05)(3.43) + (.15)(.05)(2.68) = .114$

$x' = (.05)(.05)(3.25) + (.05)(.15)(3.43) + (.05)(.15)(2.68) = .054$

Continuing on in the same fashion for generation after generation the differences between succeeding genotype frequencies will eventually become negligible and the frequencies will be those given in Table 2. It is interesting to note that this particular system is isoplethic. It is further worthwhile to note that if the frequencies given in Table 1 under the heading "Second Run" had been used instead, the same final results would have been obtained.

TABLE 2

Final Results of Example System

Genotypes	Final Frequencies		Phenotypes
aabb	.00	.33	Phenotype 1
aabb'	.06		
aa'bb	.03		
aa'bb'	.24		
aab'b'	.05	.33	Phenotype 2
aa'b'b'	.28		
a'a'bb	.02		
a'a'bb'	.14	.33	Phenotype 3
a'a'b'b'	.17		

A final remark about this example concerns the recursion formulae at isoplethy. When all the phenotypic frequencies are equal (one-third), the weight factors all become 3.00. Even so it will be noted that the recursion formulae do not simplify.

Clearly these equations are not in general linear, and solving them algebraically for fixed points (i.e., for values of p, q, r, etc. where p' = p, q' = q, r' = r, etc.) is not practical. The equations are, however, ideally suited for fixed point iteration on a computer. One simply programs the equations into the computer, reads in some initial values for the genotype frequencies, and lets the computer calculate succeeding values until either there is no appreciable change in the values or a pre-set maximum number of iterations has been made.

An inherent problem with such computations is that initial values must be supplied and there is no **a priori** assurance that the final results would be the same with different sets of initial values.* For this reason two different sets of initial values were used. (Table 1.)

The example given above is only one of many ways in which the nine genotypes can be assigned to the three phenotypic classes. It is perhaps not obvious that producing a different system by permuting the phenotype names or the allele symbols or the loci symbols will not alter the results. For instance, if in the above example those genotypes constituting Phenotype 1 were instead considered to constitute Phenotype 2 and vice versa, then no real difference should occur in the final results. Likewise, if "a" is interchanged with " a' " in all the genotypes, the relative results would still not be changed. In other words, two such systems are genetically equivalent; only the symbols used are different.

*As will be seen, systems with more than one equilibrium point are not uncommon. In even the simplest of such systems, complex problems arise in relating the possible initial values to the various possible equilibria points. See Cormack (1964) for a mathematical discussion of the problems involved.

The problem of counting and listing the genetically distinct systems constitutes an interesting sidelight. It was primarily solved by having a computer generate all possible systems, discarding those that are genetically equivalent to any system already listed. A total of 439 systems was generated. A further discussion on generating the systems and the verification of the total number is presented as an appendix.

The main program was written in Fortran IV and run on an IBM 360/50 and a CDC 3600. Generalized recursion formulae similar to those given above were used, and a convergence accelerator was employed. The latter consisted of passing to the limit of a geometric series whenever the changes in the genotypic frequencies over three generations appeared to be following the pattern of a convergent geometric series. Each system was run twice; once with each of the two sets of initial conditions. A maximum of 150 generations (iterations) was allowed for each run. With the aid of the convergence accelerator all but three systems converged on at least one run. Convergence was defined as occurring when the absolute change in every genotypic frequency was less than 10^{-6}.

GENERALIZED RESULTS

It was anticipated from Finney's work that the results obtained from the simulator would be diverse. They were. Of the 439 different systems run, three failed to converge on either of the two runs; 85 others converged on only one of the two runs. Of the remaining 351 systems, 187 had both runs converge to the same genotypic frequencies. The 787 <u>runs</u> that converged may be classified by their final phenotypic frequencies:

1. 410 runs were isoplethic.
2. 131 runs degenerated into dimorphic systems.
3. 61 were balanced-anisoplethic.
4. 185 went to some other result.

Balanced-anisoplethy is used here to denote the situation where the final frequences were .33; .33 - x; and .33 + x, where $.01 \leq x \leq .07$. There is a degree of doubt as to the integrity of this group of results since the pragmatic criteria used to determine when a run had converged to equilibrium does not necessarily imply mathematical convergence. However, several arbitrarily selected systems were allowed to run for four or five thousand generations. Their final genotypic frequencies were within ± .01 of what they had been after only 100 generations, so we have a certain amount of confidence in our results.

Some order and insight may be gained by considering the results in terms of the diagram presented in Figure 1. In the diagram the nine genotypes are arranged in a three-by-three array. The homozygotes are located in the corners and the double heterozygote in the center. The three genotypes lying on any given edge of the diagram all are homozygous for one of the four alleles. A set of three such genotypes is referred to as a Homozygous Allele Set.

A second class of sets is of interest. A set composed of any homozygote, its two adjacent heterozygotes and the double-heterozygote has the property that there is one gamete that is produced by every genotype in the set; moreover, it is produced only by members of the set. Such a set is called a Common Gamete Set. If the double-heterozygote is excluded from the set it is referred to as a Restricted Common Gamete Set.

In terms of the two types of sets just described a broad generalization of the simulator results can be made. A large

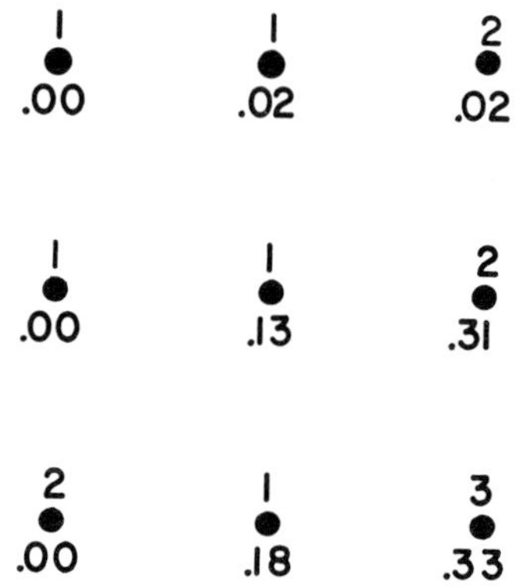

FIGURE 1

majority of the runs can be described as going to an equilibrium state in which one of the four Common Gamete Sets is decisively dominant and one of the opposing Homozygous Allele Sets is highly reduced or even completely eliminated. Dominance and reduction are used here to denote relative frequencies. Each genotype in a dominant Common Gamete Set generally has a final frequency in the range of .11 to .33, while one in a reduced Homozygous Allele Set generally has a final frequency in the range of .001 to .01.

There are frequent exceptions to this generalized, qualitative outcome. One of the major exceptions is symmetry. If the genotypes are assigned to phenotypes in a manner that is symmetric when viewed in terms of the diagram, the genotypic equilibrium frequencies may also be symmetric and not fit the generalized description. Other exceptions are exhibited by systems such as ones in which (1) all genotypes other than one Restricted Common Gamete Set are eliminated; (2) only part of a Common Gamete Set dominates; and (3) a Homozygous Allele Set dominates. This last case is rare.

GENOTYPE ELIMINATION

One of the first questions to need answering is when and why are certain genotypes completely eliminated in some systems? Analysis of the diagram in Figure 1 provides the key to this question and makes it possible to state certain conditions necessary for a genotype to be eliminated.

The lines connecting the various genotypes in Figure 1 represent paths along which gene flow can occur in passing from one generation to the next. A single path length is defined as a line connecting two genotypes with no genotype lying in between. Gene flow occurs from parent to offspring and can never travel more than one path length at a time. Generally the possible genotypes of a particular mating are those that are one or less path lengths removed from each parental genotype. For instance, in a cross of aabb with a'a'bb' the possible offspring are aa'bb and aa'bb'. The only exception to this generalized rule is that a Homozygous Allele Set is closed, which means that the only possible offspring from a mating of two members of the set must also be in the set, in addition to being not over a single path length removed from each parent.

If a genotype's neighborhood is defined as being the set of all genotypes that are not over one path length away from the given genotype, then it is clear that a given genotype can be produced only by parents in its neighborhood. When the requirements of a self-incompatible mating system are considered, this observation implies the following conclusion: a genotype will be eliminated and remain at a frequency of zero whenever all non-zero genotypes in the neighborhood are members of the same phenotypic class.

In the case of a homozygous genotype the neighborhood is precisely the Common Gamete Set to which it belongs. If the entire set belongs to just one phenotypic class, the homozygote will be eliminated. This is trivial since in this case the gamete for which the homozygote is homozygous is produced only by members of the one phenotype, and can never meet itself in any compatible mating. The situation in which a genotype belongs to one phenotypic class and all other non-zero members of its neighborhood belong to a second phenotypic class is not trivial. The only compatible crosses that will produce the genotype in question are crosses involving the genotype itself. Since there must be a third phenotype that is not in the neighborhood under consideration, there is at least one cross involving the given genotype that will not "plow back" gene flow into itself. Furthermore, since the double heterozygote belongs to every neighborhood, mendelian considerations dictate that there is at least one cross which will give less than total plow back. Therefore, unless (1) the system degenerates into a dimorphic situation and (2) the double heterozygote is eliminated, the frequency of such an isolated genotype will be monotonically decreasing from generation to generation until it is eventually eliminated.

A corollary to the neighborhood requirement for genotype elimination is that the remaining set of non-eliminated genotypes must be closed in the sense defined above for Homozygous Allele Sets. There are three possible kinds of closed sets. As previously mentioned, any Homozygous Allele Set is closed regardless of the system assigning the genotypes to phenotypes. A second case of a closed set is the set left after the elimination of an isolated genotype as discussed above. This case is dependent upon the system; it can only occur in a system that has an isolated genotype. The third kind of closed set is a Restricted Common Gamete Set in which the homozygote belongs to one phenotype and the two heterozygotes belong to a second. Again this case is dependent upon the system, since the two heterozygotes must be in the same phenotypic class.

A final consideration on closed sets is the fact that the second case given above can be extended to the situation where more than just one genotype is isolated. In particular, if the three genotypes that lie along the center column (or the center row) of Figure 1 are all in the same phenotypic class, then the two adjacent Homozygous Allele Sets are isolated from each other. This particular situation creates an unstably balanced system. Clearly both of the Homozygous Allele Sets cannot be eliminated, for then only the monomorphic center column (row) would remain. Therefore, the determination of which one is eliminated and which one is not, is dependent upon the configuration of the rest of the system and upon the initial frequencies of the genotypes.

ILLUSTRATIVE EXAMPLE

To add clarity and appreciation to the above principles an illustrative system is presented in Figure 2. It was not chosen as being "ideal" or "typical," rather it is a composite system that exemplifies most of the principles just discussed. The genotypes are represented by dots in the same relative positions as in Figure 1. The integers above each genotype indicate the phenotypic class to which it is assigned by the system. The decimal numbers below each genotype are the final equilibrium frequencies attained on the first run.

One of the primary items to be noted is that the Common Gamete Set containing a'a'b'b' is strongly dominant; another is that aabb was trivially eliminated; aab'b' is isolated and was also eliminated. With the elimination of these two homozygotes, aabb' was then left isolated and it too was eliminated. In fact, the entire Homozygous Allele Set was isolated to start with in the manner described at the end of the above section.

The final significance of this example is that it illustrates the effect of strongly unbalanced initial frequencies. By comparing Table 1 and Figure 1 it will be noted that on the second run the Restricted Common Gamete Set containing aab'b' contained 97 per cent of the initial frequencies. In this particular system that set is closed. As a result, on the second run the system eliminated all genotypes not in the set, degenerating into a dimorphic situation in 15 generations: aab'b' had a final frequency of .50 and aabb' and aa'b'b' had frequencies of .20 and .30, respectively.

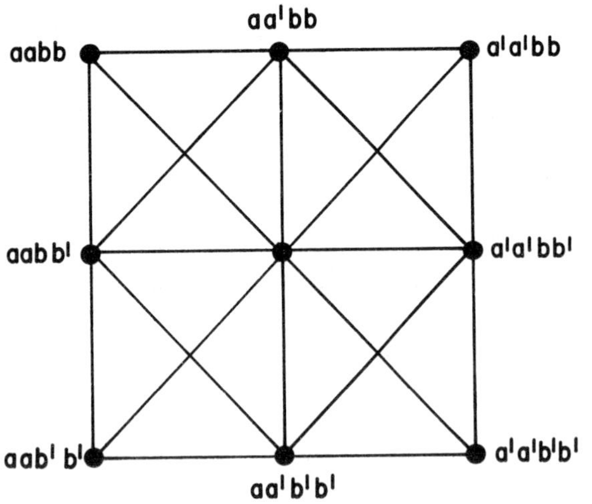

FIGURE 2

SUMMARY

The study presented in this paper deals with the equilibrium states of trimorphic self-incompatible mating systems determined by two loci with two alleles per locus. In the introduction, considerable emphasis has been placed on the theoretical challenge of self-incompatible mating systems in general. Using the formulae for trimorphic incompatibility systems derived by Finney (1952), a computer simulation program was written to generate the equilibrium states of all possible two locus-two allele systems A second program enumerated all the genetically distinct ways by which the nine genotypes can be assigned to three non-empty phenotypic classes. Each of the 439 such systems was run through the simulator twice, using different initial genotypic frequencies for the two runs. With the aid of a convergence accelerator, all but three systems converged to equilibrium within 150 generations.

A comparison of the equilibria revealed that isoplethy, equality of the three phenotypic classes, was reached in roughly half of the runs. In slightly over fifty per cent of the systems, the choice of initial frequencies made a difference in the equilibria attained. Figure 1 presents a schematic representation of gene flow between the genotypes and can be used to provide an understanding of the relative distribution of weight among the genotypes at equilibrium. It further provides a basis for stating necessary conditions for the complete elimination of one or more genotypes.

The concepts presented here constitute merely a preliminary, qualitative insight into the mechanisms and conditions governing equilibrium states in self-incompatibility systems. The complete requirements necessary for isoplethy are still under investigation.

Acknowledgement

This work was supported by NIH grant 2T1GM373 and NSF grant GB7767. The authors wish to acknowledge the generosity of the Commonwealth Scientific and Industrial Research Organization in making available its CDC computer for this work.

REFERENCES

Cormack, R. M. 1964. A boundary problem arising in population genetics. Biometrics 20:785-793.

Cotterman, C. W. 1953. Regular two-allele and three-allele phenotype systems. I. Amer. J. Hum. Gen. 5:193-235.

Finney, D. J. 1952. The equilibrium of a self-incompatible polymorphic species. Genetica 26:33-64.

Fisher, R. A. 1930. The Genetical Theory of Natural Selection. Clarendon Press, Oxford.

Fisher, R. A. 1941. The theoretical consequence of polyploid inheritance for mid style form of Lythrum salicaria. Ann. Eugen. 11:31-38.

Hartl, D. L. and T. Maruyama, 1968. Phenogram enumeration: the number of genotype-phenotype correspondences in genetic systems. J. Theoret. Biol. 20:129-163.

Moran, P. A. P. 1962. The Statistical Processes of Evolutionary Theory. Clarendon Press, Oxford.

APPENDIX I
The Generation of Genetically Unique Systems

The first step that had to be taken before it was possible to make a comparative study of all possible trimorphic self-incompatible mating systems based upon two loci with two alleles per locus was the enumeration of all the genetically distinct ways in which the nine genotypes can be arbitrarily assigned to one of the three phenotypic classes. At first glance there appear to be 3^9 or 19,683 such arrangements; fortunately, however, the requirement of genetic distinctness imposes several restrictions which reduce this number to the more workable one of 439. This latter figure was first obtained by the program to be described and was subsequently verified mathematically. In addition the program provided a listing of the 439 systems, which was then used as input for the simulator program designed to seek out equilibrium states.

The first restriction is that all three phenotypes be represented; that is, the system must be trimorphic. The second is that the phenotypic grouping be unique, and not a simple permutation of some other grouping. The third, and most worrisome, is that any given system should not be the genetic equivalent of some other system already considered. The nature of these problems and the simple but effective way in which they were handled will become clearer as the program itself is considered.

Let the genotypes be ordered arbitrarily as in Table 1. Next let a given genetic system be represented by a nine digit "system number" in which the digits correspond, in order from left to right, with the nine genotypes, and the value of each digit indicates which phenotype the corresponding genotype produces. For instance, the system number 111122233 represents the system in which the first four genotypes (in Table 1) produce Phenotype 1; the next three, Phenotype 2; and the last two, Phenotype 3. Therefore, the 3^9 total is given by all numbers from 111111111 to 333333333 in which each digit may take on the value 1, 2, or 3. Clearly all possible numbers within these limits can be enumerated one at a time by beginning with the lower limit and successively adding 1 to the rightmost digit, with the stipulation that 1 plus 3 gives 11 (i.e., the next digit to the left is also incremented by 1). For instance, adding 1 to the right digit of 111111123 gives 111111131. Note that this method will cause the system numbers to be generated in ascending numerical order.

The first restriction demands that any acceptable system number have all three values represented at least once; thus the first acceptable number generated would be 111111123. The second restriction implies that a permutation of the values should not make it possible to recreate a previous system. Thus 111111132 is redundant because it is genetically equivalent to 111111123. This restriction is easily programmed, since it is satisfied by the requirement that the first digit of the system number always have the value 1 and the first digit with a different value have a 2. A system number that satisfies this requirement is referred to as being "ordered." It follows directly that an "unordered" system number is equivalent to an ordered one that may be obtained by a suitable permutation of the values. This requirement greatly reduces the amount of work since the highest ordered system number is 123333333.

The third and most troublesome restriction arises from the ambiguity inherent in the genetic formulation. Since the assignment of specific alleles and loci is arbitrary, any given assignment must be identical to another one in which the genic symbolism is switched at the first locus (a becomes a' and vice versa), or at the second locus (b becomes b' and vice versa), or the first locus is interchanged with the second (a becomes b, a' becomes b' and vice versa), or any combination of two or more of these interchanges is made.

An alternate way of stating the problem involved is to say that two systems are genetically equivalent if one can be obtained from the other by any permutation of the allelic symbols that preserves the integrity of the loci. That is, if after permuting the symbols, two symbols represent two alleles at the same locus if, and only if, they represented two alleles (not necessarily the same ones) which were both at one locus before permuting, then the permuted system is genetically equivalent to the system from which it was obtained. This way of stating the problem is a direct extension of Cotterman's (1953) definition of "permutationally equivalent" systems.

In Table A1 it can be seen what the effect of interchanging a with a' is. In Part a of the table the effect is shown in terms of standard genotypic notation; in Part b it specifies how the interchange permutes the various digits of the system number. That is, it shows the position each of the digits would occupy in the equivalent system number that would be produced if the gene symbol a were interchanged with a'. Using Table A1, consider two specific examples. First: the system number 123332112 is transformed into 112332123. Second: the number 112321331 is transformed into 331321112. However, this last number is unordered; reordering produces the system number 112132223, which must be genetically equivalent to the original system number.

TABLE A1

Permutation of Genotypes by Interchanging a with a'

a: In Terms of Genotypes			b: In Terms of Digit Position		
aabb	----→	a'a'bb	1	----→	7
aabb'	----→	a'a'bb'	2	----→	8
aab'b'	----→	a'a'b'b'	3	----→	9
aa'bb	----→	aa'bb	4	----→	4
aa'bb'	----→	aa'bb'	5	----→	5
aa'b'b'	----→	aa'b'b'	6	----→	6
a'a'bb	----→	aabb	7	----→	1
a'a'bb'	----→	aabb'	8	----→	2
a'a'b'b'	----→	aab'b'	9	----→	3

By constructing other tables similar to Table A1, it is easily verified that the three symbol interchanges given above, plus all combinations of them, produce a total of seven distinct permutations of the system numbers' digits. It is a small matter to program these permutations. So, as each system number is generated, the program performs each of the seven permutations, checking to insure that the third restriction is met. There are only two critical points to the procedure. First, from Example 2 above it is evident that the result of a permutation may be unordered. Therefore, a routine must be supplied to do reordering. Second, there must be a way to actually check that the third restriction is in fact met. This is actually quite simple. The key to it is the fact that the program generates the system numbers in ascending numerical order. Therefore, if the system number under consideration is equivalent to another numerically smaller system number, then it follows that the restriction is not met and the number is discarded.

In conclusion, therefore, it is possible to program a computer to generate all of the 19,683 conceivable ways in which two loci with two alleles at each locus can specify three phenotypes. By performing appropriate checks to insure that three critical restrictions are met, the computer can enumerate the reduced total of 439 genetically distinct systems, each of which can then be simulated to find equilibrium values. The number of such systems produced--439--is in agreement with the theoretical results of Hartl and Maruyama (1968), who have presented a very general method for counting the genetically distinct ways an arbitrary number of phenotypes can be produced by any given set of genotypes.